ADVANCES IN DIAGNOSIS AND MANAGEMENT OF PHYTONEMATODES

ADVANCES IN DIAGNOSIS AND MANAGEMENT OF PHYTONEMATODES

Dr. P. PARVATHA REDDY

Former Director
Indian Institute of Horticultural Research
Bangalore

CRC Press
Taylor & Francis Group
Boca Raton London New York

CRC Press is an imprint of the
Taylor & Francis Group, an **informa** business

-EPH-

First edition published 2024
by CRC Press
4 Park Square, Milton Park, Abingdon, Oxon, OX14 4RN

and by CRC Press
2385 NW Executive Center Drive, Suite 320, Boca Raton FL 33431

British Library Cataloguing-in-Publication Data
A catalogue record for this book is available from the British Library

Print edition not for sale in India

ISBN: 9781032689708 (hbk)
ISBN: 9781032689760 (pbk)
ISBN: 9781032689777 (ebk)

DOI: 10.4324/9781032689777

Typeset in Adobe Caslon Pro
by Elite Publishing House, Delhi

–EPH–

Contents

Preface   xvii
About the Author   xix

1. Diagnosis and Management of Phytonematodes: An Overview   1

    1.1. Introduction
    1.2. Economic importance
        1.2.1. International scenario
        1.2.2. Indian scenario
    1.3. Emerging nematode problems
        1.3.1. Root-knot and white tip nematodes on rice
        1.3.2. Root-knot nematode on groundnut
        1.3.3. Root-knot nematode on acid lime
        1.3.4. Root-knot nematode on pomegranate
        1.3.5. Root-knot nematode on guava
        1.3.6. Root-knot nematode on mulberry
        1.3.7. Cyst nematodes on potato
        1.3.8. Foliar nematode on tuberose
        1.3.9. Nematode problems on polyhouse crops
    1.4. Recent advances in plant nematology
        1.4.1. Diagnosis and identification
        1.4.2. Management
    1.5. Nematode diagnostics
        1.5.1. Morphology-based diagnostics
        1.5.2. DNA-based methods
        1.5.3. Protein-based methods
    1.6. Nematode management
        1.6.1. Biofumigation?
        1.6.2. Seed bio-priming
        1.6.3. Bioprotectants
        1.6.4. Protected cultivation
        1.6.5. Precision nematode management
        1.6.6. Nanotechnology
        1.6.7. New molecules of chemicals
        1.6.8. Avermectins
        1.6.9. Vegetable grafting
        1.6.10. Biotechnological approaches

1.6.11. Biointensive integrated nematode management
1.6.12. Climate change
1.7. Conclusion

2. Molecular Approaches for Nematode Identification   **22**

2.1. Introduction
2.2. Protein-based diagnostics
2.3. DNA-based diagnostics
    2.3.1. Restriction fragment length polymorphisms
    2.3.2. Ribosomal DNA polymerase chain reaction (rDNA-PCR)
    2.3.3. Real-time PCR
    2.3.4. Microarrays
    2.3.5. DNA sequencing
2.4. Conclusion

3. Soil solarization: An Eco-Friendly Strategy for Nematode Management   **34**

3.1. Introduction
3.2. Soil solarization
3.3. Effects of solarization
    3.3.1. Increased soil temperature
    3.3.2. Improved soil physical and chemical features
    3.3.3. Control of soil-borne pathogens
    3.3.4. Enhancement of beneficial soil organisms
    3.3.5. Increased plant growth and yield
    3.3.6. Increased availability of nutrients
    3.3.7. Decomposition of organic matter
3.4. Adaptability of solarization
    3.4.1. Protected cultivation
    3.4.2. Nursery production
    3.4.3. Open field production (Annual crops)
    3.4.4. Open field production (Permanent crops)
    3.4.5. Non-conventional users
3.5. Advantages and limitations
    3.5.1. Advantages
    3.5.2. Limitations
3.6. Nematode management
3.7. Case studies
    3.7.1. Fruit crops
    3.7.2. Vegetable crops
    3.7.3. Plantation crops
    3.7.4. Spice crops
    3.7.5. Field crops
3.8. Integration of solarization with other management methods

3.8.1. Solarization and biofumigation

3.8.2. Solarization and chemical controls

3.8.3. Solarization, amendments and fertilizers

3.8.4. Solarization and biological controls

3.9. Mechanisms of action

3.9.1. Physical mechanisms

3.9.2. Chemical mechanisms

3.9.3. Biological mechanisms

3.10. Strategies to enhance efficacy of soil solarization

3.10.1. Two transparent films

3.10.2. Transparent over black double film

3.10.3. Improved films

3.10.4. Sprayable films

3.11. Conclusion

4. Biofumigation: Opportunities and Challenges for Nematode Management   57

4.1. Introduction

4.2. What is biofumigation?

4.3. Advantages

4.4. Modes of utilization

4.4.1. Green manuring

4.4.2. Crop rotation/intercropping

4.4.3. Processed plant products

4.5. Biofumigation crops

4.6. Biofumigation approaches

4.6.1. Brassica biofumigation

4.6.2. Non-Brassica biofumigation

4.7. Modes of action

4.8. Nematode management

4.8.1. Root-knot nematodes, *Meloidogyne* spp.

4.8.2. Cyst nematodes, *Heterodera* spp., *Globodera* spp.

4.8.3. Citrus nematode, *Tylenchulus semipenetrans*

4.8.4. Grapevine dagger nematode, *Xiphinema index*

4.8.5. Banana nematodes

4.8.6. Lesion nematode, *Pratylenchus penetrans*

4.9. Maximizing biofumigation potential

4.9.1. Enhancing GSL profiles

4.9.2. Improving efficacy in field

4.9.3. Increasing ITC production using plant stress

4.10. Cultural practices that impact the efficacy of biofumigation

4.10.1. Soil characteristics

4.10.2. Method, rate, and timing

4.11. Integrated nematode management

4.12. Conclusion

5. Precision Agriculture-The-state-of-Art Nematode Management   75

5.1. Introduction
5.2. Precision nematode management
5.3. Precision tools
    5.3.1. Global positioning system (GPS)
    5.3.2. Geographic information system (GIS)
    5.3.3. Remote sensing (RS)
    5.3.4. Variable rate technologies (VRT)
    5.3.5. Yield monitors (YM)
5.4. Decision support system
5.5. Benefits of precision farming
5.6. Nematode management
    5.6.1. Cotton root-knot nematode
    5.6.2. Cotton reniform nematode
    5.6.3. Potato root-knot nematode
5.7. Conclusion

6. Nanobiotechnology-Based Nematode Management   89

6.1. Introduction
6.2. Dimensions of nanomaterials
6.3. Nanotechnology applications
6.4. Benefits
6.5. Nanonematicides
6.6. Green synthesis methods
    6.6.1. Plant extracts
    6.6.2. Microorganisms
6.7. Detection of nematodes and nematicide residues
    6.7.1. Detection of nematodes
    6.7.2. Detection of nematicide residues
6.8. Smart delivery of plant protection products
    6.8.1. Nanoencapsulation
6.9. Nematode management
6.10. Phytosynthesized nanoformulations for nematode management
    6.10.1. Laboratory studies
    6.10.2. Glasshouse studies
    6.10.3. Field studies
6.11. Mechanism of action of AgNPs
    6.11.1. Suppression in nematode reproduction
    6.11.2. Suppression in nematode development
6.12. Conclusion

7. Role of Enriched Vermicompost in Nematode Management   108

    7.1. Introduction
    7.2. Nutritional composition
        7.2.1. C/N ratio
        7.2.2. Nitrogen
        7.2.3. Phosphorus
        7.2.4. Iron (Fe)
        7.2.5. Magnesium (Mg)
        7.2.6. Manganese (Mn)
        7.2.7. Zinc (Zn)
    7.3. Role of vermicompost in plant growth promotion
    7.4. Bacterial diversity associated with earthworms
    7.5. Enrichment with beneficial microorganisms
        7.5.1. Enrichment with bacteria
        7.5.2. Enrichment with fungi
        7.5.3. Method for enrichment of vermicompost with bioagents
    7.6. Nematode management
    7.7. Management of disease complexes
    7.8. Integrated nematode management
    7.9. Modes of action
    7.10. Conclusion

8. Seed Biopriming for Management of Nematodes   119

    8.1. Introduction
    8.2. What is bio-priming or biological seed treatment?
    8.3. Bioagents used
        8.3.1. Bacteria
        8.3.2. Fungi
        8.3.3. Arbuscular mycorrhizal fungi
    8.4. Procedure
    8.5. Nematode management
        8.5.1. *Bacillus spp.*
        8.5.2. *Pasteuria* spp.
        8.5.3. *Pseudomonas fluorescens*
        8.5.4. *Glomus mosseae*
        8.5.5. *Purpureocillium lilacinum*
        8.5.6. *Trichoderma harzianum*
        8.5.7. Several bioagents
        8.5.8. Consortium of *Pseudomonas fluorescens* and *Bacillus subtilis*
    8.6. Management of disease complexes
    8.7. Mechanisms of action
    8.8. Conclusion

9. Avermectins: Promising Solution for Phytonematode Management   **131**

    9.1. Introduction

    9.2. Distinguishing characteristics of *streptomyces avermectinius*

    9.3. Abamectin

        9.3.1. Environmental aspects

        9.3.2. Bio-efficacy on nematodes

    9.4. Emamectin benzoate

        9.4.1. Environmental aspects

        9.4.2. Bio-efficacy on nematodes

    9.5. Mode of action

        9.5.1. Gaba antagonists

    9.6. Commercial products

    9.7. Nematode management

        9.7.1. Banana nematodes, *Meleidogyne javanica, Radopholus similis*

        9.7.2. Citrus nematode, *Tylenchulus semipenetrans*

        9.7.3. Potato cyst nematode, *Globodera pallida*

        9.7.4. Tomato root-knot nematode, *Meloidogyne incognita*

        9.7.5. Tomato reniform nematode, *Rotylenchulus reniformis*

        9.7.6. Brinjal and chilli root-knot nematode, *Meloidogyne incognita*

        9.7.7. Cucumber root-knot nematode, *Meloidogyne incognita*

        9.7.8. Garlic stem and bulb nematode, *Ditylenchus dipsaci*

        9.7.9. Carnation and gerbera root-knot nematode, *Meloidogyne incognita*

        9.7.10. Tobacco root-knot nematode, *Meloidogyne incognita*

        9.7.11. Cotton root-knot nematode, *Meloidogyne incognita*

        9.7.12. Sugar beet cyst nematode, *heterodera schachtii*

        9.7.13. Cereal cyst nematode (CCN) on wheat, *Heterodera avenae*

        9.7.14. Maize lesion nematode, *Pratylenchus zeae*

    9.8. Integrated nematode management

    9.9. Conclusion

10. Natural Genetic and Induced Nematode Resistance   **147**

    10.1. Introduction

    10.2. PPN life cycles

        10.2.1. Ectoparasitic nematodes

        10.2.2. Endoparasitic nematodes

        10.2.3. Semi-endoparasitic nematodes

    10.3. Natural genetic resistance

        10.3.1. Plant genetic resistance against nematodes

        10.3.2. Durability of resistance

        10.3.3. Assessment of crop resistance

    10.4. Induced resistance to nematodes by chemicals

    10.5. Resistance mechanisms

        10.5.1. Secretion of anti-nematode enzymes into the apoplast

10.5.2. Production of anti-nematode compounds

10.5.3. Reinforcement of cell wall as a physical barrier

10.5.4. Reactive oxygen species (ROS)

10.5.5. Nitric oxide (NO) and protease inhibitor-based immunity

10.5.6. HR-cell death-based inhibition of nematode development

10.6. Research needs

10.7. Conclusion

11. Vegetable Grafting for Improved Nematode Resistance   **164**

11.1. Introduction

11.2. Objectives

11.3. Area under selected grafted vegetables seedlings

11.4. Grafting methods

11.4.1. Hole insertion grafting (HIG)/terminal or top insertion grafting

11.4.2. Tongue approach grafting (TAG)

11.4.3. Splice grafting /tube grafting /one cotyledon splice grafting

11.4.4. Cleft grafting (CG)

11.4.5. Pin grafting (PG)

11.5. Post-graft care

11.6. Nematode management

11.6.1. Cucurbitaceae

11.6.2. Solanaceae

11.7. Advantages and disadvantages

11.8. Robotic grafting

11.9. Conclusion

12. Strategies for Transgenic Nematode Resistance   **185**

12.1. Introduction

12.2. Transgenic nematode resistance

12.2.1. Plant natural resistance genes

12.2.2. Proteinase inhibitor coding genes

12.2.3. Nematicidal proteins

12.2.4. Utilization of RNA interference to suppress nematode effectors

12.2.5. Other strategies

12.3. Nematode resistant GM crops

12.3.1. Cotton root-knot nematode

12.3.2. Cotton reniform nematode

12.3.3. Soybean cyst nematode

12.3.4. Pineapple reniform nematode

12.3.5. Potato cyst nematode

12.3.6. Sugar beet cyst nematode

12.3.7. Tomato root-knot nematode

12.3.8. Tobacco root-knot nematode

12.3.9. Banana nematodes

12.3.10. Rice nematodes

12.3.11. Wheat cereal cyst nematode

12.4. Conclusion

13. Genome Editing - New Tool to Combat Phytonematodes   **204**

13.1. Introduction

13.2. Genome editing

13.3. Genome editing techniques

13.3.1. Restriction enzymes: The original genome editor

13.3.2. Zinc finger nucleases (ZFNs): Increased recognition potential

13.3.3. TALENs gene editing: Single nucleotide resolution

13.3.4. CRISPR-Cas9 gene editing: Genome editing revolutionized

13.4. Classes of CRISPR-based genome editing

13.4.1. Nucleases

13.4.2. Base editors

13.4.3. Prime editors

13.4.4. Transposases/recombinases

13.5. Application of CRISPR/Cas9 in plants

13.6. Future outlook

13.7. Conclusion

14. Bio-intensive Integrated Nematode Management   **216**

14.1. Introduction

14.2. Integrated nematode management

14.3. Bio-intensive Integrated nematode management

14.4. Bio-intensive INM strategies

14.4.1. Planning

14.4.2. Pest identification

14.4.3. Monitoring

14.4.4. Economic injury and action levels

14.4.5. Record-keeping: "Past is prologue"

14.5. BINM options

14.5.1. Proactive options

14.5.2. Reactive options

14.6. Proactive options

14.6.1. Healthy, biologically active soils (increasing below-ground diversity)

14.6.2. Crop diversity (increasing above-ground biodiversity)

14.6.3. Resistant crop cultivars

14.7. Reactive options

14.7.1. Biological control

14.7.2. Mechanical and physical controls

14.7.3. Chemical controls

14.8. Low input sustainable BINM strategy
    14.8.1. Summer
    14.8.2. Monsoon
    14.8.3. *Kharif*
    14.8.4. *Rabi*
14.9. Case studies
    14.9.1. Rice white tip nematode
    14.9.2. Black pepper foot rot and nematodes disease complex
    14.9.3. Banana burrowing nematode
    14.9.4. Potato nematodes
    14.9.5. Tomato root-knot nematode
    14.9.6. Brinjal root-knot nematode
    14.9.7. Okra root-knot and reniform nematode
    14.9.8. Sweet potato root-knot nematode
    14.9.9. Carrot root-knot nematode
    14.9.10. Tobacco root-knot nematode
    14.9.11. Strawberry lesion nematode
14.10. Future strategies
14.11. Conclusion

15. Integrated Nematode Management under Protected Cultivation   **235**
15.1. Introduction
15.2. Advantages and limitations
    15.2.1. Advantages
    15.2.2. Limitations
15.3. Area under protected cultivation
    15.3.1. International scenario
    15.3.2. Indian scenario
15.4. Nematode problems
15.5. Reasons for nematode flare up under protected cultivation
    15.5.1. Moisture
    15.5.2. Temperature
    15.5.3. Continuous cultivation of susceptible hosts
15.6. Nematode management
    15.6.1. Physical methods - soil solarization
    15.6.2. Cultural methods
    15.6.3. Biological methods
    15.6.4. Chemical methods
    15.6.5. Grafting on nematode-resistant rootstocks
    15.6.6. Integrated methods
    15.6.7. IIHR schedule for INM
15.7. Nematode management in vegetable crops
    15.7.1. Tomato root-knot nematodes, *Meloidogyne* spp.

15.7.2. Bell pepper root-knot nematodes, *Meloidogyne* spp.

15.7.3. Cucumber root-knot nematodes, *Meloidogyne incognita, M. javanica*

15.7.4. Cruciferous vegetable crops root-knot nematodes, *Meloidogyne* spp.

15.7.5. Lettuce root-knot nematodes, *Meloidogyne* spp.

15.8. Nematode management in flower crops

15.8.1. Carnation and gerbera root-knot nematode, *Meloidogyne incognita*

15.8.2. Carnation and gerbera spiral nematode, *Helicotylenchus dihystera*

15.8.3. Chrysanthemum foliar nematode, *Aphelenchoides ritzemabosi*

15.8.4. Gladiolus root-knot nematodes, *Meloidogyne* spp.

15.8.5. Lilies root-knot, lesion and foliar nematodes

15.8.6. Anthuriums burrowing nematode, *Radopholus similis*

15.9. Nematode management in fruit crop

15.9.1. Strawberry root-knot nematode, *Meloidogyne hapla*

15.10. Conclusion

16. Climate Change Adaptation and Mitigation Strategies for Phytonematodes   259

16.1. Introduction

16.2. Effects of climate change

16.2.1. Elevated temperature

16.2.2. Enrichment CO_2 levels

16.2.3. Fluctuating precipitation patterns

16.2.4. Combined changes in temperature and moisture regimes

16.2.5. Severe droughts

16.2.6. Air pollutants

16.3. Impacts of climate change

16.3.1. Expansion of geographical distribution

16.3.2. Breakdown of nematode resistance

16.3.3. Additional generations per season

16.3.4. Impact on overwintering

16.3.5. Reduced effectiveness of nematicides

16.3.6. Reduced effectiveness of bioagents

16.4. Prediction modeling

16.5. Adaptation and mitigation

16.5.1. Physical methods

16.5.2. Cultural methods

16.5.3. Biological methods

16.5.4. Host resistance

16.6. Conclusion

17. A Vision of the Future Outlook   283

17.1. Introduction

17.2. New paradigms for nematode management

17.3. Proposed approaches

17.3.1. Organic agriculture
17.3.2. Ecosystems
17.3.3. Land use changes
17.4. Future outlook
17.5. Conclusion

References **291**
Subject index **313**

Preface

The two major challenges facing agricultural scientists today include providing food and nutritional security for the growing world population which is expected to reach 10 billion by 2050 and protecting the environment from degradation. In order to ensure food security for the growing population, food production has to be doubled by 2050. Despite of Green revolution during 1960s which reduced poverty and hunger, nearly 800 million people still continue to suffer from hunger and malnutrition around the world. Prospects of increasing area under cultivation are limited. Hence, there is a need to increase crop productivity (yield/unit area).

Reducing yield losses caused by nematode pathogens of tropical agricultural crops is one measure that can contribute to increased food production. The destructive nematode diseases are posing a formidable challenge in the successful cultivation of crop plants. In India, plant parasitic nematodes cause 21.3% crop losses amounting to ₹102 billion (US$ 1577 million) annually. The enormous losses caused by these enemies of farmers have drawn the attention of agriculturists and policy makers. Recent years have witnessed an upsurge in the incidence and accentuation of nematode problems of crop plants that are attributable to changing cropping patterns, intensive cultivation, shift towards water saving techniques, and climate change. The nematode problems have manifested themselves in utmost severity. Dissemination of nematodes through infected planting materials, especially in horticultural crops, is contributing towards interception of newer nematode problems in geographically unknown areas. Nematode diagnosis and management is therefore, important for high yields and quality that are required by the high cost of modern crop production.

The purpose of book is to provide information on various aspects of diagnosis and eco-friendly nematode management. This information is very much scattered and there is no book at present which comprehensively and exclusively deals with the above aspects. The present book on "**Advances in Diagnosis and Management of Phytonematodes**" gives the detailed description of nematode diagnosis (morphological, biochemical and molecular) and management strategies (regulatory, physical, cultural, chemical, biological, host resistance and integrated management methods).

The book is divided into seventeen chapters. The First Chapter provides an overview of diagnosis and management of plant parasitic nematodes. A comprehensive account of molecular approaches for nematode diagnosis is discussed in Chapter Two. In Chapter Three, the role of soil solarization for nematode management are discussed. The use of biofumigation with soil incorporation of cruciferous and other plant material for nematode management are reviewed in Chapter Four. In Chapter Five, the nematode management using precision agriculture are discussed in detail. Aspects of nanobiotechnology-driven management of phytonematodes are envisaged in Chapter Six. In Chapter Seven, the use of vermicompost enriched with biological control agents are reviewed. Nematode management using seed priming with biocontrol agents are dealt in Chapter Eight. In Chapter Nine, the use of avermectins derived from the actinomycete, *Streptomyces avermictinius* are dealt with. Host plant resistance to nematode pathogens through natural genetic and induced systemic resistance is dealt in Chapter Ten. Chapter Eleven deals with management of nematode pests using grafting of vegetables on resistant rootstocks. Host plant resistance to nematode pathogens using genetically modified crops are dealt in Chapter Twelve. Chapter Thirteen deals with development of nematode resistant crop varieties using genome editing technologies. Biointensive integrated nematode management strategies using eco-friendly components such as botanicals, biocontrol agents and arbuscular mycorrhizal fungi are discussed in Chapter Fourteen. Chapter Fifteen reviews nematode management under protected cultivation using organic amendments enriched with biological control agents. Nematode management aspects under climate change using adaptation and mitigation strategies are dealt in Chapter Sixteen. The Final Chapter outlines a vision of future lines of research and conclusion.

This book will be of immense value to scientific community involved in teaching, research and extension activities related to diagnosis and management of phytonematodes. The material can be used for teaching both undergraduate and post-graduate courses in the field of plant protection. The book can also serve as a very useful reference to policy makers and practicing farmers. Suggestions to improve the contents of the book are most welcome (E-mail: reddypp42@gmail.com). The publisher, Elite Publishing House, New Delhi, deserves commendation for their professional contribution.

P. Parvatha Reddy

About the Author

Dr. P. Parvatha Reddy obtained his Ph. D. degree jointly from the University of Florida, USA, and the University of Agricultural Sciences, Bangalore.

Dr. Reddy served as the Director of the prestigious Indian Institute of Horticultural Research (IIHR) at Bangalore from 1999 to 2002 during which period the Institute was honored with "ICAR Best Institution Award". He also served as the Head, Division of Entomology and Nematology at IIHR and gave tremendous impetus and direction to research, extension and education in developing bio-intensive integrated pest management strategies in horticultural crops. These technologies are being practiced widely by the farmers across the country since they are effective, economical, eco-friendly and residue-free. Dr. Reddy has about 34 years of experience working with horticultural crops and involved in developing an F1 tomato hybrid "Arka Varadan" resistant to root-knot nematodes.

Dr. Reddy has over 250 scientific publications to his credit, which also include 45 books. He has guided two Ph.D. students at the University of Agricultural Sciences, Bangalore.

Dr. Reddy is serving as Senior Scientific Advisor, Dr. Prem Nath Agricultural Science Foundation, Bangalore. He had also served as Chairman, Research Advisory Committee (RAC), Indian Institute of Vegetable Research, Varanasi; Member, RAC of National Centre for Integrated Pest Management, New Delhi; Member of the Expert Panel for monitoring the research program of National Initiative on Climate Resilient Agriculture (NICRA) in the theme of Horticulture including Pest Dynamics and Pollinators; Member of the RAC of the National Research Centre for Citrus, Nagpur; and the Project Directorate of Biological Control, Bangalore. He served as a Member, QRT to review the progress of the Central Tuber Crops Research Institute, Trivandrum; AICRP on Tuber Crops; AICRP on Nematodes; and AINRP on Betel vine. He is the Honorary Fellow of the Society for Plant Protection Sciences, New Delhi; Fellow of the Indian Phytopathological Society, New Delhi; and Founder President of the Association for Advancement of Pest Management in Horticultural Ecosystems (AAPMHE), Bangalore.

Dr. Reddy has been awarded with the prestigious "Association for Advancement of Pest Management in Horticultural Ecosystems Award", "Dr. G.I. D'souza Memorial Award", "Prof. H.M. Shah Memorial Award" and "Hexamar Agricultural Research and Development Foundation Award" for his unstinted efforts in developing sustainable, bio-intensive and eco-friendly integrated pest management strategies in horticultural crops.

Dr. Reddy has organized "Fourth International Workshop on Biological Control and Management of *Chromolaena odorata*", "National Seminar on Hi-Tech Horticulture", "First National Symposium on Pest Management in Horticultural Crops: Environmental Implications and Thrusts", and "Second National Symposium on Pest Management in Horticultural Crops: New Molecules and Biopesticides".

Chapter - 1

Diagnosis and Management of Phytonematodes: An Overview

1.1. INTRODUCTION

The two major challenges facing agricultural scientists today include providing food and nutritional security for the growing world population which is expected to reach 10 billion by 2050 and protecting the environment from degradation. In order to ensure food security for the growing population, food production has to be doubled by 2050. Despite of Green revolution during 1960s which reduced poverty and hunger, nearly 800 million people still continue to suffer from hunger and malnutrition around the world. Prospects of increasing area under cultivation are limited. Hence, there is a need to increase crop productivity (yield/unit area).

The destructive nematode diseases are posing a formidable challenge in the successful cultivation of crop plants. The enormous losses caused by these enemies of farmers have drawn the attention of agriculturists and policy makers. Recent years have witnessed an upsurge in the incidence and accentuation of nematode problems of crop plants that are attributable to changing cropping patterns, intensive cultivation, and shift towards water saving techniques. The nematode problems have manifested themselves in utmost severity. Dissemination of nematodes through infected planting materials, especially in horticultural crops, is contributing towards interception of newer nematode problems in geographically unknown areas. Reducing yield losses caused by pathogens of tropical agricultural crops is one measure that can contribute to increased food production. Meagre representation of crop protection specialists in extension set-up, both in public and private sector, has impacted the proper diagnosis and management of nematode problems.

1.2. ECONOMIC IMPORTANCE

1.2.1. International scenario

According to Thorne (1961), *"Each year these minute organisms exact an ever- increasing toll from almost every cultivated acre in the world: a bag of rice in Burma, a pound of tea in Ceylon, a ton of sugar beets in Germany, a bag of potatoes in England, a bale of cotton in Georgia, a bushel of corn in Iowa, a box of apples in New York, a sack of wheat in Kansas, or a crate of oranges in California".*

On a worldwide basis, the 10 most important genera of plant parasitic nematodes were reported to be as follows (Sasser and Freckman, 1987):

- » *Meloidogyne*
- » *Pratylenchus*
- » *Heterodera*
- » *Ditylenchus*
- » *Globodera*
- » *Tylenchulus*
- » *Xiphinema*
- » *Radopholus*
- » *Rotylenchulus*
- » *Helicotylenchus*

Estimated overall average annual yield loss of the world's major crops due to damage by plant parasitic nematodes is 12.3% (Table 1.1) (Sasser and Freckman, 1987). For the 20 crops (left-hand column) that stand between man and starvation (life sustaining crops), an estimated annual yield loss of 10.7% is reported. For the 20 crops (right-hand column) that represent a miscellaneous group important for food or export value were reported to have an estimated annual yield loss of 14%.

Table 1.1. Estimated annual yield losses due to damage by plant parasitic nematodes – World basis (Sasser and Freckman, 1987).

Life sustaining crops	Loss (%)	Economically important crops	Loss (%)
Banana	19.7	Cocoa	10.5
Barley	6.3	Citrus	14.2
Cassava	8.4	Coffee	15.0
Chickpea	13.7	Cotton	10.7
Coconut	17.1	Cowpea	15.1
Corn	10.2	Eggplant	16.9
Field bean	10.9	Forages	8.2
Millet	11.8	Grapes	12.5
Oat	4.2	Guava	10.8
Peanut	12.0	Melons	13.8
Pigeon pea	13.2	Misc. other	17.3
Potato	12.2	Okra	20.4
Rice	10.0	Ornamentals	11.1

Rye	3.3	Papaya	15.1
Sorghum	6.9	Pepper	12.2
Soybean	10.6	Pineapple	14.9
Sugar beet	10.9	Tea	8.2
Sugarcane	15.3	Tobacco	14.7
Sweet potato	10.2	Tomato	20.6
Wheat	7.0	Yam	17.7
Average	10.7 %	Average	14.0 %
Overall average – 12.3%			

Monetary losses due to nematodes on 21 crops, 15 of which are life sustaining, were estimated at US $ 77 billion annually based on 1984 production figures and prices. These figures are staggering and the real figure, when all crops are considered, probably exceeds US $ 100 billion annually. The losses are 5.8% greater in developing countries than in developed countries (Sasser and Freckman, 1987).

Crop losses due to nematodes are difficult to calculate accurately, with global estimates varying considerably from US$ 80 billion (101) to US$ 157 billion per year (Abad *et al.*, 2008). On a global scale, annual economic losses due to nematode infection of crops have been estimated at $173 billion (Elling, 2013). In the United Kingdom, the potato cyst nematode (PCN) problem (*Globodera rostochiensis* and *G. pallida*) accounts for an estimated ~$US70 million or 9% of UK potato production (Nicol *et al.*, 2011).

1.2.2. Indian scenario

In India, the crop losses by phytonematodes were estimated at about ₹ 2,100 million annually (Jain *et al.*, 2007).

Based on data generated through AICRP on Nematodes over the years, a critical analysis has been made on losses in different crops (Table 1.2). Overall, plant parasitic nematodes cause 21.3% crop losses amounting to ₹ 102039.79 million (US$ 1577 million) annually; the losses in 19 horticultural crops were assessed at ₹ 50224.98 million, while for 11 field crops it was estimated at ₹ 51814.81 million. Rice root-knot nematode, *Meloidogyne graminicola* was economically most important causing yield loss of ₹ 23272.32 million in rice. Citrus (₹ 9828.22 million) and banana (₹ 9710.46 million) among fruit crops; and tomato (₹ 6035.2 million), brinjal (₹ 3499.12 million), and okra (₹ 2480.86 million) among the vegetable crops suffered comparatively more losses. The details of crop losses in different crops are provided in Table 1.2 (Walia and Chakraborty, 2018).

Table 1.2. Estimated losses due to economically important plant parasitic nematodes to various crops in India (2014-15) (Walia and Chakraborty, 2018).

Crop	Nematode	Production (million tons)	Yield loss (%)	Price per metric ton (₹)	Monetary loss (₹ in million)
Fruit crops					
Banana	*Meloidogyne incognita*	02.92 (29.22)	15	22170	9710.46
Citrus	*Tylenchulus semipenetrans*	01.16 (11.65)	27	31380	9828.22
Grapes	*M. incognita*	00.28 (02.82)	30	46910	3940.44
Guava	*Meloidogyne* spp.	00.40 (03.99)	28	20990	2350.88
Papaya	*M. incognita* + *Rotylenchulus reniformis*	00.49 (04.91)	30	17120	2516.64
Pomegranate	*Meloidogyne* spp.	00.18 (01.78)	23	73030	3023.44
Mean yield loss in fruit crops-25.5%		**Total monetary loss in fruit crops - ₹ 31370.08 million**			
Vegetable crops					
Bitter gourd	*M. incognita*	00.08 (00.77)	13.5	23410	252.82
Bottle gourd	*M. incognita*	00.19 (01.82)	22	9660	403.78
Brinjal	*Meloidogyne* spp.	01.25 (12.58)	21	13330	3499.12
Capsicum	*Meloidogyne* spp.	00.02 (00.18)	10	26460	52.92
Carrot	*Meloidogyne* spp.	00.10 (00.96)	34	22180	754.12
Chilli	*Meloidogyne* spp.	00.20 (01.99)	15	24830	744.90
Cucumber	*Meloidogyne* spp.	00.07 (00.67)	12	13150	110.46
Okra	*Meloidogyne* spp.	00.57 (05.70)	19.5	22320	2480.86
Potato*	*Globodera* spp.	0.032 (0.032)	26	15270	127.04
Tomato	*Meloidogyne* spp.	01.64 (16.38)	23	16000	6035.20
Mean yield loss in vegetable crops -19.6%		**Total monetary loss in vegetable crops – ₹ 14461.22 million**			
Spice crops					
Ginger *M. incognita*		00.08 (00.76)	29-33	75170	1894.28
Black pepper	*Radopholus similis*	0.006 (00.06)	24	605450	871.84
Turmeric	*M. incognita*	00.08 (00.83)	33	61650	1627.56
Mean yield loss in spice crops - 29.5%		**Total monetary loss in spice crops - ₹ 4393.68 million**			
Cereal crops					
Maize	*Heterodera zeae*	02.41 (24.17)	12	13100	3788.52
Rice	*M. graminicola*	10.54 (105.48)	16	13800	23272.32

Wheat**	*H. avenae*	02.17 (21.71)	28.5	14500	8967.52
Mean yield loss in cereal crops - 18.80%		**Total monetary loss in cereal crops – ₹ 36028.36 million**			
Pulse crops					
Black gram	*M. incognita*	00.19 (01.96)	19	43500	1570.35
Chickpea	*M. incognita*	00.73 (07.33)	21	31750	4867.27
Green gram	*M. incognita*	00.15 (01.50)	29	46000	2001.00
Mean yield loss in pulse crops -23.00%		**Total monetary loss in pulse crops - ₹ 8438.62 million**			
Oilseed crops					
Castor	R. reniformis	00.18 (01.87)	15	40103	1082.78
Groundnut***	*M. arenaria*	00.22 (02.20)	4.5	40000	396.00
Sunflower	*M. incognita*	00.04 (00.43)	16	37500	240.00
Mean yield loss in oilseed crops – 11.80%		**Total monetary loss in oilseed crops – ₹ 1718.78 million**			
Fiber crops					
Cotton	*M. incognita/ M. javanica*	00.59 (05.91)	20.5	39000	4717.05
Jute	*Meloidogyne* spp.	00.20 (02.00)	19	24000	912.00
Mean yield loss in fiber crops-19.75%		**Total monetary loss in fiber crops - ₹ 5629.05 million**			
SUMMARY					
Mean yield loss in field crops – 18.23		**Total monetary loss in field crops - ₹ 51814.81 million**			
Mean yield loss in horticultural crops-23.03%		**Total monetary loss in horticultural crops- ₹ 50224.98 million**			
OVERALL MEAN YIELD LOSS -21.30%		**GRAND TOTAL MONETARY LOSS – ₹ 102039.79 million**			

Figures on production have been taken from Anonymous (2016) and AGMARKETNET Portal (2014).

Loss estimations are based mainly on work done at various Centers of AICRP on Nematodes.

Only ten per cent area of total production infested with phytonematodes has been considered for estimation of national losses.

Figures in parentheses are total production of the crops in country.

* *Globodera* spp. is widespread only in Nilgiri Hills (Tamil Nadu), so the area, production of Nilgiri Hills only has been considered.

** Production in Rajasthan and Haryana states only taken into consideration.

*** Production in Gujarat state only taken into consideration.

1.3. EMERGING NEMATODE PROBLEMS

The changing cropping patterns, introduction of new crops, crop diversification, and agronomic practices *etc.* change the spectrum of pests and pathogens, including plant parasitic nematodes. For example, the adoption of water saving techniques like System of *Rice* Intensification (SRI) in rice and drip irrigation in horticultural crops; diversification towards horticultural crops, particularly protected cultivation systems; and widespread and unchecked movement of planting materials from horticultural nurseries, have led to emergence of new nematode problems in newer areas and intensification of existing nematode problems. A number of new nematode problems have emerged in the intensive cropping systems.

Numerous challenges in the discipline of Nematology have opened up new opportunities to prove nematodes as important limiting factors in agricultural production: root-knot nematode problem on rice, pomegranate, guava; and vegetables and flowers grown under protected cultivation; cyst nematodes problem on potato; floral malady in tuberose; impact of climate change; threat of introduction of new nematode pests under globalization of agricultural produce; lack of management strategy in standing crops, *etc.* are some of the challenges in near future.

1.3.1. Root-knot and white tip nematodes on rice

The root-knot nematode, *Meloidogyne graminicola* prevalent in eastern, north-eastern and southern states of India has emerged as a serious threat for the successful rice production in nurseries, uplands, deep water, and irrigated fields due to rice cropping intensification and increasing scarcity of water. In India, the losses due to *M. graminicola* have been estimated to be 16-32% and yield loss due to poorly filled kernels to be 17-30% (Prasad and Somasekhar, 2009). The widespread detection of white tip disease of rice, *Aphelenchoides besseyi* in the southern and eastern states of India is emerging as a serious nematode problem causing 17-54% yield reduction in rice (Das and Khan, 2007).

1.3.2. Root-knot nematode on groundnut

Root-knot nematodes, *Meloidogyne arenaria* and *M. javanica* are emerging as important constraints to groundnut production in Gujarat which are responsible for 13-50% and 10-23% of yield losses, respectively. *M. arenaria* causes crop loss amounting to ₹ 396 million in Gujarat alone. It is evident that root-knot nematodes are important constraints to groundnut production in Rajkot, Junagadh and Kaira districts of Gujarat (Patel *et al.*, 1996).

1.3.3. Root-knot nematode on acid lime

The root-knot nematode, *Meloidogyne indica* is devastating the acid lime in Gujarat. Acid lime orchards in North Gujarat, particularly in Banaskantha district were observed to be heavily infected by *M. indica*. The citrus root-knot nematode infection was observed in Banaskantha, Mahesana, and Anand districts of Gujarat (Patel *et al.*, 1999).

1.3.4. Root-knot nematode on pomegranate

There has been an outbreak of serious root-knot nematode problems in the recent years associated with pomegranate (*Meloidogyne incognita* Race 2) disseminated through infected planting material and capable of killing the tress outright within three months after the onset of symptoms (Khan *et al.*, 2005). Besides causing direct damage, the root-knot nematode species are responsible for causing wilt disease complexes in association with soil-borne fungal plant pathogens. The root-knot nematode complex has similarly created serious situation with *Ceratocystis fimbriata* on pomegranate in Maharashtra, Karnataka, and North Gujarat states.

1.3.5. Root-knot nematode on guava

Till recently, *Meloidogyne incognita* and *M. javanica* have been reported associated with guava from various parts of the country (Ansari and Khan, 2012); however, these were not considered highly pathogenic to guava. The situation turned alarming with the interception of an exotic species of root-knot nematode, *Meloidogyne enterolobii* (syn. *M. mayaguensis*) on guava in Ayyakudi area (District Dindigal) of Tamil Nadu (Poornima *et al.*, 2016). Young guava trees witnessed heavy mortality (30-50%) within three months of first appearance of the symptoms (Poornima *et al.*, 2016; Ashokkumar and Poornima, 2019). Subsequent surveys revealed its occurrence in many districts of Tamil Nadu (Ashokkumar *et al.*, 2019) besides 10 other states (Andhra Pradesh, Telangana, Karnataka, Gujarat, Kerala, Haryana, Uttar Pradesh, Rajasthan, Uttarakhand, West Bengal of India) (AICRP-Nematodes Centers' Reports).

1.3.6. Root-knot nematode on mulberry

The root-knot nematode, *Meloidogyne incognita* is a major constraint in mulberry cultivation and plays an important role in reducing herbage yield and quality of leaves besides the life span of mulberry plants. It is responsible for 10-12% herbage yield loss and adversely affects the silk industry in Karnataka (Govindaiah *et al.*, 1991).

1.3.7. Cyst nematodes on potato

Potato cyst nematodes (*Globodera rostochiensis* and *G. pallida*), which were hither to restricted to South Indian states such as Tamil Nadu (Nilgiri Hills) and Kerala (Kodai Hills), have now been detected in North Indian state like Himachal Pradesh (Shimla, Mandi, Kullu, Chamba and Sirmaur districts), which is a serious concern with ramification on export and quarantine issues (Ganguly *et al.*, 2010). Recently, the potato cyst nematodes have also been recorded from Jammu and Kashmir and Uttarakhand states.

1.3.8. Foliar nematode on tuberose

Foliar nematode, *Aphelenchoides besseyi* is of great concern for profitable cultivation of tuberose (*Polianthes tuberosa*). It induces a typical 'floral malady' symptoms which is a prime threat for growing quality flowers particularly in major tuberose growing areas of West Bengal and Odisha in India (Das and Khan, 2007). The foliar nematode is responsible for 38% loss in spike yield and 59% loss in loose flower yield.

1.3.9. Nematode problems on polyhouse crops

Vegetable crops (tomato, bell pepper, cucumber, and okra) and flower crops (carnations and gerbera) are being grown throughout India under protected cultivation (in polyhouses/greenhouses/shade nets) which are seriously infested with nematodes such as *Meloidogyne incognita*, *M. javanica* (root-knot nematodes) and *Rotylenchulus reniformis* (reniform nematode). Nematode problems on all these crops under protected conditions have assumed alarming proportions leading to huge losses (up to 80%) in select crops. The nematode infestations exacerbate severity of fungal diseases leading to complete crop losses. *M. incognita* infection makes the plants highly susceptible to *Fusarium oxysporum* f. sp. *dianthi* attack. *Phytophthora parasitica* + *M. incognita* interact to produce a disease complex in gerbera leading to reduction in the yield by around 40–60%.

1.4. RECENT ADVANCES IN PLANT NEMATOLOGY

1.4.1. Diagnosis and identification

Accurate, reliable and rapid identification of PPNs are key to devise appropriate management strategy and to prevent the introduction of exotic nematodes through imported plant material (Hunt and Handoo, 2009). Traditionally, morphological and morphometric characters are used to identify PPNs (Luc *et al.*, 1990). However, morphology-based identification is time-consuming, prone to human error and may not be reproducible across different laboratories (Fortuner, 2013). Besides, the morphological character-based diagnosis requires skilled nematode taxonomists and suitable equipment for proper identification of the species (Abebe *et al.*, 2011).

Molecular methods are more accurate, rapid, cost-effective and do not depend on the specific life stages of nematodes (Braun-Kiewnick and Kiewnick, 2018). Several DNA-based methods, for example, restriction fragment length polymorphism (RFLP) (Han *et al.*, 2004), amplified fragment length polymorphism (AFLP) (Esquibet *et al.*, 2003), random amplification of polymorphic DNA (RAPD) (Cenis, 1993) and sequence characterised amplified region (SCAR) (Carrasco-Ballesteros *et al.*, 2007) have been extensively used in identification and description of existing and new species of PPNs associated with different crops. Genotyping by sequencing (GBS) has also recently been used to identify *Meloidogyne* species (Rashidifard *et al.*, 2018).

The development of polymerase chain reaction (PCR) and availability of genomic sequence databases has facilitated identification and diagnosis of PPNs by means of molecular methods (Carneiro *et al.*, 2017). PCR based detection became popular because of its specificity, sensitivity, speed, ease and cost-effectiveness as compared to the other diagnostic protocols. The conserved regions of ribosomal DNA, including internal transcribed spacer regions (ITS1 and ITS2), intergenic spacer regions (IGS1 and IGS2), and specific regions from mitochondrial DNA have been used as markers for detection and diagnosis of PPNs (Madani *et al.*, 2019).

1.4.2. Management

Presently practiced nematode management measures mainly orient towards chemical control. The excessive dependence on chemical nematicides leads to the development of resistance in nematode pathogens, outbreaks of biotypes and occurrence of residues in food chain. The chemical control also has other limitations such as high cost, low cost benefit ratio; poor availability; selectivity; temporary effect; efficacy affected by physico-chemical and biological factors; health hazards; toxicity towards plants, animals and natural enemies; environmental pollution, *etc*. Because of highly intensive agricultural practices and chemicalization of agriculture, the age-old environment friendly pest management practices like sanitation, crop rotation, mixed cropping, adjustment of date of planting, fallowing, summer plowing, green manuring, composting, *etc*, to combat plant nematodes are not being practiced in Indian agriculture. The pace of development and durability of resistant varieties had been slow and unreliable.

Considering these limitations, there has been a growing awareness and increasing demand for novel and improved nematode diagnosis and management approaches to guarantee effective and sustainable food production which offers new opportunities for crop protection research. The most recent advances include precision nematode management, nano-formulations of nematicides, avermectins, biofumigation, biotechnological approaches, bio-priming of seeds, vegetable grafting, nematode management under protected cultivation, biointensive integrated nematode management and others.

1.5. DIAGNOSTICS

Proper identification of nematode pests of crop plants is still a great problem at this time. It has also drawn the attention from farmers and agricultural extension workers in general and scholars of the crop protection streams, in particular. Wrong identification is responsible for the improper nematode management recommendations.

Reliable diagnostic tests allow these nematode pests to be detected, identified and/or monitored in a timely fashion. The correct identification of nematode pests can hardly be over-emphasized since it is fundamental to management. Rapid detection is of the essence and may be the key to success in containment and control.

Reliable diagnostics serve as the basis for the cultivation and trade of healthy crop plants. Healthy crops in turn are essential for safe and sustainable agriculture, contributing to wholesome nutrition and a better quality of life.

1.5.1. Morphology-based diagnostics

Traditionally, identification is based on characteristics such as body length, morphology of sexual organs, mouth and tail parts, and other physical characters. This morphology-based classification can prove inadequate due to lack of clear variation among closely related taxa and the need for highly skilled taxonomists, whose number is on the decline (De Oliveira *et al.*, 2011). Morphology-based identification is also a demanding endeavour, especially when large numbers of samples are involved.

Important morphological identification characters in nematodes include shape of head, number of annules, body length, length of stylet, shape of stylet knob, structure of lateral fields, presence/absence and shape of spermatheca, shape of female tail terminus, shape and length of spicule and gubernaculum (Handoo *et al.*, 2008). Measurements of these characteristics and processing of samples for this purpose requires skilled taxonomists, whose number is on the decline (De Oliveira *et al.*, 2011). Morphology may also be altered due to variation in geographic location, host plant, nutrition, and other environmental factors as is observed among some free-living and plant parasitic nematodes. Concisely, it can be difficult for non-specialists to identify a nematode species with a high level of confidence based on morphology alone (Carneiro *et al.*, 2017).

1.5.2. DNA-based methods

Many forms of DNA-based methods have been developed for the identification of nematodes (Semblat *et al.*, 998; Correa *et al.*, 2013). These can be broadly categorized into fingerprint- and nucleotide-based methods. Fingerprint-based methods may include Restriction Fragment Length Polymorphism (RFLP), Amplified Fragment Length Polymorphism (AFLP), Random Amplification of Polymorphic DNA (RAPD) and the use of species-specific primers, which relies on the presence/absence of a PCR amplification product. Except for RFLP, where PCR may not be needed, all fingerprint-based methods involve PCR followed by electrophoresis. The resulting DNA fingerprint, i.e., the pattern of resolution of the DNA fragments, is used for identification and/or phylogenetic analyses of the nematode taxa considered. On the other hand, nucleotide-based methods involve PCR amplification, specific probe hybridizations and sequencing of a region(s) of the DNA, which is then used in phylogenetic analyses. Each of these methods has its own advantages and/or disadvantages compared to other nematode identification methods, DNA-based or otherwise. However, it is notable that nematode sequences have greatly altered our understanding of the evolutionary relationships between taxa (Blaxter *et al.*, 1998).

1.5.3. Protein-based methods

Like DNA-based methods, protein sequences, mass-to-charge ratios, and immunological techniques focus on using unique protein composition and structures to delineate nematode species. Proteins provide a reduced vocabulary compared to DNA due to redundancy of the genetic code; however, the alphabet used is vastly more complex, utilizing 20 plus characters compared to the four DNA bases. Additionally, protein structure and post-translational modifications increase the potential diversity available to define nematode species and facilitate identification. Nonetheless, the requisite specialization in protein-based techniques is often a significant deterrent.

1.6. NEMATODE MANAGEMENT

1.6.1. Biofumigation

As originally defined, the term "biofumigation" demonstrates the suppressive effects of Brassicaceae plant family on noxious soil-borne pathogens and is specifically attributed to the release of biocidal isothiocyanates (ITCs) due to the hydrolysis of glucosinolates (GSLs, thioglucosides) present in crop residues, catalysed by myrosinase (MYR, β-thioglucoside glucohydrolase) isoenzymes (Matthiessen and Kirkegaard, 2006). Stapleton *et al.* (1998) defined biofumigation as an agronomic practice that release volatile biotoxic compounds into the soil atmosphere during the decomposition of organic amendments. Cultivation and incorporation of cover/rotation crops, especially Brassicaceae plants, occasionally suppresses soil-borne diseases, including nematodes. Utilization of volatile toxic compounds, such as isothiocyanates generated from the glucosinolates in such crops, for soil-borne disease control is generally termed biofumigation (Kirkegaard *et al.*, 1993).

Biofumigation is a sustainable agronomic practice using naturally produced plant compounds to manage soil pests, including plant-parasitic nematodes (PPNs). This practice primarily relies on volatile organic compounds (VOCs) when they or their by-products are incorporated into soil. Many plants produce VOCs; however, the biofumigation practice is dominated by the use of glucosinolates (GSLs) that are hydrolyzed into isothiocyanates (ITCs) capable of killing or driving away PPNs. Biofumigation efforts concentrate on the Brassicaceae family, with mustards and radish at the forefront of success (Daneel *et al.*, 2018). Plant material is grown as a cover crop and tilled into the soil at maturity, where it breaks down to release ITCs. Biofumigation plants or by-products may also be incorporated as seed meal or applied as extractions added to water (Estupiñan-López *et al.*, 2017).

The use of certain crops as biological fumigants ahead of crop production to manage soil-borne pests is receiving considerable interest in recent times. The crops that have shown the potential to serve as biological fumigants include plants in the mustard family (such as mustards, radishes, turnips, and rapeseed) and Sorghum species (Sudan grass, sorghum-Sudan grass hybrids). The crops from the mustard family show some promise to reduce soil-borne pests by releasing naturally occurring compounds called glucosinolates in plant tissue (roots and foliage). When chopped plant tissues are incorporated in soil, they are further broken down by enzymes (myrosinase) to form chemicals (glucosinolates) that behave like fumigants. Isothiocyanates are the breakdown products of glucosinolates, which are the same chemicals that are released from metam-sodium (Vapam) and metam-potassium (K-Pam), commonly used as chemical fumigants.

Biofumigation research focuses on the most economically damaging PPNs, including root-knot nematodes (RKNs) (*Meloidogyne* spp.), cyst nematodes (CNs) (*Heterodera* spp.), and lesion nematodes (LNs) (*Pratylenchus* spp.) (Jones *et al.*, 2013).

1.6.2. Seed bio-priming or biological seed treatment

Biological seed treatments for control of seed and seedling diseases offer the grower an alternative to chemical fungicides. While biological seed treatments can be highly effective, it must be recognized that they differ from chemical seed treatments by their utilization of living microorganisms. Storage and application conditions are more critical than with chemical seed protectants and differential reaction to hosts and environmental conditions may cause biological seed treatments to have a narrower spectrum of use than chemicals. Conversely, some biocontrol agents applied as seed dressers are capable of colonizing the rhizosphere, potentially providing benefits to the plant beyond the seedling emergence stage (Nancy *et al.*, 1997).

Seed treatment with bio-control agents along with priming agents may serve as an important means of managing many of the soil and seed-borne diseases, the process often known as 'bio-priming'. The bio-priming seed treatment combines microbial inoculation with pre-plant seed hydration. Bio-priming involves coating seed with bacterial and fungal biocontrol agents such as *Pseudomonas aureofaciens* AB254, *Trichoderma harzianum, Purpureocillium lilacinum, etc.* and hydrating for 20 hr under warm (23 °C) conditions in moist vermiculite or on moist germination blotters in a self-sealing plastic bag. The seeds are removed before radical emergence. The biocontrol agents may multiply substantially on seed during bio-priming (Callan *et al.*, 1990).

PGPR application through seed bio-priming involves soaking the seeds for pre-measured time in liquid bacterial suspension, which starts the physiological processes inside the seed while radicle and plumule emergence is prevented (Anitha *et al.*, 2013) until the seed is sown. The start of physiological process inside the seed enhances the abundance of PGPR in the spermosphere (Taylor and Harman, 1990). This proliferation of antagonist PGPR inside the seeds is 10-fold than attacking pathogens which enables the plant to survive those pathogens (Callan *et al.*, 1990) increasing the use of biopriming for biocontrol too.

Bio-priming process had potential advantages over simply coating seed with *P aureofaciens* AB254. Seed priming often results in more rapid and uniform seedling emergence and may be useful under adverse soil conditions. Sweet corn seedling emergence in pathogen infested soil was increased by *Pseudomonas aureofaciens* AB254 at a range of soil temperatures, but emergence at 10 °C was slightly higher from bio-primed seeds than from seeds coated with the bacterium (Mathre *et al.*, 1994).

Induction of defense-related enzymes by biocontrol agents and chemicals, osmotic priming of seeds using poly ethylene glycol (PEG) solution is known to improve the rate and uniformity of seed germination in several vegetable crops (Smith and Cobb, 1991). The observed improvements were attributed to priming induced quantitative changes in biochemical content of the sweet corn seeds (Sung and Chang, 1993).

A successful antagonist should colonize rhizosphere during seed germination (Weller, 1983). Priming with PGPR increase germination and improve seedling establishment. It initiates the physiological process of germination, but prevents the emergence of plumule and radical. Initiation of physiological process helps in the establishment and proliferation of PGPR on the spermosphere (Taylor and Harman, 1990). Bio-priming of seeds with bacterial antagonists increase the population load of antagonist to a tune of 10-fold on the seeds thus protected rhizosphere from the ingress of nematode pathogens (Callan *et al.*, 1990).

1.6.3. Bioprotectants

The expanding hazardous effects of nematicides have led to check for eco-friendly biological control of PPNs, and hence biological nematicides are becoming more popular and adding as major component of biomanagement practices for PPNs. Bio-products containing fungi and bacteria antagonists rank high among the bio-nematicides used against PPNs (Askary, 2015a, b; Eissa and Abd-Elgawad, 2015). Their effect on root-knot and cyst nematode, has been well demonstrated in previous studies (Stirling, 1991; Meyer, 2003).

Biopesticides are the substances extracted from the natural materials such as living organisms (natural enemies) or their products (microbial products, phytochemicals) or they may be their by-products (semi-chemical), which are used in crop protection and management (Dybas, 1989). These biopesticides include broad area of microbial pesticides and biochemicals derived from the natural sources and microorganisms (bacteria, fungi, and nematodes). Biopesticides have a significant role in control and management of nematode population under the economic thresholds. They are released in masses to control pest and pathogens. In the integrated pest management programs, biopesticides are an important component and are considered as effective and eco-friendly compared to synthetic materials, and also biopesticides do not persist in high number in the crop environment.

Microbial pesticides include microbes that are bacterium, fungus, virus, and protozoans, which act as biological control for plant, which are specific to pest species. Many microorganisms colonize in the rhizoplane and rhizosphere of plants. Plant growth promoting microbes (PGPM) produce plant growth promoting substance as well as antibiotics, which are having capability to protect the plants from nematode disease (Siddiqui and Mahmood, 1996). Most of the research has been carried out among the nematode-antagonistic organisms including nematophagous fungi and bacteria (Askary and Martinelli, 2015). Among the nematophagous fungi, *Purpureocillium lilacinum* has antagonistic effects on root-knot nematodes (Sharon *et al.*, 2001). In cotton and some vegetables crops grown by seed treatment, *Streptomyces avermitilis* is used as biopesticide. Abamectin, which is a mixture of macrocyclic lactone metabolites of fungus *S. avermitilis*, is used to control plant-parasitic nematode. Abamectin is active against root-knot and reniform nematodes (*M. incognita* and *R. reniformis*) and lesion nematodes (*Pratylenchus* spp.) (Faske and Starr, 2007). Use of *Trichoderma* spp. for management of plant-parasitic nematode has been confirmed (Haseeb and Khan, 2012). Also, few nematophagous fungi can be used as potential biological control agents, for example, *Pochonia chlamydosporia* for *M. incognita* in vegetable crops. In a root-knot nematode susceptible tomato, the use of *P. chlamydosporia*, along with crop rotational methods, demonstrated a reduction in nematode levels (Atkins *et al.*, 2003).

Another parasitic bacterium, *Pasteuria* spp., is found to be effective on 323 nematode species (both plant-parasitic nematodes and free-living), which adds a plus point for its use as biopesticide (Chen and Dickson, 1998). For example, in a greenhouse study conducted on cucumber, treatment with three species of *Pasteuria* reduced gall caused by *M. incognita* and also reduced the number and reproduction of nematodes. Bacteria belonging to *Bacillus* spp. have shown greater potential in nematode management. *Bacillus cereus* strain S2 treatment in *M. incognita* resulted in a mortality of 90.96% (Gao *et al.*, 2016).

1.6.4. Protected cultivation

Among the productivity enhancing technologies, protected cultivation has a tremendous potential to increase the crop yields several folds. Growing of vegetables (Tomato, bell pepper, cucumber, gherkins, muskmelon, watermelon) and flowers (carnations, roses, gerbera and anthuriums) crops under protected cultivation is receiving utmost attention and gaining popularity among farming community across the country.

The ideal conditions provided by protected cultivation by way of continuous availability of the host plant round the year often result in high population build-up of soil borne pathogens including plant parasitic nematodes. The root-knot nematodes (*Meloidogyne* spp.), reniform nematode (*Rotylenchulus reniformis*), lesion nematodes (*Pratylenchus* spp.) and spiral nematode (*Helicotylenchus dihystera* on carnation and gerbera in North Karnataka) are the major nematode problems. The nematodes are responsible for about 25% yield loss in vegetables and flower crops under protected cultivation. Sometimes the yield losses have reached up to 80%. Hence, there is an urgent need to develop prophylactic as well as curative measures to check the build-up of nematode populations under polyhouse conditions.

The following nematode management practices have been recommended:

»   Incorporation of bio-pesticide (*Pseudomonas fluorescens* + *Trichoderma harzianum* + *Purpureocillium lilacinum*) enriched FYM @ 2 kg/sq. m or bio-pesticides enriched vermicompost @ 500 g/sq. m in top 18 cm of soil in the beds.

»   Application of 100 g of neem / pongamia / mahua cake or 250 g of vermicompost enriched with *P. fluorescens* + *T. harzianum* + *P. lilacinum* on 1 sq. m. beds or around the rhizosphere of the standing crop.

Farmers who adopted the above technology developed at Indian Institute of Horticultural Research, Bangalore, reduced the use of agro-chemicals to 40 to 45% and obtained 30 to 35% increased yields in capsicum, gerbera and carnations with benefit: cost ratio of 3.

1.6.5. Precision nematode management

Uniform application of nematicides across the field and on an area-wide basis brings in habitat homogeneity which is congenial for nematode outbreaks. Similarly, eradication of all non-crop plants in the field and its vicinity by herbicide applications may make it easier for nematodes to locate crop plants and deprives beneficial predators and parasitoids of important food sources (Shelton and Edwards, 1983). Hence, there is a need to improve the efficacy of nematode-control strategies and to minimize adverse environmental impact by responding to the spatial variability in nematode occurrence and habitats. Crop diversity through the use of multi-line cultivars, varietal mixtures and intercropping enhances natural enemy populations that reduce the density of nematode populations through predation, parasitism and competitive interactions. The nematode population outbreaks can be retarded by fine-tuning of nematicide applications that are made only at 'hot-spots' where nematode densities reach action thresholds, which would create a mosaic of communities that differ in species composition and inter-specific interactions. The release of toxic chemicals into the environment would be greatly reduced by agro-ecologically based nematode management strategy i.e., precision nematode management (Weisz *et al.*, 1996; Brenner *et al.*, 1998).

Application of broad-spectrum nematicides is responsible for killing of natural enemies of the target nematode which can also lead to the resurgence of nematode populations (Hardin *et al.*, 1995). Site-specific applications through precision farming would minimize the exposure of predators and parasitoids to nematicides, and favor biological control which subsequently reduces nematode damage. Site-specific nematode management is responsible for the buffering effect of inter-patch dispersal and would make nematode population outbreaks less likely because it links sink and source populations. Sampling procedures are considerably influenced by the alteration of nematode distribution patterns in the field (approaches random or even distribution) requiring less sampling effort.

The environmental nematicide load is significantly reduced by precision farming leading to the reduction of material costs, as the necessary nematicide amount is 8-10% lower (calculated in active ingredient) than in case of traditional treatment. Takács-György *et al.* (2014) estimated that the amount of pesticides saved on the level of EU-25 countries is 31.7-84.5 thousand tons in case 15% of farms apply precision farming, 63.4-169.1 thousand tons in case 25% of them introduce it, while in the most favorable case (40%) it is 126.8-338.1 thousand tons.

The information needed to make soil and crop management decisions that fit the specific conditions found within each field is provided by precision nematode management that combines the best available technologies. It enables to take more informed management decisions by using GPS, GIS, and RS to revolutionize the way data are collected (at resolutions of 1 to 5 m) and analyzed. Precision nematode management has the potential to have detailed records covering every phase of the crop production process, thus enhancing sound nematode management decisions.

The agricultural crop production costs and crop and environmental damage can be potentially reduced by following precision nematode management.

1.6.6. Nanotechnology

Nowadays, the rapid development of nanotechnology presents a new way to improve the performance of conventional pesticide formulations through the construction of nanotechnology-based agricultural systems such as drug carriers and a controllable drug targeting and releasing system (Nakamura *et al.*, 2017). Many nano-products or nanomaterials were used against plant-parasitic nematodes such as nano-silver and nano-sulfur (Hardman 2006).

Nanonematicides are formulations of active ingredient of a nematicide in nanoform that have slow degradation, targeted delivery, and controlled release of active ingredient for longer period that make them environmentally safe and less toxic in comparison with conventional chemical nematicides. Several studies have reported an enhancement in the efficacy of certain biological substances on nematodes and a reduction of losses due to physical degradation through encapsulation of these substances in nanoparticulate systems.

Use of nanoparticles in nematode management is a novel and fancy approach that may prove very effective in future with the progress of application aspect of nanotechnology (Ladner *et al.*, 2008). The effect of applications of nanoparticles in nematode management can be divided into two perspectives; direct effect of nanoparticles on nematodes and use of nanoparticles in formulating nematicides. Due to their ultra-sub microscopic size, nanoparticles gain the high degree of reactivity and sensitivity and thus have potential to prove very useful in control of plant parasitic nematodes (PPNs), in addition, nematicidal residue analysis. As chemical and physical properties of nanoparticles vary greatly as compared to larger form, thus it has become imperative to evaluate the effect of nanoparticles on nematodes to harness the beneficial effects of this technology in plant protection, especially against PPNs. Ultra-small size and very high reactivity will affect the activity of PPNs (Gatoo *et al.*, 2014).

An indigenous bacterial strain isolated from indigenous gold mines identified as *Bacillus* sp. GPI-2 was found to convert $AuCl_4$ into AuNP of size 20 nm with the spherical shape. These bio gold AuNPs were evaluated for their ability to kill nematodes (Siddiqi and Husen, 2017). Gold and silver NPs possess nematicidal activity that may provide an alternative to high-risk synthetic nematicides or inconsistent biological control agents. High frequency (biweekly) and high application doses (90.4 mg/m^2) of gold and silver NPs may be required to achieve effective field efficacy for root-knot nematode (Kalishwaralal *et al.*, 2008). Combining AgNP with an irrigation system such as fertigation or tank-mixture with compatible chemicals that supplement the AgNP nematicidal effect may increase applicability of AgNP. Further understanding of the mechanism in the nematicidal action of Silver NP also warrants improvement of gold and silver NPs efficacy (Ganesh Babu and Gunasekaran, 2009).

Silver nanoparticle (AgNP) has shown evidence of being a potentially effective nematicide (Roh *et al.*, 2009), and its toxicity is associated with induction of oxidative stress in the cells of targeted nematodes (Lim *et al.*, 2012). Ag-nanoparticles of *Urtica urens* extracts concomitant with rugby were effective in the management of *M. incognita*, since it increased nematicidal activity 11-fold more than the least toxic extract against eggs (Nassar, 2016). The toxicity of three nanoparticles, silver, silicon oxide and titanium oxide, to the root-knot nematode, *M. incognita*, was recorded in laboratory and pot experiments (Ardakani, 2013).

1.6.7. New molecules of chemicals

1.6.7.1. Nimitz [Fluensulfone (Adama)]: Nimitz is a systemic fluoroalkenyl compound that recently received U.S. Environmental Protection Agency (EPA) registration for use in vegetable crops. In vegetable production systems, Nimitz has received much interest as a methyl bromide alternative in part because it causes mortality of target nematodes within 24 to 48 hours of product application. This nematicide controls several types of plant-parasitic nematodes (e.g., root-knot, stubby root, sting, lesion, and needle nematodes) resulting in improved fruit quality and yield of several vegetable crops, including cucumbers, eggplants, tomatoes, and peppers. Nimitz can be broadcast or applied through drip irrigation seven days before seeding or transplanting vegetables. Nimitz has shown the greatest efficacy in the control of plant-parasitic nematodes when used as pre-plant application. When applied on plant foliage, the active ingredient of Nimitz moves from the application point downwards into the roots, where it might affect parasitic nematodes. To prevent phytotoxicity and subsequent crop losses, foliar application of Nimitz is not recommended.

1.6.7.2. Velum Prime [Fluopyram (Bayer CropScience)]: Velum Prime is manufactured as a liquid formulation and is one of the few available non-fumigant nematicides with systemic properties labeled for control of multiple species of plant-parasitic nematodes in the Southern U.S. This nematicide can be broadcast or applied through soil drench and drip irrigation at planting of vegetables. Velum Prime is registered for vegetable crops including potatoes and sweet potatoes; cucurbit vegetables such as cucumbers, pickling cucumbers, squash, watermelons, and cantaloupes; fruiting vegetables such as tomatoes, eggplants, okra, and peppers; and brassica vegetables such as cabbage and broccoli.

1.6.8. Avermectins

Excitement is being generated by the results of a new class of insecticide - nematicide which have demonstrated a broad spectrum of anthelminthic activity. Promising results for the control of root-knot nematodes have been reported.

Avermectins (abamectin and emamectin) can be utilized to control plant parasitic nematodes because of their chemical and biological properties, as well as relative safety. Avermectins have short half-lives and their residues can be eliminated easily through different food processing methods. Both abamectin and emamectin are very effective nematicides which proved capability of reducing PPNs significantly in various crops.

The aqueous solution of avermectins (250 ml of 0.001%/m^2 nursery bed) significantly reduced root galls of tomato seedlings raised in root-knot infested nursery beds as compared to carbofuran (10 g a.i. /m^2 nursery bed) or neem cake (1 kg/m^2 nursery bed). The treated tomato nursery beds yielded robust and healthy seedlings with no root-knot nematode infection (Parvatha Reddy and Nagesh, 2002).

1.6.9. Vegetable grafting

Under continuous cropping, soil-borne nematode problems are likely to increase. Since soil sterilization can never be complete, grafting has become an essential technique for the production of repeated crops of fruit-bearing vegetables.

Grafting as a technique is gaining wide attention throughout the world, especially for greenhouse cultivation of vegetable crops, mainly the Solanaceous (tomato, eggplant and sweet pepper) and Cucurbitaceous (cucumber, watermelon and musk melon) ones, from the view point of resistance against the soil-borne nematode pathogens in addition to obtaining better yield and quality.

Many rootstocks having distinctive characteristics are available (Yamakawa, 1983), and growers select the rootstocks, they think are the most suitable for their growing season, cultivation methods (field or greenhouses), soil type, and the type of crops and cultivars (Lee, 1989).

The vigorous roots of the rootstock exhibit excellent tolerance to serious soil-borne diseases, such as those caused by *Fusarium, Verticillium, Ralstonia*, and root-knot nematodes, even though the degree of tolerance varies considerably with the rootstock. The mechanism of disease resistance, however, has not been intensively investigated. The disease tolerance in grafted seedlings may be entirely due to the tolerance of stock plant roots to such diseases.

However, in actual plantings, adventitious rooting from the scion is very common (Lee, 1989). Plants having the root systems of the scion and rootstock are expected to be easily infected by soil-borne diseases.

In Morocco and Greece, grafting is used to control root-knot nematodes (*Meloidogyne* species) in both tomatoes and cucurbits. Researchers have proposed using grafted plants instead of methyl bromide to manage soil-borne nematode diseases in these regions of the world.

In cucurbits, resistance to *Meloidogyne incognita* was identified in *Cucumis metuliferus, C. ficifolius*, and bur cucumber (*Sicyos angulatus*) (Fassuliotis, 1970; Gu *et al.*, 2006). Using *C. metuliferus* as a rootstock to graft RKN-susceptible melons, led to lower levels of root galling and nematode numbers at harvest (Siguenza *et al.*, 2005). Moreover, *C. metuliferus* showed high graft compatibility with several melon cultivars (Trionfetti Nisini *et al.*, 2002). Cucumbers grafted on the bur cucumber rootstock exhibited increased RKN resistance (Zhang *et al.*, 2006). Promising progress has also been made in developing *M. incognita*-resistant germplasm lines of wild watermelon (*Citrullus lanatus* var. *citroides*) for use as rootstocks (Thies *et al.*, 2010). However, at present, cucurbit rootstocks with resistance to RKN are not commercially available (Thies *et al.*, 2010).

In Solanaceae, the *Mi* gene, which provides effective control against RKN in tomato, has been introgressed into cultivated tomatoes and rootstock cultivars (Louws *et al.*, 2010). Grafting of susceptible tomato cultivars on RKN-resistant rootstocks was effective in controlling RKN in fields naturally infested with RKN (Rivard *et al.*, 2010). However, as a result of temperature sensitivity of the *Mi* gene, such resistance may not be uniformly stable (Cortada *et al.*, 2009).

Wild brinjal, *Solanum torvum* identified as resistant against root-knot nematode was used as rootstock to graft with scions of promising tomato varieties Kashi Aman, Kashi Vishesh and Hisar Lalit. Grafted plants were compatible between rootstock and scion and also showed significant resistance against root-knot nematode by reducing soil population, reproduction and gall index (Gowda *et al.*, 2017).

1.6.10. Biotechnological approaches

Due to an increasing awareness of potential adverse effects of pesticides on the safety of foods and on the environment, the development of new and safer pest control methods, including alternatives to widely used chemical pesticides, has become a serious priority for the entire agri-food industry.

Hence, in the next generation agriculture practices, there is a need to do more with less and increase the yield by optimizing the available resources. Therefore it is essential to adopt modern innovative crop protection methods to ensure more optimized and make productive usage of the resources to harness the growth potential of this sector. Prof. Swaminathan (2000) emphasized the need for 'Evergreen Revolution' keeping in view the increase in population. The challenge before the plant protection scientist is to do this without harming the environment and resource base.

1.6.10.1. Recombinant DNA technology: The genetic engineering leading to transgenic plants harboring nematode resistance genes has demonstrated its significance in the field of plant nematology. The strategies include natural resistance genes, cloning of proteinase inhibitor coding genes, anti-nematodal proteins and use of RNA interference to suppress nematode effectors. Furthermore, the manipulation of expression levels of genes induced and suppressed by nematodes has also been suggested as an innovative approach for inducing nematode resistance in plants.

In the past, several transgenic strategies have been used for enhancement of nematode resistance in plants. The resistance genes from natural resources have been cloned from numerous plant species and could be transferred to other plant species, for instance, *Mi* gene from tomato for resistance against *Meloidogyne incognita*, *Hs1*$^{pro-1}$ from sugar beet (*Beta vulgaris*) against *H. schachtii*, *Gpa*-2 from potato against *Globodera pallida* and *Hero A* from tomato against *G. rostochiensis* (Fuller *et al.*, 2008). The overexpression of different protease inhibitors (PIs) such as cowpea trypsin inhibitor (CpTI), PIN2, cystatins, and serine proteases has been used for producing nematode resistant plants (Lilley *et al.*, 1999). Another main strategy was the targeted suppression of important nematode effectors in plants using RNA interference (RNAi) approach. Unlike these strategies, some recent researches have suggested that nematode resistance could be enhanced in plants by modifying the expression of particular genes in syncytia (Klink and Matthews, 2009; Ali, 2012; Ali *et al.*, 2013a, b).

1.6.10.2. Genome editing: The emergence of genome manipulation methods promises a real revolution in biotechnology and genetic engineering. Targeted editing of the genomes of living organisms not only permits investigations into the understanding of the fundamental basis of biological systems but also allows addressing a wide range of goals towards enhancing productivity including conferring resistance to various biotic stresses. A major area of application of genome editing approaches in plant breeding is to create varieties resistant to various pathogens and/or pests. These methods have been used for the modification of the key plant immunity stages at different levels in several crops. This goal can be achieved by modifying:

- » Susceptibility genes (S-genes)
- » Resistance genes (R-genes)
- » Genes regulating the interaction between the effector and target
- » Genes regulating plant hormonal balance (Andolfo *et al.*, 2016)

The Clustered Regularly Interspaced Short Palindromic Repeats associated Cas9 (CRISPR/ Cas9) system opened up its vast area of applications as a promising component of genome editing termed as RNA-guided engineered nucleases (RGENs), which were used as sequence specific nucleases for precise genetic modifications (Fichtner *et al.*, 2014; Liang *et al.*, 2015). RGENs are developed as programable nucleases composed of two components, which must be expressed in cells to perform genome editing; the Cas9 nuclease and an engineered single guide RNA (sgRNA). The sgRNA has 20 nucleotides at the 5′ end that directs Cas9 to the complementary target site. Any DNA sequence of the form N20-NGG can be targeted by altering the first 20 nucleotides of the sgRNA for novel genome editing applications (Sander and Joung, 2014).

1.6.11. Biointensive integrated nematode management

Integrated nematode management (INM) is an economically viable and socially acceptable approach to crop protection. INM is a system that, in the context of the associated environment and the population dynamics of the nematode species, utilizes all suitable techniques and methods in as compatible manner as possible and maintains the nematode populations below those causing economically unacceptable damage or loss. The criteria for judging the effectiveness of INM are productivity, stability, sustainability and equity. Sound knowledge of nematodes, cropping systems and the environment are important for the implementation and success of INM.

Biointensive integrated nematode management (BINM) is a systems approach to pest management based on an understanding of nematode ecology. It begins with steps to accurately diagnose the nature and source of nematode problems, and then relies on a range of preventive tactics and biological controls to keep nematode populations within acceptable limits. Reduced-risk nematicides are used if other tactics have not been adequately effective, as a last resort, and with care to minimize risks.

An important difference between conventional and biointensive INM is that the emphasis of the latter is on proactive measures to redesign the agricultural ecosystem to the disadvantage of a nematode and to the advantage of its parasite and predator complex

1.6.11.1. Proactive options: Proactive options, such as crop rotations and creation of habitat for beneficial organisms, permanently lower the carrying capacity of the farm for the nematode. The carrying capacity is determined by the factors like food, shelter, natural enemies complex and weather, which affect the reproduction and survival of a nematode species. Cultural control practices are generally considered to be proactive strategies. Proactive practices include crop rotation, resistant crop cultivars including transgenic plants, disease-free seeds and plants, crop sanitation, spacing of plants, altering planting dates, mulches, *etc*.

The proactive strategies (cultural controls) include:

» Healthy, biologically active soils (increasing below-ground diversity).
» Habitat for beneficial organisms (increasing above-ground biodiversity).
» Appropriate plant cultivars.

1.6.11.2. Reactive options: The reactive options mean that the grower responds to a situation, such as an economically damaging population of nematodes, with some type of short-term suppressive action. Reactive methods generally include inundative releases of biological control agents, mechanical and physical controls, botanical pesticides, and chemical controls.

1.6.12. Climate change

Climate change will cause alterations in the spatial and temporal distribution of nematodes and consequently the control methods will have to be altered to suit these new situations. Assessments of the impact of climate change on nematode infestations and in crops provide a basis for revising management practices to minimize crop losses as climate conditions change (Ghini *et al.*, 2008).

Global warming resulting in elevated carbon dioxide (CO_2) and temperature in the atmosphere may influence plant pathogenic nematodes directly by interfering with their developmental rate and survival strategies and indirectly by altering host plant physiology. Available information on effect of global warming on plant pathogenic nematodes though limited, indicate that nematodes show a neutral or positive response to CO_2 enrichment effects with some species showing the potential to build up rapidly and interfere with plant's response to global warming. These findings underline the importance of understanding the impact of climate change on soil nematodes and its implications to crop production while developing mitigation and adaptation strategies to address impact of climate change on agriculture.

Prediction of changes in geographical distribution and population dynamics of pests will be useful to adapt the pest management strategies to mitigate the adverse effects of climate change on crop production. Pest outbreaks might occur more frequently, particularly during extended periods of drought, followed by heavy rainfall. Some of the components of pest management such as host plant resistance, bio-pesticides, natural enemies, and synthetic chemicals will be rendered less effective as a result of increase in temperatures and UV radiation, and decrease in relative humidity. Climate change will also alter the interactions between the pests and their host plants. As a result, some of the cultivars that are resistant to pests, may exhibit susceptible reaction under global warming. Adverse effects of climate change on the activity and effectiveness of natural enemies will be a major concern in future pest management programs. Rate of pest multiplication might increase with an increase in CO_2 and temperature. Therefore, there is a need to have a concerted look at the likely effects of climate change on crop protection, and devise appropriate measures to mitigate the effects of climate change on food security.

Crop protection professionals routinely develop and deploy strategies and tools based on well-established principles to manage plant nematodes and many may also be applicable under climate change when projected changes, processes and interactions are factored in. Therefore research to improve adaptive capacity of crops by increasing their resilience to nematodes may not involve a totally new approach, although managing plant nematodes may have the added advantage of mitigating rising CO_2 levels

Recent observations suggest that nematode pressure on plants may increase with climate change (Ghini *et al.*, 2008). As a result, there may be substantial rise in the use of nematicides in both temperate and tropical regions to control them. Non-chemical nematode management methods (green manuring, crop rotation, mulching, application of organic manures, *etc.*) assume greater significance under changing climate scenario.

A major shift in nematode-management strategies is occurring, from almost exclusive reliance on soil-applied nematicides, to the use of combinations of alternative strategies such as crop rotation, host plant resistance, cultural manipulations and biological control (Ferris *et al.*, 1992). Greater understanding of nematode genetic response and adaptation to abiotic factors will be important in optimizing the design of cultural management tactics such as manipulations of planting and harvest times, wet or dry fallow, and soil solarization. Wet or dry fallowing starve nematodes while active (wet fallow) or die from extreme moisture stress (dry fallow). Soil solarization involves natural heating of soil under plastic cover to attain the thermal death point of nematodes. Avoidance may include changes in planting and harvest dates, such as delaying planting in the fall to avoid infection activity, and early planting or late harvest of crops to avoid additional nematode generations. Soil amendments such as green manures and various bio-proprotectans show promise for nematode suppression in some systems. Some potential biological control agents are known to give effective management of nematodes. Current research is addressing many areas of nematode management, including: development and deployment of nematode-resistant plants; crop rotation to reduce population densities of these pathogens; cover crops, trap crops and soil amendments including green manures to reduce population densities; the role of weed hosts in bringing about phenotypic changes in nematode populations; characterization of resistance genes and resistance responses; the development of molecular diagnostic protocols for nematode identification and the reference databases necessary for their implementation. Efforts to develop improved techniques for nematode diagnostics, understanding of nematode diversity and genetic variability, processes of nematode fitness and adaptation, and the incorporation of this knowledge into the design and analysis of improved nematode management strategies needs to be intensified.

1.7. CONCLUSION

Considering the importance of economic losses caused by nematodes and the fact that the restrictions governing the use of chemical nematicides are elevating, it is clear that there is need for the development of new environmentally benign strategies. It is also crucial to continue improving the current green methods in search of making them more efficient. For example, some questions to be addressed for antagonistic fungi/bacteria which include what are the optimum rate, timing, frequency, and method of application for biocontrol agents especially under field conditions?

It is critical to consider the development and improvement of multidisciplinary management strategies for nematodes such as combining microbial strategies using both bacterial and fungal agents with other cultural control practices or host resistance. Both biocontrol and application of soil amendments have been studied to some extent against PPN, but with the recent advances in technology there is room for deeper studies on how these two strategies can be synergistically used.

It is thus necessary to support a cognitive approach to study ecology of nematode communities which will aim to include/understand assemblage rules, their impacts on populations in communities (life trait evolution, adaptation), and the consequences on system functions and management.

Considering the whole nematode diversity, according to an ecological approach, makes it possible to consider more suitable strategies for sustainable management of these parasites in agriculture, based on an auto-regulation of the global pathogenic effect of the nematode communities. It is consequently critical to focus control practices on agronomic methods because they involve a great impoverishment of the inputs generated by the agro-industrial research (pesticides, biopesticides, and resistant seeds), conferring an acute brittleness in terms of sustainability at the same time.

Integrated pest management should hence flank the development of nematodes more than killing them. We must be convinced that both biological balances in soil and environmental factors are fundamental and non-avoiding helps for plant-parasitic nematode management. Every strategy processed in order to eradicate plant parasites would hence invariably prove to be unsustainable.

Biotechnology-, molecular biology-, precision agriculture, and nanotechnology-based approaches have added a new dimension to nematode disease diagnosis and management. Identification of genes that reduce nematode's ability to reproduce has allowed the breeding of nematode-resistant plants. Marker-assisted selection, genetic engineering, and RNA interference to confer resistance in crop plants, nematode suppression using host plant proteinase inhibitors, and genome-editing technologies have helped tremendously in developing management strategies for plant-parasitic nematodes.

In conclusion, future studies should focus on environmentally benign approaches which are based on multidisciplinary strategies that can fill the gaps of single sided management methods. Such approaches will also reduce the chance of resistance as the complexity of different nematicidal components would make resistance highly improbable. Whatever strategy is devised, future attempts should focus on important factors such as synergism between nematode antagonists, environmental conditions, sustainability, studying the effect of new treatments on non-target organisms, and association of individual plants with nematode antagonists of interest. In summing up, a sustainable management of plant-parasitic nematodes is feasible when two or more compatible tactics are applied concurrently while appraising environmental protection.

Chapter - 2

Molecular Approaches for Nematode Identification

2.1. INTRODUCTION

Nematode identification is crucial for nematologists, diagnosticians and policy-makers. Due to the nematodes small size, life cycle and different habitats, scientists have been struggling to find morphological differences among species that would differentiate them. Nematode identification and differentiation can provide accurate decisions for the control of parasitic nematodes and the conservation of non-parasitic nematodes. Many misidentifications due to morphological errors resulted in huge economic impacts around the world. The accurate identification of species is essential for the effective implementation of non-chemical management strategies (Hyman, 1990).

Diagnostic laboratories that provides testing for plant parasitic nematodes have been increasing in recent years due to increased occurrence, damage, and dissemination of plant parasitic nematodes, lack of proper control management strategies, and high population density of key nematode pests in agricultural systems (Lima *et al.*, 2015). The current withdrawal of most chemical nematicides from the market is direct consequence of their toxicity and side effects to environment and human health. Alternative means in controlling plant parasitic nematodes for a sustainable cropping system include the use of resistant cultivars, the use of non and poor hosts, crop rotation, crop succession, and biological control (Hunt and Handoo, 2009). However, accurate and fast identification of nematodes to species and subspecies levels is mandatory not only to be successful in choosing a proper management strategy but also for studying their genetic and biological variability or to avoid global spread of exotic and quarantine pathogens (Blok, 2005; Blok and Powers, 2009; Castagnone-Sereno, 2011). Currently, new methods and tools using biochemical and molecular approaches have been successfully used as diagnostic for plant parasitic nematodes (Blok, 2005; Blok and Powers, 2009; Castagnone-Sereno, 2011; Esbenshade and Triantaphyllou, 1990).

In nematology, species identification has been based primarily on light microscopy observations and measurements of morphological and morphometrical features mainly of females and males (Coomans *et al.*, 1978). Many nematode genera, especially plant-parasitic nematode genera, exhibit little morphological diversity. Intraspecific variation of the features important for distinguishing species, the possibility of observational and interpretative mistakes, among other factors, make the precise and reliable identification of nematode species a formidable task even for well-qualified taxonomists

(Coomans, 2002). Correct identification up to the species level is crucial to the prevention, locally and internationally, of the spread of pathogenic nematodes and the success of effective nematode management strategies.

During the past two decades, plant nematologists have increasingly used molecular techniques, both protein- and DNA-based, to confirm the validity of existing nematode species and to assist in the identification and description of new species. There is a need to examine the involvement of tropical nematologists in the development and use of molecular techniques, and review to what extent species of tropical plant-parasitic nematodes were used in molecular studies.

A huge step towards nematode identification has been the use of the most effective, precise and fast PCR-based methods with the use of DNA which has provided solutions in several identification problems. The internal transcribed spacer (ITS) has proven to be a useful DNA region from which universal or species-specific primers are used in PCR reactions. The ITS regions are considered to be the most widely used for identification purposes by nematologists (Powers *et al.*, 1997). The use of PCR technology enables nematologists to diagnose nematode diseases rapidly and accurately. Furthermore, the use of PCR is adopted by the European and Mediterranean Plant Protection Organization (EPPO) and used in standardized protocols (EPPO, 2013).

2.2. PROTEIN-BASED DIAGNOSTICS

Protein electrophoresis was the first molecular technique to be applied in plant nematology. Esbenshade and Triantaphyllou (1985) used polyacrylamide-gel electrophoresis to define the isozyme phenotypes of 16 *Meloidogyne* species from approximately 300 populations originating from 65 countries from various continents. They found esterase and malate dehydrogenase to be the most useful isozymes for the identification of the most economically important species. A few years later, simplification and miniaturization of the electrophoretic procedures led to the development of commercially available automated electrophoretic systems using precast slab gels. Identification of most common and some rare *Meloidogyne* species became routine based on isozymes from the soluble protein extract of a single young egg-laying female (Chen *et al.*, 1998; Esbenshade and Triantaphyllou, 1990; Karssen *et al.*, 1995; Molinari, 2001; Molinari *et al.*, 2005) or from galled root pieces harboring a female (Ibrahim and Perry, 1993).

Esterase patterning has been used for diagnosing *Meloidogyne* spp. from a wide range of samples and has been proved to be species-specific for a number of species (Carneiro and Almeida, 2001; Carneiro and Cofcewics, 2008; Esbenshade and Triantaphyllou, 1985). *Meloidogyne* spp., isozyme electrophoresis patterning has discriminated all of these otherwise cryptic species, however, this technique is restricted to females (Carneiro and Cofcewics, 2008). Examples of esterase patterning for major *Meloidogyne* spp. are shown in Fig. 2.1.

One of the earliest examples of the use of isozyme phenotypes to distinguish *Meloidogyne* spp. was given by Esbenshade and Triantaphyllou (1985), who reported esterase patterns for 16 *Meloidogyne* species, with the most common phenotypes being A2 and A3 (*M. arenaria*), H1 (*M. hapla*), I1 (*M. incognita*), and J3 (*M. javanica*). In landmark surveys for *Meloidogyne* spp. using isozyme (Esbenshade and Triantaphyllou, 1990; Esbenshade and Triantaphyllou, 1985) study, approximately 300 populations originate from 65 countries and several continents. In later surveys, Carneiro *et al.* (2000) found 18 esterase phenotypes among 111 populations of *Meloidogyne* spp. from Brazil and other South American countries. Isozymes continue to be widely used for diagnosis of *Meloidogyne* spp. despite some limitations. Nonetheless, isozyme phenotyping has been used for a large number of species (Blok and Powers, 2009). Schematic diagrams of isozyme patterns based on surveys, including those conducted in the international *Meloidogyne* project have been published (Esbenshade and Triantaphyllou, 1990; Carneiro *et al.*, 2000; Esbenshade and Triantaphyllou, 1985) and provide important references.

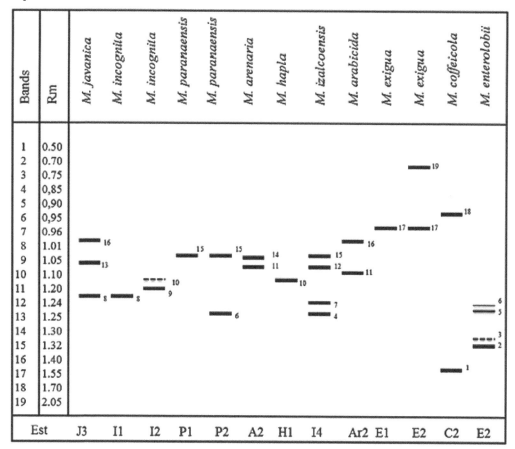

Fig. 2.1. Esterase phenotypes (Est) of major *Meloidogyne* spp. associated with coffee. Rm = ratio of migration in relation to the fast band of *M. javanica*. Dotted lines indicate weak bands (Christoforou *et al.* 2014).

Several isozyme systems have been used, nonetheless, carboxylesterase/esterase EST proved to be the most useful in discriminating *Meloidogyne* species. Others, such as malate dehydrogenase (MDH), are also often included to confirm species identification (Esbenshade and Triantaphyllou, 1985). Phenotypes with the same number of bands are differentiated by small letters (Esbenshade and Triantaphyllou, 1985; 1990). Enzyme patterns are usually compared with a known standard, with *M. javanica* being frequently used to determine migration distances among bands. Isozymes are used primarily with female egg-laying stage, using single individuals. Miniaturization and automation of the electrophoresis systems and the use of precast polyacrylamide gels (i.e., PhastSytem, Pharmacia Ltd, Uppsala, Sweden) have made isozyme phenotyping a widely used technique in most labs (Carneiro *et al.*, 2000; Esbenshade and Triantaphyllou, 1985). Classical electrophoresis methods using vertical and horizontal systems were also described in details in References (Carneiro and Almeida, 2001; Esbenshade and Triantaphyllou, 1985), respectively.

Although isozyme electrophoresis is currently one of the best methods for *Meloidogyne* spp. diagnosis, it seems likely that DNA-based methods and tools will soon usurp this method for many applications where finer resolution, particularly of intraspecific variation, is paramount (EPPO, 2013). Nonetheless, the use of an integrative diagnosis, combining more than one approach, such as morphology, morphometrics, biochemical, and molecular data is less prone to error and could be used when possible.

Extensive characterization of isozyme phenotypes has subsequently been carried out for other plant-parasitic nematode genera (Andres *et al.*, 2000; 2001; Fallas *et al.*, 1996; Fox and Atkinson, 1988; Ibrahim *et al.*, 1994; Ibrahim and Rowe, 1995; Mokabli *et al.*, 2001). For many genera, these studies revealed a wide variation in isozyme phenotypes between populations of the same species with the exception of *Meloidogyne*, in which limited intraspecific variation was detected.

In the relatively few studies where protein-based diagnostics have been used, nematologists have obtained a much better insight into the diversity of *Meloidogyne* species present and the frequency of occurrence and abundance of the individual species.

Protein-based diagnostics of *Heterodera* species were carried out in India on cyst nematodes associated with graminaceous and leguminous host plants, such as *H. avenae, H. cajani, H. filipjevi, H. graminis, H. sorghi*, and *H. zeae* (Bishnoi *et al.*, 2004; Ganguly *et al.*, 1990; Ibrahim and Rowe, 1995; Meher *et al.*, 1998; 2004; Singh *et al.*, 1998; Venkatesan *et al.*, 2004). These studies were also mainly based on differences among esterase and malate dehydrogenase phenotypes. Differences in esterase banding patterns also allowed the separation of *H. elachista, H. oryzae, H. oryzicola*, and *H. sacchari* that attack rice (Nobbs *et al.*, 1992).

Andres *et al.* (2000) used differences in isozyme banding patterns obtained from isoelectric focusing to differentiate and establish the genetic relatedness among 40 nematode populations comprising *Radopholus similis* and *Pratylenchus* species from broad host and geographic origins, including some important tropical *Pratylenchus* species. Ibrahim *et al.* (1994) studied the usefulness of alpha and beta esterase banding patterns in three populations of *Aphelenchoides besseyi* and one population each of *A. arachidis, A. bicaudatus, A. fragariae*, and *A. hamatus*, two undescribed species of *Aphelenchoides* from rice, and *Ditylenchus angustus* and *D. myceliophagus* using native and SDS-PAGE electrophoresis. Certain enzyme bands were common between the species whereas other bands were specific.

2.3. DNA-BASED DIAGNOSTICS

Since the development of polymerase chain reaction (PCR) and the vast amount of genetic data generated with DNA sequencing, molecular-based detection tools have been widely developed and successfully used for the diagnosis of plant parasitic nematodes. Molecular-based detection tools have the following advantages as compared with other methods:

- » Can be used in a high throughput manner.
- » DNA information can be acquired easily with the vast number of databases and sequencing information.
- » Cheap, fast, and accurate.
- » DNA markers are independent of phenotypic variation and developmental stage of the nematode (Powers and Harris, 1993).

» Able to exclude the effects of environmental and developmental variation (Powers, 2004; Subbotin and Moens, 2006).

DNA-based markers have been proved reliable and have allowed diagnosis and description of new species for several groups of nematodes, including key genera such as *Meloidogyne, Pratylenchus, Globodera*, and *Heterodera* (Harris *et al.*, 1990; EPPO, 2013; MacMillan *et al.*, 2006; Castagnone-Sereno *et al.*, 1995; Fullaondo *et al.*, 1999; Nakhla *et al.* 2010; Devine *et al.*, 2001; Nocker *et al.*, 2007). DNA-based detection tools make excellent methods for nematode diagnosis since they are simple, accurate, and fast (EPPO, 2013; MacMillan *et al.*, 2006) and can be used with a wide range of sample types, including host tissue, eggs, egg masses, soil extracts, and fixed samples (Powers *et al.*, 2001).

Nowadays, most labs worldwide are commonly using molecular methods to diagnose nematodes since cost associated with reagents and equipment are affordable and there has been a crescent interest in molecular taxonomy by young scientists (Castagnone-Sereno *et al.*, 1995; Powers *et al.*, 2001). These methods have been used ordinary and are sensitive enough to detect individual nematodes from complex types of samples, including soil samples and species mixtures in the field (Bates *et al.*, 2002; Bae and Wuertz, 2009; Vesper *et al.* 2008; Blok and Powers, 2009). Some limitations of molecular-based detection tools include problems associated with optimization and validation of tools and methods, DNA extraction protocols, conditions of samples (i.e., quarantine specimens), amount of target DNA in a sample, cross contamination, false positive and negative results, which overall should be used carefully as to not compromise the ultimate result of diagnosis (Powers *et al.*, 2001).

PCR-based methods involve the extraction of DNA from single or numerous juveniles, nematode cysts or complex soil samples. The PCR-based molecular diagnostic tools used for nematode identification and quantification are restriction fragment length polymorphisms (RFLPs), ribosomal DNA (rDNA) PCR, mitochondrial DNA amplification, microsatellite DNA fragment analysis, real-time PCR, microarrays, sequence-characterized amplified regions (SCARs) and next-generation sequencing (NGS).

2.3.1. Restriction fragment length polymorphisms

Curran *et al.* (1986) differentiated the *Meloidogyne* population on the level of race and strains by using total genome analysis from washed eggs. The egg DNA was purified and digested with EcoRI and electrophoresed in an agarose gel and visualized (Curran *et al.*, 1986). Due to the large number of specimens and thus the high amount of DNA needed for RFLPs analysis, the technique was improved in the early 1990s with the use DNA hybridization (Schnick *et al.*, 1990; Dalmasso, 1993) and finally PCR (Harris *et al.*, 1990). The combination of amplification and digestion (PCR-RFLP) of a single DNA strand has been found useful for DNA comparisons among individual nematodes (Powers *et al.*, 1997). Various PCR products during restriction endonuclease digestion lead to differences in fragment length within the restriction site yielding different RFLP profiles. To obtain a desirable result, different digestive enzymes participate. Nonetheless, the digestive enzymes used in RFLP do not separate all species within a genus, an issue that will be overcome with the use of species-specific primers. The specificity of RFLP could be used for the examination of a broad range of isolates from different sites around the world and thus confirm the general applicability of the RFLP method (Powers and Harris, 1993). Nevertheless, as a diagnostic tool, PCR-RFLP could eliminate much of the ambiguity involved in morphological identification of nematode specimens since differences in RFLP can be presented as the existence of differences in restriction sites in the ITS sequence (Fig. 2.2) (Subbotin *et al.*, 2000; Powers *et al.*, 2001). Nowadays, PCR-RFLP is still used when species-specific primers are absent.

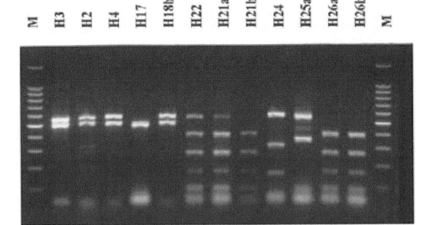

Fig. 2.2. Restriction fragments of amplified ITS regions of cyst-forming nematodes digested by Tru9I. M: 100 bp ladder and H: *Heterodera* **species (Subbotin** *et al.*, **2000).**

2.3.2. Ribosomal DNA polymerase chain reaction (rDNA-PCR)

PCR brought the evolution in molecular diagnostics of nematodes since the early 1990s. Primers were designed to produce large DNA products from which species-specific primers were then designed for producing unique products of each species. By the late 1990s, species- specific primers were designed for quarantine species such as *Globodera pallida* and *G. rostochiensis* (Bulman and Marshall, 1997; Fullaondo *et al.*, 1999). Nematode PCR products were derived from the 18S, 28S, 5, 8S coding genes and the ITS regions. The ITS region is considered a variable area of DNA that has been repeatedly examined for molecular differences among species. Mulholland *et al.* (1996) presented a multiplex PCR technique based on the use of species-specific primers, able to identify potato cyst nematodes (PCN) at the species level and without the use of restriction endonuclease digestion.

The PCR method requires DNA extracted from specimens, two pairs of 12–24-bp oligonucleotides named primers, which are complimentary to the 3'end of each strand in a specific binding site of the DNA region that will be amplified, a DNA polymerase (Taq DNA polymerase), four deoxynucleotides (dATP, dCTP, dGTP and dTTP) and a buffer-containing MgCl2 . The steps of the PCR method contain the activation of the Taq DNA polymerase (usually above 90°C), the denaturation of the DNA chain into two separated strands (usually above 90°C), the annealing of the primers (between 45 and 65°C) and the extension of the new strands, which involves the attachment of the Taq enzyme on the primers 3'end and the moving of the enzyme downstream along the DNA template, incorporating the free dNTPs on the new strand. The extension process is usually done at 72°C. PCR method usually uses around 35–40 cycles in a PCR thermocycler. The PCR product is mixed with a fluorescent dye and then transferred into separated wells of agarose gel, with the first well having a DNA ladder used as molecular weight marker. The loaded agarose gel is placed in a tray with buffer (the same buffer with which the gel was prepared) and plugged with electrodes (– electrode in the wells side and + electrode in the other site of the tray) at 100 volts. The higher the voltage, the faster the DNA moves but the heat increases and thus decreases resolution. Agarose gel is then visualized in UV light and photographed. Fleming *et al.* (1998), used the PCR method for diagnosing and estimating population levels of PCN. They demonstrated a correlation between the number of viable juveniles hatched from a cyst with the amount of DNA that could be extracted from them in a quantitative manner (Fleming *et al.*, 1998). A multiplex PCR was presented by Bulman and Marshal (Fig. 2.3), when species-specific primers were used and combined with mixed populations of PCNs (Bulman and Marshall, 1997). A few years later, the PCR method was named conventional PCR (CoPCR) due the appearance of quantitative real-time PCR (qPCR) (Bates *et al.*, 2002).

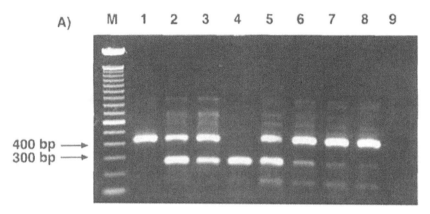

Fig. 2.3. Polymerase chain reaction (PCR) differentiation of the potato cyst nematode (PCN) species, *Globodera rostochiensis* and *G. pallida*, with various concentrations of DNA. A: multiplex PCR with primers Plp4, Plr3 and ITS% upon DNA from Ro1 Lincoln and Pa2/3 Lincoln. M: ladder, Lane 1, Ro1 1:20 H$_2$O, Lane 2, Ro1 1:20 Pa2/3, Lane 3, Ro1 1:1 Pa2/3 (1:20 H$_2$O each), Lane 4, Pa2/3 1:20 H$_2$O, Lane 5, Ro1 10:1 Pa2/3, Lane 6, Ro1 20:1 Pa2/3, Lane 7, Ro1 50:1 Pa2/3, Lane 8, Ro1 100:1 Pa2/3, Lane 9, no DNA control (Bulman and Marshall, 1997).

A complete list of species-specific primers along with references is presented in Table 2.1.

Table 2.1. Species-specific primers for diagnosis of selected plant parasitic nematodes.

Nematode species	Target region	Method	References
Meloidogyne spp.			
M. arabicida, *M. izalcoensis*	SCAR*	PCR	Correa *et al.*, 2013
M. arenaria	SCAR	PCR	Zijlstra *et al.*, 2000
M. chitwoodi	IGS	PCR	Petersen *et al.*, 1997
	SCAR	PCR	Zijlstra, 2000
M. exigua	SCAR	PCR	Randig *et al.*, 2002
M. enterolobii	mtDNA	PCR	Blok *et al.*, 2002
	SCAR	PCR	Tigano *et al.*, 2010
M. ethiopica	SCAR	PCR	Correa *et al.*, 2014
M. fallax	IGS	PCR	Petersen *et al.*, 1997
	SCAR	PCR	Zijlstra, 2000
M. graminis	ITS	PCR	Ye *et al.*, 2015
M. hapla	satDNA	PCR	Piotte *et al.*, 1995

	SCAR	PCR	Zijlstra, 2000
	IGS	PCR	Wishart *et al.*, 2002
M. incognita	SCAR	PCR	Zijlstra *et al.*, 2000
	SCAR	PCR	Randig *et al.*, 2002
M. javanica	SCAR	PCR	Zijlstra *et al.*, 2000
	SCAR	PCR	Zijlstra *et al.*, 2004
M. marylandi	28S D2-D3	PCR	Ye *et al.*, 2015
M. naasi	ITS	PCR	Zijlstra *et al.*, 2004
	28S D2-D3	PCR	Ye *et al.*, 2015
M. paranaensis	SCAR	PCR	Randig *et al.*, 2002
Other parasitic nematodes			
Bursaphelencus xylophilus	satDNA	PCR	Castagnone *et al.*, 2005
	satDNA	qPCR	François *et al.*, 2007
	Heat shock protein	qPCR	Leal *et al.*, 2007
Ditylenchus destructor *D. dipsaci*	rDNA	PCR/qPCR	Jeszke *et al.*, 2015
Heterodera glycines	rDNA	qPCR	Goto *et al.*, 2009
	SCAR	PCR	Ye, 2012
H. schachtii	ITS	qPCR	Amiri *et al.*, 2002
Pratylenchus penetrans	rDNA	qPCR	Sato *et al.*, 2017

*SCAR—sequence characterized amplified region; IGS—intergenic spacer region; ITS—internal transcribed spacer; mtDNA—mitochondrial DNA; satDNA—satellite DNA; PCR—polymerase chain reaction; qPCR—quantitative real-time PCR.

2.3.3. Real-time PCR

While conventional PCR was used worldwide for identification purposes, there was a need for more rapid, sensitive and cost-efficient method for identifying nematodes. As the genome analysis was heading deeper and deeper, more and more sequence data became available which made nematode identification and species discrimination more rapid and accurate (Bates *et al.*, 2002). Real-time PCR provides simultaneous amplification of the DNA target sequence and direct analysis of the PCR products by incorporating fluorescent probes or dyes into the reaction mix and thus the need for gel electrophoresis is avoided (Zarlenga and Higgins, 2001). In real-time PCR, the fluorescent molecule (probe or dye) reports the amount of DNA as it is multiplied in each cycle as the fluorescent signal increases proportionally. The two types of fluorescent molecules used in real-time PCR bind on DNA as DNA-binding dyes or fluorescently labelled specific primers or probes and specialized thermal cyclers detect, monitor and measure the fluorescence which reflects the amount of the amplified products in each cycle, in real time.

Quantitative real-time PCR is used for the detection and quantification of DNA present in a sample which is reflected by the number of nematodes present in the sample. For the quantification of nematodes using qPCR, a standard curve is needed (Fig. 2.4) (Christoforou *et al.* 2014). Standard curves are constructed by plotting the Ct values against the logarithm of the DNA amount isolated from different amounts of nematode eggs and juveniles. The amplification efficiency (E) is calculated from the slope of the standard curve using the following formula $E = 10^{-[1/slope]} - 1$ (Pfaffl, 2001). qPCR is used for the quantitative detection, species identification and discrimination in plant and in veterinary parasitic nematodes (Papayiannis *et al.*, 2013; Zarlenga and Higgins, 2001; Christoforou *et al.* 2014; Madani *et al.*, 2005; Berry *et al.*, 2008; Nowaczyk *et al.*, 2008; Madani *et al.*, 2008; Nakhla *et al.* 2010).

Although quantification of nematodes was a step forward for estimating population levels of parasitic nematodes in a sample, the stability of DNA from dead specimens in samples especially those extracted from cysts appears to be an obstacle (Madani *et al.*, 2005). In the case of PCN, it is very common for dead juveniles to be present within a cyst (in-egg mortality) (Devine *et al.*, 2001) and their DNA intact, while in soil, dead juveniles' DNA can be degraded in a short time. The DNA of *Phasmarhabditis hermaphrodita* was degraded in unpasteurized soil within 6 days as the dead juveniles were in direct contact to soil microflora (MacMillan *et al.*, 2006). Christoforou *et al.* (2014) reported the detection and amplification of nematodes DNA in a 34-year-old cyst stored at room temperature using PCR (Fig. 2.5) and qPCR with Taqman probes.

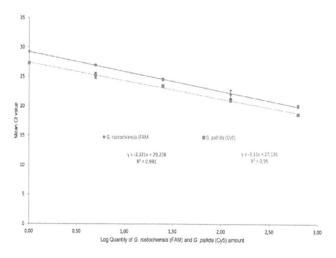

Fig. 2.4. PMA-qPCR method for the detection and quantification of viable potato cyst nematodes. A: standard curves generated by duplex real-time PCR using DNA isolated from standard PCN solutions containing 1, 5, 25, 125 and 625 live eggs or juveniles (J2), respectively. The mean Ct values corresponding to the PCR cycle number are plotted against the logarithmic quantity of nematodes DNA used in triplicate as standards. The error bars represent standard deviations of three samples (Christoforou *et al.*, 2014).

Fig. 2.5. PCR-amplified products at 465 bp of genomic DNA from non-PMA and PMA-treated cysts (A1–A4: 1976, A1 and A2 non-PMA and A3 and A4 PMA; B1–B4: 1990, B1 and B2 non-PMA and B3 and B4 PMA; C1–C4: 2007, C1 and C2 non-PMA and C3 and C4 PMA; D1–D4: 2010, D1 and D2 non-PMA and D3 and D4 PMA) [Christoforou *et al.*, 2014].

Although the use of DNA appears to be the best approach for live/dead specimen differentiation, its stability outside cell membranes allows the amplification of outbound DNA from dead cells as well, thus introducing inaccuracies in live nematode quantification. Recently, a new chemical dye propidium monoazide (PMA) has been used for selective detection of viable bacteria, fungi and nematodes, in combination with qPCR (Christoforou *et al.* 2014; Nocker *et al.*, 2007; Bae and Wuertz, 2009; Vesper *et al.* 2008). PMA is a photoreactive DNA-intercalating dye which renders exposed DNA of dead cells, is unable to amplify and thus, only DNA from viable/intact cells is PCR amplified and detected. Christoforou *et. al.* (2014) presented a qualitative estimation of viable PCN inocula using species-specific primers and Taqman probes designed by Papayiannis *et al.* (2013), in a PMA-qPCR method which was developed for the two PCN species. The PMA-qPCR method successfully discriminates dead from living specimens in heat-treated samples as also the eggs from old and newly formed cysts.

qPCR method proves to be very useful for routine identification and discrimination of nematode species from field samples. The optimization of the qPCR and DNA extraction methods is essential for the specificity, sensitivity and accuracy of the procedure. Madani *et al.* (2005) described a real-time PCR method using SYBR green-I dye with melting curve analysis for the detection and quantification of PCN species and mentioned the dependence of nematode quantification on the efficacy of DNA-extraction methods. Papayiannis *et al.* (2013) evaluated five DNA extraction methods (silica columns, magnetic-based surface, Chelex resin, chloroform- based and disruption in TE) and compared them for their preparation time, cost and technical difficulty as well as the limit of detection between PCR and qPCR assays for all extraction methods. Another important factor for an accurate qPCR assay is the primers' specificity and the limitations in detecting nematodes when species are mixed in a sample. When three plant parasitic nematodes (PPNs), *Meloidogyne javanica*, *Pratylenchus zeae* and *Xiphinema elongatum*, were tested for identification and quantification in a mixture of species and primers, competition between the DNA of *M. javanica* with *P. zeae* and *X. elongatum* was found (Berry *et al.*, 2008).

2.3.4. Microarrays

Microarrays show high potential for discriminating nematodes in multi-complex samples since many targets can be identified simultaneously due to the specificity of the microarray method to detect unique sequences for each target species (Blok and Powers, 2009; Ahmed *et al.*, 2011). Microarrays are composed of complementary DNAs (cDNAs) that can be detected due to a fluorescence bind on the cDNA, on microscope slides or silicon chips, which contain specific synthesized known DNA after hybridization of the cDNA. Ahmed *et al.* (2011) mentioned the potential of using the microarrays to identify gastrointestinal nematodes. Besides the high prospective of microarrays as diagnostic tools for identifying nematodes, it still has not been achievable. The high cost, the amplification of unknown sequences in mixed samples and the better hybridization of mismatched targets rather than the perfectly matched targets lead to the limited use of the microarray method as a diagnostic tool for nematodes (Blok and Powers, 2009).

2.3.5. DNA sequencing

DNA sequencing or DNA barcoding is referred to many nematode-related publications and has been the main driving force in studies, and as availability of instrumentation increases while cost is constantly reduced, it is apparent that it will be the dominating approach. The Sanger method or NGS approaches accumulate a substantial amount of genetic data with sufficient, if not to say overwhelming, information on sequence divergence, which may be often characterized as erroneous due to sample or analysis limitations.

For diagnostic purposes, most studies have targeted two main genomic regions for sequence divergence. These regions are the nuclear ribosomal RNA genes and their transcribed and untranscribed spacers and the mitochondrial cytochrome oxidase I (COI) gene. These regions are highly conserved but sufficiently divergent and occur in multiple copies in the genome, thus made easily amplifiable by PCR. A key element of this approach is the use of standardized markers and a relatively standardized experimental approach not introducing significant subjectiveness. On the other hand, this methodology builds taxonomic reference libraries where all submitted sequences from different organisms can be compared. As a result, unidentified organisms can be determined according to the level of DNA homology (Taberlet *et al.*, 2012). Results can be acquired in as fast as 8–12 h, making the method competent to be used in control of pest movement within trade activities and border control (Powers, 2004). rDNA genes are preferred over COI gene in most studies due to the availability of sequences and the level of conservation in order to design universal primers even though COI is capable of discriminating between species at a better level. Porazinska *et al.* (2009) had shown that the use of SSU and LSU genes together improves resolution.

With the development of NGS approaches, similarly to metagenomics, a term that has been used solely for microorganisms, DNA metabarcoding, is rapidly evolving. Bulk DNA deriving from environmental samples (water, soil) but same approach can be applied elsewhere (i.e. infected plant tissues, animal gut, blood samples), can uncover the entire hidden microcosm (Taberlet *et al.*, 2012). This approach can be used both for ecology studies, including soil quality and health, and for plant/animal diagnostics.

Limitations of high-throughput DNA barcoding still exist and are mainly the following:

» Efficiency of DNA recovery is an issue but experimentation and protocol development studies will soon address this,

» Identification of a suitable marker to provide good taxonomic coverage and species resolution.

» Formation of chimeras (artefacts of PCR when an incompletely extended DNA fragment from one cycle anneals to a template of an unrelated taxon and gets copied to completion in subsequent cycles). Bioinformatic tools are trained to identify and discard such sequences.

The second limitation referred to above can be indirectly resolved by taking advantage of the high throughput of NGS technology, where multiple genes can be simultaneously sequenced and analysed with relatively low cost. Genetic information will be easily acquired and the main technical obstacle is the vastness of information in genetic repositories of sequences, that storage and computing capacities require constant upgrade to convey bio- logical and taxonomic meanings to scientists.

It is worth referring the most recent achievement of DNA sequencing, using third-generation sequencing technology and providing whole genome analysis that has used the portable device MinION. Tyson *et al.* (2017) have reported performing the whole genome and assembly of a *Caenorhabditis elegans* genome with complex genomic arrangements. Two astonishing elements of this study are the USB type and sized instrument of MinION and the long reads that the technology offers. This second attribute improves immensely the NGS technology for *de novo* sequencing of complex genomes, in part due to repeat regions that nematodes as metazoans have in common. The flowcell of MinION is currently able to provide 5–10 Gb of sequence, which is a sufficient performance for a 100-Mb genome of a nematode with long reads for an unambiguous assembly of the chromosome.

2.4. CONCLUSION

Molecular diagnostics are used as tools for the identification of parasitic and free-living nematodes since the early 1990s. Currently, most of the plant protection laboratories use molecular tools for the identification, discrimination and quantification of important parasitic nematodes for common everyday diagnostic activities. From all the molecular tools and methods mentioned in the literature and in this review, only few are used in routine protocols. These selected ones are highly correlated with the reliability, the time and cost effectiveness as well as the expertise necessary for applying the methods.

From the methods reviewed in this chapter, real-time PCR is currently the fastest, most-sensitive and accurate method. Taqman PCR assay could detect, identify and quantify nematodes, reaching 100% accuracy. Real-time PCR methodologies can be of use in field applications with the use of a mobile qPCR instrument that is able to operate in field conditions along with easy-to-perform kits like DNA extraction and PCR reaction chemistries. For more analytical protocols and methodologies, DNA barcoding is fast progressing as DNA sequencing tools develop. However, the DNA barcoding based on NGS technologies and proteomic analysis based on mass spectrometry will soon dominate the market and offer low-budget, kit-type applications even for mobile diagnostic laboratories.

Chapter - 3

Soil Solarization: An Eco-Friendly Strategy for Nematode Management

3.1. INTRODUCTION

The destructive nematode diseases are posing a formidable challenge in the successful cultivation of crop plants. The enormous losses caused by these enemies of farmers have drawn the attention of agriculturists and policy makers. Recent years have witnessed an upsurge in the incidence and accentuation of nematode problems of crop plants that are attributable to changing cropping patterns, intensive cultivation, and shift towards water saving techniques. The nematode problems have manifested themselves in utmost severity. Dissemination of nematodes through infected planting materials, especially in horticultural crops, is contributing towards interception of newer nematode problems in geographically unknown areas. Meagre representation of crop protection specialists in extension set-up, both in public and private sector, has impacted the proper diagnosis and management of nematode problems.

Soil-borne pests and pathogens, including weed propagules, nematodes, insects, fungi, bacteria and certain other agents, can be limiting factors in the production of horticultural and other crops. One of the principal strategies used by the growers of high-value horticultural crops to combat these organisms is pre-plant soil disinfestation, using pesticides or other physical or biological methods. Soil fumigants are the most effective soil disinfestation chemicals, and methyl bromide (MB) is the most important soil fumigant chemical used by growers around the world. It is a broad spectrum pesticide with excellent activity against most potential soil pests.

However, MB was identified as a risk to the stratospheric ozone layer in 1992 and targeted for worldwide phase out in 1997 by means of the Montreal Protocol, an international treaty. Scientists have been continuously working to develop usable alternatives for soil disinfestation. Among the potential alternative control method being touted to replace methyl bromide is soil solarization which is the most useful non-chemical soil disinfestation methods.

3.2. SOIL SOLARIZATION

The use of clear polyethylene film to cover moistened soil and trap lethal amounts of heat from solar radiation was first reported by Katan and colleagues in Israel in the mid-1970s (Katan, 1987).

DeVay and associates at University of California, Davis began an intensive research program on the promising technique shortly thereafter, and the term "soil solarization" was soon coined to describe the process by co-operators in the San Joaquin Valley. Researchers found that solarization could be a useful soil disinfestation method, especially in areas with hot and arid conditions during the summer months, such as the Central Valley and southern deserts. In certain cases, the treatment has also been effective, primarily for weed management, in cooler coastal areas (Elmore *et al.*, 1993). The pesticidal activity of solarization was found to stem from a combination of physical, chemical and biological effects, as described in several comprehensive reviews (Katan 1987; Chen *et al.*, 1991; DeVay *et al.*, 1991; Stapleton, 1998, 2000).

Soil solarization technique consists of covering moist soil with a plastic film during periods of intense sunshine and heat (Fig. 3.1), thereby capturing radiant heat energy from the sun, causing physical, chemical, and biological changes in the soil. Linear Low Density Polyethylene (LLDPE) clear films were efficient to manage root knot nematode incidence.

Fig. 3.1. Plastic sheets laid by machines in strips for soil solarization.

Although solarization can provide excellent soil disinfestation under suitable conditions, it has significant limitations and should not be considered a cure-all or universal replacement for MB. For example, solarization is most effective close to the surface of the soil under climatic and weather conditions of high air temperature and long days for soil heating. It will not control all pest organisms, may require that land be taken out of production for 3-to-6- week treatments during the summer months, and requires disposal of used plastic film. Therefore its practical value to the user must be assessed by several factors, including extent and predictability of pesticidal efficacy, effect on crop growth and yield, economic cost/benefit and personal pest management philosophy (Stapleton, 1997).

3.3. EFFECTS OF SOLARIZATION

3.3.1. Increased soil temperature

The heating effect of soil solarization is greatest at the surface of the soil and decreases with depth. The maximum temperature of soil solarized in the field is usually from 42° to 55°C at a depth of 5 cm and from 32° to 37°C at 45 cm. Control of soil pests is usually best in the upper 10-30 cm of soil. Higher soil temperatures and deeper soil heating may be achieved inside greenhouses or by using a double layer of plastic sheeting. Soil solarized in greenhouses may reach 60°C at a depth of 10 cm and 53°C at 20 cm. Soil solarized in black plastic nursery sleeves under a single or double layer of clear plastic can exceed 70°C.

3.3.2. Improved soil physical and chemical features

Solarization initiates changes in the physical and chemical features of soil that improve the growth and development of plants. It speeds up the breakdown of organic material in the soil, resulting in the release of soluble nutrients such as nitrogen (NO_3^-, NH_4^+), calcium (Ca^{++}), magnesium (Mg^{++}), potassium (K^+), and fulvic acid, making them more available to plants. Improvements in soil tilth through soil aggregation are also observed.

3.3.3. Control of soil-borne pathogens

Repeated daily heating during solarization kills many plant pathogens, nematodes, and weed seeds and seedlings. The heat also weakens many organisms that can withstand solarization, making them more vulnerable to heat-resistant fungi and bacteria that act as natural enemies. Changes in the soil chemistry during solarization may also kill or weaken some soil organisms.

Although many soil pests are killed at temperatures above 30° to 33°C, plant pathogens, weeds, and other soil-borne organisms differ in their sensitivity to soil heating. Some pests that are difficult to control with soil fumigants are easily controlled by soil solarization. Other pests are also affected but cannot be consistently controlled by solarization. These may require additional control measures.

Soil solarization can be used to control many species of nematodes including lesion, root-knot, reniform, cyst, sting, ring, stubby root and dagger nematodes in different crops. However, soil solarization is not always as effective in controlling nematodes as it is in controlling fungal diseases and weeds because nematodes are relatively mobile and can recolonize soil rapidly. Nematode management may therefore require yearly treatment. Control by solarization is greatest in the upper 30 cm of the soil. Nematodes deeper in the soil profile may survive solarization and damage plants with deep root systems.

Nematode control by solarization is usually adequate to improve the growth of shallow-rooted, short-season plants. It is particularly useful for organic gardeners and home gardeners. Solarization may also be a beneficial addition to an integrated nematode management system. For example, excellent control of root knot nematode (*Meloidogyne incognita*) was obtained in the San Joaquin Valley by combining solarization with the application of composted chicken manure (Gamliel and Stapleton, 1993b).

3.3.4. Enhancement of beneficial soil organisms

Fortunately, although many soil pests are killed by soil solarization, many beneficial soil organisms are able to either survive solarization or recolonize the soil very quickly afterwards. Important among these beneficials are the mycorrhizal fungi and fungi and bacteria that parasitize plant pathogens and aid plant growth. The shift in the population in favor of these beneficials can make solarized soils more resistant to pathogens than non-solarized or fumigated soil.

3.3.4.1. Earthworms: The effect of soil solarization on earth worms has not received much attention, but it is thought that they retreat to lower depths and escape the effects of soil heating.

3.3.4.2. Fungi: Beneficial fungi, especially *Trichoderma*, *Talaromyces*, and *Aspergillus* spp., survive or even increase in solarized soil. Mycorrhizal fungi are more resistant to heat than most plant pathogenic fungi. Their populations may be decreased in the upper soil profile but studies have shown that this is not enough to reduce their colonization of host roots in solarized soil.

3.3.4.3. Bacteria: Populations of the beneficial bacteria *Bacillus* and *Pseudomonas* spp. are reduced during solarization but recolonize the soil rapidly afterward. Populations of *Rhizobium* spp., which fix nitrogen in root nodules of legumes, may be greatly reduced by solarization and should be reintroduced by inoculation of leguminous seeds. Soil-borne populations of other nitrifying bacteria are also reduced during solarization. Population levels of actinomycetes are not greatly affected by soil solarization. Many members of this group are known to be antagonistic to plant pathogenic fungi.

During solarization of soil, populations of oxidized negative fluorescent pseudomonads and gram positive bacteria, including *Bacillus* species, may be reduced by 78 to 86 percent compared with non-solarized soil (Stapleton and DeVay, 1984); whereas, populations of Actinomycetes may be reduced from 45 to 58 percent in solarized soil (Stapleton and DeVay, 1984). Surprisingly, after solarization, *Pseudomonas* species quickly recolonize the soil and their populations reach high levels. Of great significance is the change in populations of *Bacillus* species during solarization; the percentage of colonies in solarized soil which exhibited antibiosis to *Geotrichum candidum* increased nearly 20-fold when compared with non-solarized soil (Stapleton and DeVay, 1984). These bacteria are among those which are rhizosphere competent and are believed to contribute to the increased growth response of plants grown in solarized soil (Katan, 1987). Although initial populations of *Bacillus* species are greatly reduced, they are spore formers and are a major component of the soil microflora.

In contrast to the studies in California and Israel, studies in Western Australia showed that solarization increased the total numbers of bacteria and actinomycetes in soil. However, as in the California study (Stapleton and DeVay, 1984) where there was an increase in the proportion of antagonistic gram positive bacteria in solarized soil. The Western Australia study showed that the proportion of bacteria (actinomycetes) antagonistic to *Fusarium oxysporum*, *F. solani*, and *Rhizoctonia solani* was increased compared with non-solarized soil. In other studies, *Actinomyces scabies* was controlled by soil solarization.

3.3.5. Increased plant growth and yield

Plants often grow faster and produce both higher and better-quality yields when grown in solarized soil. This can be attributed, in part, to improved disease and weed control; but increases in plant growth are still seen when soil apparently free of pests is solarized. A number of factors may be involved. First, minor or unknown pests may also be controlled. Second, the increase in soluble nutrients improves plant growth. Third, relatively greater populations of helpful soil microorganisms have been documented following solarization, and some of these, such as certain fluorescent pseudomonad and *Bacillus* bacteria, are known to be biological control agents.

Solarization increased okra biomass; the longer the duration of solarization, the greater the increase in okra biomass (based on comparison among 2-, 4-, and 6-week solarization periods). The positive yield response indicated that solarization did not impair organic matter decomposition and subsequent release of plant nutrients. Both of these studies (Ozores-Hampton *et al.* 2004; Seman-Varner *et al.* 2008) suggest that solarization does not interfere much with beneficial soil organisms that decompose organic matter. Solarization may also result in an increased growth response (as evidenced by increased trunk diameters) and yield in orchard trees, by increasing the availability of plant nutrients and changes the soil microflora to favor biological pest control. Improved growth of rice and wheat plants was observed following solarization. Soil inorganic nitrogen levels were consistently higher as was extractable manganese.

3.3.6. Increased availability of nutrients

In general, availability of some nutrients can be expected to increase with solarization, because the heat generated under clear plastic will encourage accelerated decomposition of organic matter. In a field study in southwest Florida, soil nutrients were not affected by solarization compared to conventional treatment with methyl bromide (Ozores-Hampton *et al.* 2004). Solarization coupled with compost increased soil nutrient levels more than treatments with methyl bromide or solarization alone, both of which were combined with inorganic fertilizer. Seman-Varner *et al.* (2008) measured nutrient concentration in the soil and plant tissue of an okra crop following different durations of solarization. While soil potassium (K) and manganese (Mn) were higher following solarization, Copper (Cu) and zinc (Zn) were lower. In addition, soil pH was slightly decreased by solarization. Soil phosphorus (P), magnesium (Mg), calcium (Ca), and iron (Fe) were not affected by solarization. Nutrients supplied to the crop were exclusively provided by chopped cowpea hay. Okra tissue concentrations of K, N, Mg, and Mn were higher when grown on solarized plots. In contrast, concentrations of P and Zn were lowered by solarization.

The increased availability of mineral nutrients following soil solarization are particularly those tied up in organic fraction, such as NH_4-N, NO_3-N, P, Ca, and Mg, as a result of the death of the microbiota. Extractable P, K, Ca, and Mg sometimes have been found in greater amounts after soil solarization. The liberation of N compounds (vapor and liquid) is a component of the mode of action, increased concentration of reduced N would then nitrify after termination of soil solarization to provide NO_3 for increased crop growth.

Soils covered with plastic film mulches usually retain a higher level of soluble minerals. Constant moisture content, higher temperature and better aeration of the soil all tend to favor higher microbial populations in the soil thus ensuring more complete nitrification.

Plastic mulch prevents leaching of nutrients, particularly nitrogen. The dominant advantage of using polyethylene is that it aids in the retention of nutrients within the root zone, thereby permitting more efficient nutrient utilization by the crop.

3.3.7. Decomposition of organic matter

Solarization increases the levels of available mineral nutrients in soils by breaking down soluble organic matter and increasing bioavailability. Soil solarization also speeds up the breakdown of organic material in the soil, often resulting in the added benefit of release of soluble nutrients such as nitrogen ($NO3^-$, $NH4^+$), calcium (Ca^{++}), magnesium (Mg^{++}), potassium (K^+), and fulvic acid, making them more available to plants.

3.4. ADAPTABILITY OF SOLARIZATION

Solarization has considerable versatility, being adaptable to various agricultural production applications.

3.4.1. Protected cultivation

Worldwide, probably the major commercial use of solarization is in conjunction with greenhouse/glasshouse/plastic-house culture, especially in regions where the protected crops are grown only in the winter. The empty structures can be closed in the heat of the summer and plastic film laid on the soil for solarization. This method of double insulation provides a still-air chamber above the solarized soil for added heating and greater efficacy. Greenhouse solarization is used primarily in Japan, the Near East and other Mediterranean countries (Stapleton, 1997).

3.4.2. Nursery production

Solarization without combination with a chemical pesticide is not suitable for open field nursery production, due to zero tolerance for surviving pathogen propagules deeper in the soil. However, in containerized nursery production, solarization can provide complete eradication of pests. A "double-tent" solarization technique, which functions in much the same way as the greenhouse treatment just described, was approved by the California Department of Food and Agriculture in 1999 for nematode-free production of container-, flat- and frame-grown nursery stock. Among other mandated details, the approved treatment specifies solarization of soil at a minimum of 70°C for at least 30 contiguous minutes (Stapleton *et al.*, 1999).

3.4.3. Open field production (Annual crops)

Open field cultivation of row crops is the setting in which solarization was first discovered and tested. Soil to be treated should be thoroughly moistened either by pre-irrigation or by drip irrigation beneath the clear plastic film. Soil can be solarized in fields by either complete coverage or by covering only the planting beds. This technique works best with shallow-rooted, late-season crops in warmer locations so that the solarization can be done between crops in the summer (Elmore *et al.*, 1997).

3.4.4. Open field production (Permanent crops)

Solarization can be modified for use in managing certain weed, disease and nematode pests before or during establishment or replanting of orchard and vineyard crops, especially in warm areas such as the Central Valley. The method is most useful for managing shallowly distributed pests. It would not be expected to effectively control nematodes, fungi and other pests deep in the soil. When using solarization in conjunction with growing plants, treatment with black, rather than clear, plastic film may be preferable. For example, early studies indicated that solarization with clear plastic film beginning shortly after planting killed almond and apricot trees due to excessive heat. However, similar treatment with black film controlled the disease *Verticillium* wilt without damaging the trees. Other benefits included conservation of soil moisture and reducing humidity in the crop canopy (Elmore *et al.*, 1997). This type of solarization is mainly used on a commercial basis in the Central Valley for *Prunus* spp. and citrus.

Although solarization can be an effective soil disinfestant, the stand alone process, which largely depends on passive solar heating, has inherent limitations. Fortunately, it is compatible with other physical, chemical and biological methods of soil disinfestation to provide more efficacious and/or predictable treatment through integration. As MB is phased out, many current users will turn to other pesticides for soil disinfestation. Combining these pesticides (perhaps at lower dosages) with solarization (perhaps for a shorter treatment period) may be more effective. When using solarization in conjunction prove to be the most popular option for users in warm climatic areas who want to continue using chemical soil disinfestants.

3.4.5. Non-conventional users

Solarization is presently used on a relatively small scale in conventional agriculture, but its use will probably increase as MB becomes unavailable. On the other hand, solarization has become a widespread practice for organic growers, home gardeners and other users who cannot or will not use chemical soil disinfestants.

3.5. ADVANTAGES AND LIMITATIONS

3.5.1. Advantages

» Non-pesticidal and simple.

» No health or safety problems associated with use.

» No registration is required.

» Crops produced are pesticide-free and may command a higher market price.

» Controls multiple soil-borne diseases and pests.

» Selects for beneficial microorganisms.

» Tends to increase soil fertility.

» Increases soluble NO_3, NH_4, Ca, Mg, K and soluble organic matter.

» May improve soil tilth.

» Can speed up in-field composting of green manure.

3.5.2. Limitations

» Is restricted to areas with warm to hot summers.

» May be less effective in cooler coastal areas.

» Land must be taken out of production for 4 to 6 weeks during the summer.

» May not fit in with some cropping cycles.

» May be difficult for those using a small amount of land intensively.

» Limited number of retail outlets for UV-inhibiting plastics.

» Disposal of plastic may be a problem.

» Large amounts of plastic cannot currently be recycled.

» Some pests are not controlled or are difficult to control.

» No pest control in the furrows between strips (if applied in strip coverage).

» High winds and animals may tear the plastic.

3.6. NEMATODE MANAGEMENT

Plant-parasitic nematodes are killed by high temperatures. Early research on *Meloidogyne javanica* (root-knot nematode) showed that movement of juveniles stopped immediately after being exposed to 50°C and there was no recovery even after returning the temperature to 25°C. Lowering the temperature resulted in longer periods required to kill juveniles; at 42°C it required 3 hours (Wallace, 1966). In Tanzania, egg masses of *M. javanica* were buried in soil (15-cm depth) and exposed to solarization. Within two to three weeks, all eggs were dead. Soil temperatures under solarization reached an average of 43°C, with maximum of 45°C (Madulu and Trudgill, 1994).

  Solarization can be effective in Florida, but may be impaired by overcast skies and rainfall during the warmest summer months. In order to counteract these effects, the solarization time could be prolonged. Increasing the length of solarization will cause mortality based on the accumulation of sub-lethal, but detrimental temperatures. McGovern *et al.* (2000) solarized soil for 41 days mainly during October in southwest Florida, which resulted in the reduction of awl (*Dolichodorus heterocephalus*) and stubby-root (*Paratrichodorus minor*) nematodes. Nematodes were exposed to average maximum temperatures of 38.4, 33.6, and 29.8°C at depths of 5, 15, and 23 cm. A recent laboratory study confirms that nematodes can be affected by an accumulation of sub-lethal temperatures (Wang and McSorley, 2008). Eggs and juveniles of *Meloidogyne incognita* were exposed to a series of temperatures from 38 to 45°C. At 44 and 45°C, juveniles were killed within one hour. For lower sub-lethal temperatures, the hours required for suppression decreased with increasing temperatures (Table 3.1). Success of solarization depends on maximum temperatures reached in the field. The higher the temperature, the shorter the duration of solarization required to kill nematodes. Temperatures over 40°C should be the goal in order to shorten the solarization period. In a 6-week solarization period (July to August in Florida), temperatures high enough to kill nematodes could be accumulated (Wang and McSorley, 2008).

Table 3.1. Hours needed to kill 100% of *Meloidogyne incognita* **eggs and juveniles.**

Temperature (°C)	Hours to kill 100%	
	Eggs	Juveniles
38	389.8	--
39	164.5	47.9
40	32.9	46.2
41	19.7	17.5
42	13.1	13.8

Several nematodes species were negatively affected by solarization. It was effective in reducing populations of *M. incognita, D.heterocephalus, P. minor, Belonolaimus longicaudatus* (sting), *Criconemella* spp. (ring), and *Roytlenchulus reniformis* (reniform) (Chellemi *et al.*, 1997; McGovern *et al.*, 2002, McSorley and McGovern, 2000, McSorley and Parrado, 1986; Ozores-Hampton *et al.*, 2004). However, Chellemi *et al.* (1997) and Chellemi (2006) reported that they were unable to reduce *Meloidogyne* spp. and *R. reniformis*. In some cases, solarization was able to suppress populations of *M. incognita* or *R. reniformis* initially, but numbers recovered at the end of the cropping season (McSorley and McGovern, 2000; McSorley and Parrado, 1986). In addition, when using strip solarization, it is possible that fast recolonization of plant-parasitic nematodes may occur, because row middles (occupying up to 50% of the field) will remain untreated, although this late-season recovery may not limit overall yield (Chellemi *et al.*, 1997).

Resurgence of certain nematode species may occur to higher levels than before solarization. The stubby-root nematode *P. minor* increased in numbers following three weeks of solarization (McSorley and McGovern, 2000). A possible explanation for resurgence could be that some nematode species have population reservoirs in deeper soil layers that are larger than those found in the upper soil layers. This unusual vertical distribution in soil often occurs with stubby-root nematodes. Although solarization reduces or eliminates these nematodes in the upper layers, recolonization can occur quickly by drawing upon a population pool from deeper soil layers. Similar resurgence of stubby-root nematodes has also been observed when these nematodes are managed with soil fumigation (Weingartner *et al.*, 1983).

Soil solarization can be used to control many species of nematodes (Table 3.2). However, nematodes deeper in the soil profile may survive solarization and damage plants with deep root systems. For example, the southern root-knot nematode, *M. incognita* is unpredictably controlled by soil solarization.

Table 3.2. Nematodes controlled by soil solarization.

Scientific name	Common name
Criconemella xenoplax	Ring nematode
Ditylenchus dipsaci	Stem and bulb nematode
Globodera rostochiensis	Potato cyst nematode
Helicotylenchus digonicus	Spiral nematode
Heterodera schachtii	Sugar beet cyst nematode
Meliodogyne hapla	Northern root knot nematode
M. javanica	Javanese root knot nematode
Paratylenchus hamatus	Pin nematode
Pratylenchus penetrans	Lesion nematode
P. thornei	Lesion nematode
P. vulnus	Lesion nematode
Tylenchulus semipenetrans	Citrus nematode
Xiphinema spp.	Dagger nematode

Solarization may also be a beneficial addition to an integrated nematode control system. For example, excellent control of root knot nematode (*Meloidogyne incognita*) was obtained in the San Joaquin Valley by combining solarization with the application of composted chicken manure (Gamliel and Stapleton, 1993b).

Soil solarization using thin transparent polyethylene mulches for six weeks caused as much as 72.6 and 88.5% decrease in the population densities of *Tylenchorhynchus vulgaris* and *Hoplolaimus indicus*, respectively in nursery beds. Application of Mahua cake before solarization usually caused slightly greater reduction in nematode population (Kamra and Gaur, 1998).

In presence of sufficient moisture, a variety of nematodes (species of *Aphelenchoides*, *Ditylenchus*, and *Anguina*) already accompanying harvested grains cause decay. Such decay, however, can be avoided if seeds are harvested when properly mature and then allowed to dry in the air or are exposed to sun. If the seeds are properly sun dried, the nematode inoculum gets inactivated. Hot air can be used until the desired moisture is reached (about 12% moisture) before storage. Subsequently, they are stored under conditions of ventilation, which do not allow build-up of moisture.

3.7. CASE STUDIES

3.7.1. Fruit crops

3.7.1.1. Banana: Covering soil (mulching) with black polythene at 20 per cent moisture depletion recorded highest yield with low population of *R. similis* (Vadivelu *et al.*, 1987). Covering soil with black polyethylene, sugarcane trash and banana trash are reported to reduce *R. similis* and *Pratylenchus coffeae* in banana roots (Bhattacharya and Rao, 1984).

Solarization of banana roots wrapped with plastic (low-density, 25-μm-thick, UV-stabilized, polyethylene mulch) suppressed nematodes (burrowing, root-knot, and spiral) infecting banana collected from the field (Fig. 3.2). There was a clear trend that 1.5 hours of solarization treatment could suppress the plant-parasitic nematodes present significantly as compared to the control (Table 3.3).

Fig. 3.2. Solarization of banana roots infected with nematodes by wrapping in transparent plastic sheet. Control is wrapped in black plastic sheet.

Table 3.3. Management of nematodes using solarisation of banana roots

Treatment	Number of nematodes/g of roots		
	Root-knot	**Spiral**	**Burrowing**
Control	446	164	93
Solarization	0	9	0

3.7.1.2. Citrus: Soil solarization is one of the effective ways of suppressing root-knot nematode populations (*Meloidogyne javanica, M. indica*) on citrus and can be employed mostly during hot weather days. The nursery area should be plowed well and leveled after breaking the clods. The prepared field may be covered with polythene sheet of 100 gauze thickness for about 7-10 days before sowing.

3.7.1.3. Peach: A 37-100% reduction in population of *Mesocriconema xenoplax* and other nematodes on peach has been achieved by soil solarization in South Africa (Barberecheck and Broembsen, 1986).

3.7.1.4. Strawberry: Solarization gave good control of nematodes (*Meloidogyne, Pratylenchus. Helicotylenchus, Tylenchorhynchus* and *Hoplolaimus* spp.) in strawberry, compared with non-solarized plots (Table 3.4).

Table 3.4. Effect of solarization for nematode control on strawberry at different localities

Localities	Solarized			Non-solarized		
	Av. no. of nemas/200 g soil		Reproduction factor (RF) RF = Pf/Pi	Av. no. of nemas/200 g soil		Reproduction factor (RF) RF = Pf/Pi
	(Pi)	**(Pf)**		**(Pi)**	**(Pf)**	
Beni-Sweif	80	3	0.04	115	130	1.13
Fayoum	125	0	0	170	135	0.80
Qalyobia	270	0	0	100	107	1.07
Ismailia	75	0	0	90	136	1.51

Pi = Initial nematode population; Pf = Final nematode population.

3.7.2. Vegetable crops

3.7.2.1. Tomato: The effectiveness of solarization, a soil disinfection technique that uses passive solar heating, to control the incidence of root-knot nematode on tomato under closed plastic greenhouse condition was accomplished by the application of 0.05-0.06 mm clear polyethylene sheets to moist soil for 4-week during the hot season which induced an increase of temperature. Maximum soil temperatures achieved under solarization were 51 °C in the alluvial soil and 52.4 °C in the heavy clay soil type at 20 cm. depth. In the non-solarization unheated greenhouse soil temperature achieved was 39 °C (Fig. 3.3).

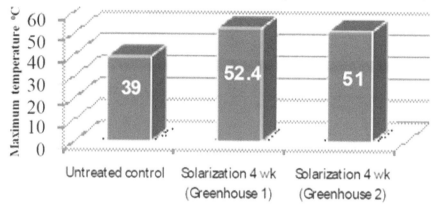

Fig. 3.3. Maximum temperatures attained during soil solarization.

The effect of soil solarization treatments on the population densities of the root-knot nematode, *M. incognita* on tomato was very clear. The larval density per 100 ml of soil was markedly reduced compared with the untreated control (Fig. 3.4). The soil solarization method had a significant influence to reduce the severity of root damage. Root galling index is the most important symptom for the root-knot nematode. This index was dramatically reduced in both solarized greenhouses compared with untreated control where the highest root galling index was found (Fig. 3.5).

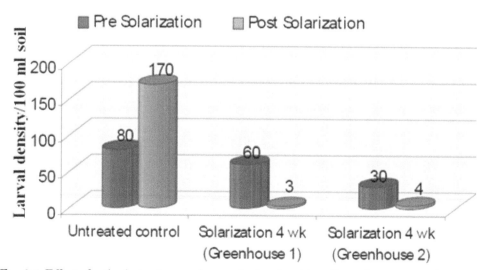

Fig. 3.4. Effect of soil solarization on the population densities of root-knot nematode in soil.

The growth of tomato plants, as indicated by visual assessment and plant height measurements was significantly improved by the soil solarization. Tomato fruit yields in solarized greenhouses were four-fold higher than those obtained in untreated control (Fig. 3.6). Such significant increases were consistent with effective control of nematode provided by solarization.

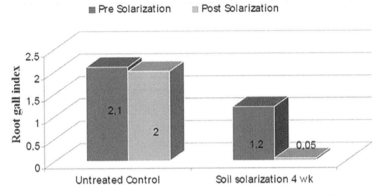

Fig. 3.5. Effect of soil solarization on root-knot nematode gall index.

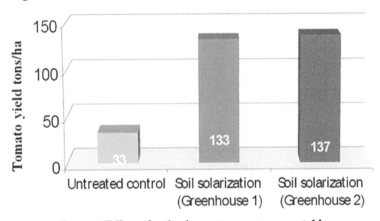

Fig. 3.6. Effect of soil solarization on tomato yield.

Solarization is also known to improve plant growth and yield through the release of nutrient induced by high soil temperatures. The soil solarization treatment also reduced nematode populations more than 90% in lettuce. The yield of lettuce was about two-fold higher in solarized soil as compared with untreated control in several experimental plots.

Soil solarization should be done during the hottest period of the year. It offers a satisfactory and environmentally friendly solution for the control of root-knot nematode. This method is easily adopted in organic, conventional, and integrated control growing system.

Soil solarization of tomato nursery beds with 100 gauge LLDPE clear plastic film during summer (in between April 15th and June 15th) for 15 days is recommended for the management of root-knot nematodes. It increased the number of transplantable seedlings by 615% giving net profit of Rs.7,661 from 1000 sq. m. (ICBR 1: 5.65) and decreased the root-knot disease and weeds by 66% and 93%, respectively (Table 3.5).

Table 3.5. Effect of soil solarization of nursery beds on root galling and number of transplantable seedlings of tomato in 1000 sq. m.

Treatment	No. of seedlings	Weight of 100 seedlings	Root-knot index	Net profit	Benefit: Cost ratio
With plastic	2,60,400	36 g	1.37	Rs. 7,661	5.65
Without plastic	36,400	26 g	3.97	---	---
% Increase (+)/decrease (-) over control	(+) 615	(+) 38	(-) 66	---	---

Red plastic mulch suppresses root-knot nematode damage in tomatoes by diverting resources away from the roots (and nematodes) and into foliage and fruits (Adams, 1997).

Tomato roots were heavily galled and rotting by the end of the growing season in control plots, while they were nearly free of nematodes in plots treated with methyl bromide. In solarized plots root infestation was intermediate because of the late nematode attack, but no rotting of the root was observed (Cartia *et al.*, 1988). Moreover, alternating methyl bromide with solarization treatments inhibited root infestation as did soil treated with methyl bromide for two consecutive years. In the same solarized plots, yield increases averaged 60 percent over the control and were similar to those obtained with methyl bromide treatments.

The reproduction factor in the solarized treatment was 0 on tomato plants for *Meloidogyne* spp., *Tylenchorhynchus* spp., and *Trichodorus* spp. The corresponding RF value in non-solarized soil was 0.89 on the same crop and the same nematodes (Table 3.6).

Table 3.6. Effect of solarization on the control of nematodes on tomato

Treatment	Solarized		
	Average number of nematodes per 200 g soil		Reproduction factor (RF) RF = Pf/Pi
	(Pi)	(Pf)	
Tomato - solarized	80	0	0
Tomato - non - solarized	156	140	0.89

Pi = Initial nematode population; Pf = Final nematode population.

In Japan and other Asian countries, several farmers growing successive crops, such as tomato susceptible to root-knot nematodes use solarization in plastic tunnels for 30 days in summer as an alternative to methyl bromide fumigation (Sano, 2002).

Plowing during summer not only leads to disturbance and instability in nematode community but also causes their mortality by exposing them to solar heat and desiccation. Three summer plowings each at 10 days interval during June (average atmospheric temperature ranging between 40-46ºC) at Hisar (Haryana) led to 96.5 % reduction in *M. javanica* population (Jain and Bhatti, 1987). Additional use of plastic sheets for covering soil either in nursery beds or in field further enhanced nematode reduction. Such an approach also helps in reducing the intensity of weeds, fungi and bacteria in the soil.

3.7.2.2. Brinjal: Generally, 2 to 3 summer ploughings each at 10 days interval during April-May (40-46ºC) have been recorded to reduce 96.5 % *M. javanica* population. Additional use of plastic sheets for covering soil either in nursery beds or in field further enhanced nematode reduction.

The maximum reduction in *M. javanica* population and increase in fruit yield was recorded in brinjal by summer plowing + covering soil with polythene sheet, followed by summer plowing + exposure to sun (Anon, 1989).

3.7.2.3. Bell pepper: Integrating summer plowing with soil solarization with polythene mulching effectively reduced *M. javanica* population in brinjal (Jain and Gupta, 1991) (Table 3.7).

Table 3.7. Management of root-knot nematodes by summer plowing/fallowing and nursery bed treatment in capsicum.

Treatment	Redn. in nema popn. (%)		Yield in kg/m^2	
	Hisar	Vellayani	Hisar	Vellayani
Plowing + Fallowing for 15 days	87.00	66.20	28.98	---
Plowing + Covering with polythene sheet for 15 days	94.90	77.00	18.84	78.00
No plowing + No covering	77.60	---	---	---

3.7.2.4. Onion: solarization resulted in reductions of nematode populations. Data in Table 3.8 show that solarization gave good control of nematodes. The RF in solarized treatments was 0, while in non-solarized plots, reductions ranged from 1.18 to 1.57.

Table 3.8. Effect of solarization on the control of nematodes on onion

Treatment	Average number of nematodes per 200 g soil*		Reproduction factor (RF) RF = Pf/Pi
	(Pi)	(Pf)	
Onion nursery - solarized	235	0	0
Onion field - solarized	65	0	0
Onion nursery - non - solarized	165	260	1.57
Onion field - non-solarized	125	147	1.18

* Pi = Initial nematode population; Pf = Final nematode population.

3.7.2.5. Garlic: In fields heavily infested with *D. dipsaci* in Israel, soil solarization protected garlic bulbs throughout the growing season resulting in a greater yield increase compared with EDB and methyl bromide applications (Siti *et al.*, 1982). In Italy, control of *D. dipsaci* in a sandy soil increased with solarization periods and only 10, 6 and 2 percent nematodes were still viable after four, six, and eight weeks of solarization, respectively (Greco *et al.*, 1985).

3.7.2.6. Beans: Root-knot nematodes can be effectively controlled by a 4-8 weeks solarization, assuming that the land will remain uncropped during summer. Data in Table 3.9 show that solarization gave good control of nematodes. The RF in solarized treatment was 0.47, while in non-solarized plots was 0.98.

Table 3.9. Effect of solarization on the control of nematodes on broad bean

Treatment	Average number of nematodes per 200 g soil*		Reproduction factor (RF) RF = Pf/Pi
	(Pi)	**(Pf)**	
Bean - solarized	85	40	0.47
Bean - non-solarized	95	93	0.98

* Pi = Initial nematode population; Pf = Final nematode population.

3.7.2.7. Melon: In Japan and other Asian countries, several farmers growing successive crops, such as melon susceptible to root-knot nematodes use solarization in plastic tunnels for 30 days in summer as an alternative to methyl bromide fumigation (Sano, 2002).

3.7.3. Plantation crops

3.7.3.1. Tea: Soil solarization in nurseries is adequate to manage root-knot (*Meloidogyne brevicauda*) and other nematodes as the plants develop complete resistance by about 9-15 months (Gnanapragasam *et al.*, 1989).

3.7.3.2. Betel vine: White polyethylene mulching for 15 days on beds prepared for planting betel vine revealed high reduction in nematode populations due to increased soil temperature of 44.1°C (Sivakumar and Marimuthu, 1987). Summer plowing, summer plowing along with solarization have been found beneficial.

3.7.4. Spice crops

3.7.4.1. Cardamom: Soil solarization for 40-45 days using 400 gauge transparent LDPE was proved ideal pre-sowing treatment for the management of nematodes and other soil-borne diseases in cardamom nurseries. Mean soil temperature during the solarization period increased by 8.7°C in solarized beds and the germination was enhanced by 25.5%, while weed growth was suppressed by 82%. *Pythium vexans* was totally eliminated, while populations of *Rhizoctonia solani*, *Phyllosticta elettaria* and nematodes were suppressed to varying levels. Solarization was also found to stimulate the growth of cardamom seedlings (Eapen, 1995).

3.7.4.2. Mustard (Rape seed): Table 3.10 clearly shows the effect of solarization on the control of nematodes on rape with RF's of 0.14 for *Tylenchorhynchus* and *Trichodorus* spp. The corresponding RF values in non-solarized soil was 1.45 on the same crops and the same nematodes.

Table 3.10. Effect of solarization on the control of nematodes on rape seed

Treatment	Av. no. of nematodes/200 g soil*		Reproduction factor (RF) RF = Pf/Pi
	(Pi)	**(Pf)**	
Rape - solarized	285	180	0.14
Rape - non-solarized	1980	2880	1.45

Pi = Initial nematode population; Pf = Final nematode population.

3.7.5. Field crops

3.7.5.1. Wheat: Deep summer plowing (2-3 times) of the infested field during May-June to expose cysts to harsh sunlight that in turn kills the cysts thereby reduces the population to 40% in each season.

Summer plowing reduces nematode population to a great extent (up to 42.4%) in the population of *H. avenae* and consequent yield increase by up to 97.5% of wheat in Rajasthan.

3.7.5.2. Rice: Sun drying of seeds for 6 hours on 4 consecutive days was found effective against seed-borne infection of *A. besseyi* on paddy seeds.

Management of root-knot nematodes using soil solarization in different crops is presented in Table 3.11.

Table 3.11. Management of root-knot nematodes in different crops using soil solarization

Crop	Root-knot nematode species	Duration	Results
Citrus	*Meloidogyne javanica, M. indica*	7-10 days	Suppress nematode population
Tomato nursery	*Meloidogyne* spp.	15 days	Decreased root-knot disease and weeds by 66% and 93%
French bean	*Meloidogyne* spp.	4-8 weeks	Nematodes effectively controlled
Cardamom nursery	*Meloidogyne* spp.	40-45 days	Nematodes suppressed
Betel vine	*Meloidogyne incognita*	15 days	High reduction in nematode population
Turmeric	*Meloidogyne* spp.	40 days	Reduce nematode population

3.8. INTEGRATION OF SOLARIZATION WITH OTHER MANAGEMENT METHODS

Under conducive conditions and proper use, solarization can provide excellent control of soil-borne pathogens in the field, greenhouse, nursery, and home garden. However, under marginal environmental conditions, with thermotolerant pest organisms or those distributed deeply in soil, or to minimize treatment duration, it is often desirable to combine solarization with other appropriate pest management techniques in an integrated pest management approach to improve the overall efficacy of treatment (Stapleton, 1997). Solarization is compatible with numerous other methods of physical, chemical, and biological pest management. This is not to say that solarization is always improved by combining with other methods.

Combining soil solarization with pesticides, organic fertilizers, and biological control agents has led to improved control of pathogens, nematodes, and weeds and may be especially useful in cooler areas, against heat-tolerant organisms, or to increase the long-term benefits of solarization.

3.8.1. Solarization and biofumigation

The primary focus of solarization research will be towards the indirect effects of solarization and the effectiveness of combining it with other pest management strategies such as biofumigation. Katan and Devay (1991) states that microbial processes, induced by solarization, may contribute to disease control beyond the physical effect of heating. Both the vulnerability of soil-borne pathogens to mycoparasitic attack and the activity of beneficial microorganisms could be enhanced by solarization.

Organic amendments may increase the effectiveness of solarization against pests (Gamliel and Stapleton 1993a, Ozores-Hampton *et al* 2004, Wang *et al.* 2008). Wang *et al.* (2008) showed that following a cowpea cover crop with solarization was more effective than solarization alone. In fact, the effectiveness of this combined treatment was comparable to methyl bromide fumigation.

3.8.2. Solarization and chemical controls

Many field trials have shown that, under the prevailing conditions, pesticidal efficacy of solarization or another management strategy alone could not be improved upon by combining the treatments (Stapleton and DeVay, 1995). However, even in such cases, combination of solarization with a low dose of an appropriate pesticide may provide the benefit of a more predictable treatment which is sought by commercial users. For example, although combining solarization with a partial dose of 1, 3-dichloropropene did not statistically improve control of northern root-knot nematode (*Meloidogyne hapla*) over either treatment alone; it did reduce recoverable numbers of the pest to near undetectable levels to a soil depth of 46 cm.

3.8.3. Solarization, amendments and fertilizers

Solarization can also be combined with a wide range of organic amendments, such as composts, crop residues, green manures, and animal manures and inorganic fertilizers to increase the pesticidal effect of the combined treatments (Ramirez-Villapudua and Munnecke 1987; Gamliel and Stapleton 1993a, b; Chellemi *et al.* 1997). Incorporation of these organic materials by themselves may act to reduce number of soil-borne pests in soil by altering the composition of the resident microbiota, or of the soil physical environment (biofumigation). Combining these materials with solarization can sometimes greatly increase the biocidal activity of the amendments. However, this appears to be an inconsistent phenomenon, and such effects should not be generalized without first conducting confirmatory research. The concentrations of many volatile compounds emanating from decomposing organic materials into the soil atmosphere have been shown to be significantly higher when solarized (Gamliel and Stapleton 1993b).

Many commercial users of solarization in California apply manures or other amendments to soil before laying the plastic. There is evidence that these materials release volatile compounds in the soil that kill pests and help stimulate the growth of beneficial soil organisms. For example, the southern root-knot nematode, which was incompletely controlled in lettuce by either solarization or application of composted chicken manure, was completely controlled by combining the two, resulting in a large yield increase (Gamliel and Stapleton 1993a).

3.8.4. Solarization and biological controls

The successful addition of biological control agents to soil before, during, or after the solarization process in order to obtain increased and persistent pesticidal efficacy has long been sought after by researchers. There have been great hopes of adding specific antagonistic and/or plant growth promoting microorganisms to solarized soil, either by inundative release or with transplants or other propagative material, to establish a long-term disease-suppressive effect to subsequently planted crops (Katan 1987; Stapleton and DeVay 1995).

It is considered that by combining solarization with biofumigation and composted amendments the re-introduction of biocontrol agents such as *Trichoderma* spp. and *Bacillus* spp. may be more effective than either treatment alone in controlling soil-borne diseases. Populations of these two microbial antagonists increase relative to other microorganisms in solarized soil.

3.9. MECHANISMS OF ACTION

The mechanisms involved in soil solarization for the management of plant nematodes are as follows:

» Accumulation of heat due to transmission of short-wave solar radiation and prevention of loss of long wave radiation in solarized soil.

» Increase in temperature due to greenhouse effect.

» Soil moisture helps in solarization process by conducting heat energy.

» Increase in microbial and physico-chemical reactions in the soil resulting in accumulations of gases, some being toxic to pathogens and others acting as a nutrient source or induce resistance to subsequent crop.

» Prolonged exposure to higher temperature resulting in increased mortality of nematodes and also making them susceptible to antagonists.

Soil solarization is a method of killing soil pests by heat. This technique consists of covering moist soil with a plastic film during periods of intense sunshine and heat, thereby capturing radiant heat energy from the sun, causing physical, chemical and biological changes in the soil. The soil temperature is increased (by 5-15°C) to levels lethal to many soil-borne fungi, bacteria, nematodes and weeds.

Although the execution of solarization is simple, the overall mode of action can be complex, involving a combination of several interrelated processes which occur in treated soil and result in increased health, growth, yield, and quality of crop plants (Katan, 1987; Stapleton and DeVay, 1995; Stapleton, 1997).

3.9.1. Physical mechanisms

Direct thermal inactivation of soil-borne pathogens and pests is the most obvious and important mechanism of the solarization process. Under suitable conditions, soil undergoing solarization is heated to temperatures which are lethal to many plant pathogens and pests. Thermal inactivation requirements have been experimentally calculated for a number of important plant pathogens and pests (Katan, 1987; Stapleton and DeVay, 1995). Although most mesophilic organisms in soil have thermal damage thresholds beginning around 39-40°C, some thermophilic and thermotolerant organisms can survive temperatures achieved in most types of solarization treatment (Stapleton and DeVay, 1995).

Because solarization is a passive solar process, soil is heated to maximal levels during the day time, then cooled at night. The highest temperatures during solarization are achieved at or near the soil surface, and soil temperature decreases with increasing depth. Typical, diurnal maximum/minimum soil temperatures during summer solarization of open field soils in the inland valleys of California might be 50/37°C at 10 cm, and 43/38°C at 20 cm with 35/20°C air temperature flux. Solarizing soil in closed greenhouses or in containers with a limited volume of soil may produce considerably higher soil temperatures. For example, solarizing soil in 3.8 liter plastic containers resting on steel pallets under low plastic tunnels constructed using two layers of transparent 1 m separated by ca. 23 cm air space ("double tent") gave maximum/minimum soil temperatures of 75/16°C with corresponding air temperatures of 38/17°C (Stapleton *et al.*, 2000). During solarization of open fields or greenhouse floors, destruction of soil-borne pest inoculum usually is greatest near the surface and efficacy decreases with increasing depth (Stapleton, 1997).

There are a number of physical factors which influence the extent of soil heating during solarization. First, solarization is dependent on high levels of solar energy, as influenced by both climate and weather. Cloud cover, cool air temperatures, and precipitation events during the treatment period will reduce solarization efficiency (Chellemi *et al.*, 1997). Solarization is commercially practiced mainly in areas with Mediterranean, desert, and tropical climates which are characterized by high summer air temperatures. In order to maximize solar heating of soil, transparent plastic film is most commonly used for solarization. Transparent film allows passage of solar energy into the soil, where it is converted into longer wavelength infrared energy. This long wave energy is trapped beneath the film, creating a "greenhouse effect". Opaque black plastic, on the other hand, does not permit passage of most solar radiation. Rather, it acts as a "black body" which absorbs incoming solar energy. A small portion of the energy is conducted into soil, but most of the solar energy is lost by reradiation into the atmosphere. Nevertheless, solarization with black, or other colors of plastic mulch is sometimes practiced under special conditions (Abu-Gharbieh *et al.*, 1991; Stapleton, 1997).

The transparent film used for mulching transmits the visible spectrum of short wave length of the solar radiation but prevents the long wave infra-red radiation from the soil. The passage of light through a semitransparent body depends on the characteristics of the material that constitutes it and on the angle of the ray, i.e. the angle formed by the luminous ray and the perpendicular to the surface of the point of penetration. When both direct and diffuse solar radiation strike the plastic tarp on the soil, they are in part reflected back towards the atmosphere, in part adsorbed by the plastic material and in part penetrate the underlying soil. The short wave incident solar radiation penetrates the polyethylene sheet but the long wave radiation is prevented from soil, thus trapping the heat resulting in thermal inactivation and production of heat shock proteins and irreversible heat injury (Fig. 3.7). The material used for tarping is therefore, the main element in the phase of capturing and storing the solar energy during the day, but it also acts as a barrier to dispersion of the heat energy during the night when the flux of solar energy has ceased.

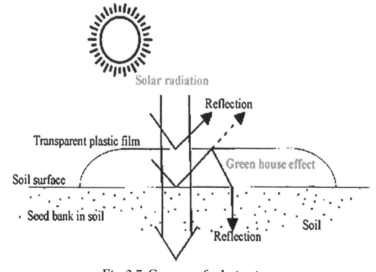

Fig. 3.7. Concept of solarization

The transfer of energy from the soil depends on its internal conductivity which is a function of the soil type and its humidity content. The choice of the optimal characteristics of thermic films aim to have a transmittance of ≥80% in the visible spectrum and ≥25% in the long IR. To protect the quality of plastics for agricultural use the Italian Institute of Plastics has introduced "marchio di qualita" which ensures the incorporation of characteristics mentioned in Table 3.12. Hence, the PVC films are most efficient in solarization but its greenhouse effect depends on the percentage of vinyl acetate which varies between 12-14% in the films available in the market.

Table 3.12. Characteristics of polyethylene types used for soil solarization.

Type of polyethylene	Total luminous transmittance	Greenhouse effect of the medium-long IR radiations
Low density polyethylene (LDPE)	88-90%	24-38%
Polyvinyl chloride (PVC)	90%	< 15
Copolymer of ethylene vinyl acetate (EVA)	80-90%	< 25

Apart from solar irradiation intensity, air temperature, and plastic film color, other factors play roles in determining the extent of soil heating via solarization. These include soil moisture and humidity at the soil/tarp interface, properties of the plastic, soil properties, color and tilth, and wind conditions. The procedure of covering of very moist soil with plastic film to produce microaerobic or anaerobic soil conditions, but without lethal solar heating, can by itself produce varying degrees of soil disinfestation (Katan, 1987; Stapleton and DeVay, 1995).

3.9.2. Chemical mechanisms

In addition to direct physical destruction of soil-borne pest inoculum, other changes to the physical soil environment occur during solarization. Among the most striking of these is the increase in concentration of soluble mineral nutrients commonly observed following treatment. For example, the concentrations of ammonium- and nitrate- nitrogen are consistently increased across a range of soil types after solarization. Results of a study in California showed that in soil types ranging from loamy sand to silty clay, NH_4-N and NO_3-N concentration in the top 15 cm soil depth increased 26-177 kg/ha (Katan, 1987; Stapleton and DeVay, 1995). Concentrations of other soluble mineral nutrients, including calcium, magnesium, phosphorus, potassium, and others also sometimes increased, but less consistently. Increases in available mineral nutrients in soil can play a major role in the effect of solarization, leading to increased plant health and growth, and reduced fertilization requirements. Increases in some of the mineral nutrient concentrations can be attributed to decomposition of organic components of soil during treatment, while other minerals, such as potassium, may be virtually cooked off the mineral soil particles undergoing solarization. Improved mineral nutrition is also often associated with chemical soil fumigation (Chen *et al.*, 1991).

3.9.3. Biological mechanisms

In addition to direct physical and chemical effects, solarization causes important biological changes in treated soils. The destruction of many mesophilic microorganisms during solarization creates a partial "biological vacuum" in which substrate and nutrients in soil are made available for recolonization following treatment (Katan, 1987; Stapleton and DeVay, 1995). Many soil-borne plant parasites and pathogens are not able to compete as successfully for those resources as other microorganisms which are adapted to surviving in the soil environment. This latter group, which includes many antagonists of plant pests, is more likely to survive solarization, or to rapidly colonize the soil substrate made available following treatment. Bacteria including *Bacillus* and *Pseudomonas* spp., fungi such as *Trichoderma*, and some free-living nematodes have been shown to be present in higher numbers than pathogens following solarization. Their enhanced presence may provide a short- or long-term shift in the biological equilibrium in solarized soils which prevents recolonization by pests, and provides a healthier environment for root and overall plant productivity (Katan, 1987; Stapleton and DeVay, 1995; Gamliel and Stapleton, 1993a).

3.10. STRATEGIES TO ENHANCE EFFICACY OF SOIL SOLARIZATION

Strategies are needed for improving the level of pest control and creating a long term effect, lasting throughout a season or over successive seasons. Some of the other reasons for improvements are cost reduction, increasing the reliability and reproducibility of the method, which is climate-dependent; shortening the period during which the soil is occupied with mulch, and making solarization possible for longer periods during the year and more acceptable.

Improvement of solarization can be achieved by either using improved plastic or modifying the application technology. For example, by solarizing shallow layers of growth substrates, the temperature can be increased to very high levels, thus leading to control of even the thermo-tolerant pathogen, *Monosporoascus cannonballus* (Pivonia *et al.*, 2002).

3.10.1. Two transparent films

One way to increase soil heating is the use of double-layer mulch, which heats the soil to higher levels than a single one. Raymundo and Alcazar (1986) achieved an increase of 12.5°C at a depth of 10 cm using a double-layer film compared to a single-layer one (60°C versus 47°C, respectively). Ben Yephet *et al.* (1987) observed a 98% reduction in the viability of *F oxysporum* f. sp. *vasinfectum* after 30 days under the double mulch compared with 58% reduction under single mulch, at a depth of 30 cm. Double-layer films form a static air space under and between the plastic layers. This construction apparently acts as an insulator for heat loss from the soil to atmosphere, especially during the night. The use of a double-layer film also offers opportunities for applying solarization in areas and climatic conditions which are not favorable for solarization when a single layer is used. In Central Italy, in an area which is climatically marginal for solarization, double-film solarization was very effective in reducing the viability of *Pythium, Fusarium* and *Rhizoctonia* in a forest nursery (Annesi and Motta 1994). Solarization in a closed greenhouse is another version of the double-layer film which improved disease control (Garibaldi and Gullino 1991).

3.10.2. Transparent over black double film

Stevens *et al* (1999) reported a 5°C increase in soil temperature when applying solarization in strips in a cloudy climate. Similarly, Arbel *et al.* (2003) achieved an increase in temperature by mulching transparent polyethylene sheet over a layer of sprayable black mulch. In field experiments, they observed that the mortality of resting structures of *F oxysporum, S rolfsii* and *R solani* was higher than in the plots which were solarized by a single plastic layer. Consequently, in the solarized plots with mulching transparent polyethylene sheet over a layer of sprayable black mulch, effective control of Fusarium crown and root rot of tomatoes and vine collapse of melons (caused by *M cannonballus*) was achieved, while solarization with regular films was not effective (Arbel *et al.* 2003).

3.10.3. Improved films

A polyethylene film which was formulated with the addition of anti-drip (AD) components prevents condensation of water droplets on the film surface, leading to a 30% increase in irradiation transmittance over regular film. Soil temperatures under AD films were 2-7°C higher than under regular film. Solarization with AD film in field experiments resulted in effective control of sudden wilt of melons, while solarization with common transparent film had no effect on disease level. Virtually impenetrable films were more effective in raising soil temperatures and killing Fusarium than regular polyethylene (Chellemi *et al.* 1997).

3.10.4. Sprayable films

Sprayable polymers also offer a feasible and cost-effective alternative to plastic tarps for soil heating. The plastic-based polymers are sprayed on the soil surface in the desired quantity and form a membrane film, which can maintain its integrity in soil and elevate soil temperatures. Nevertheless, the formed membrane is porous and allows overhead irrigation. Stapleton and Gamliel (1993) achieved effective soil heating and a reduction in the viability of *Pythium* propagules. In Israel, a sprayable polymer product, 'Ecotex', was developed together with the technology to apply it economically on soils for various purposes (Skutelsky *et al.* 2000). Soil coating with this technology with a black polymer formulation resulted in a membrane film that could raise soil temperatures close to solarization levels. Soil heating with sprayable mulch is faster than that with plastic film, but the soil also cools down to lower temperatures at night. Overall, soil temperatures under sprayable mulch are lower than those obtained under plastic film. The thickness of the sprayed coat is critical to obtaining effective heating (Skutelsky *et al.* 2000). Soil heating using sprayable mulches was effective in controlling Verticillium wilt and scab in potato (Gamliel *et al.* 2001), at a level matching that achieved by solarization using plastic films.

3.11. CONCLUSION

The biggest challenge in crop protection sciences is to effectively control soil-borne pests, while avoiding environmental hazards and degradation of natural resources. Soil solarization is an additional tool for achieving this task, when it is used in appropriate situations. The integration of pest management methods, rather than relying on one powerful control agent, is not only desirable but also the only feasible solution for coping with our need for methods of controlling soil-borne pests. There are many challenges awaiting the further development of soil solarization: improvements in implementation technology and control effectiveness, thereby shortening the mulching period and extending the period for solarization; a better understanding of control mechanisms which can lead to more effective disturbance of pathogens' life cycles.

Soil solarization is a method of soil-disinfestation based on its solar heating by mulching a soil with a transparent polyethylene during the hot season, thereby controlling soil-borne pests including plant parasitic nematodes. Nematode control is attributed to microbial, chemical, and physical processes in addition to the thermal killing. These occur in the soil during the solarization treatment and even after its termination. Frequently, a beneficial microbial shift is created in the solarized soil, resulting in soil suppressiveness. Soil solarization can be combined with other control measures for an integrated approach, thus improving its performance. The uses of soil solarization have expanded beyond soil disinfestation including structure disinfestation, sanitation, controlling pathogens, and more.

Future perspectives for the use of stand-alone solarization will be probably represented by application in greenhouse cropping systems, where high crop values and environmental benefits highly enhance economic convenience of this technique. Based on similar considerations, a great potential for solarization application can also be expected for disinfestation of seedbeds and planting substrates in nurseries (Chaube and Dhananjay, 2003), or for pre-plant disinfestation from nematodes in greenhouse or fruit orchards (Jensen and Buszard, 1988; Stapleton *et al.*, 1989; Duncan *et al.*, 1992; Rieger *et al.*, 2001). Moreover, soil solarization can also be a valuable soil disinfestation tool for irrigated agriculture in field conditions, when specific crop pest problems do not allow use of pesticides.

An interesting approach, suggested by Grinstein and Ausher (1991), is the use of soil solarization as a 'cleaning tool' for infested soils, in the framework of crop rotation. Since solarization frequently has a long-term effect, it can be applied in field every 3-5 years, before the field becomes heavily infested and prior to planting a cash crop, to provide the most benefit from solarization. This procedure also reduces cost of solarization per crop and maintains low infestations, provided that a proper crop rotation is practiced.

Chapter - 4

Biofumigation: Opportunities and Challenges for Nematode Management

4.1. INTRODUCTION

One of the principal strategies used by the growers of high-value horticultural crops to combat soil-borne pests and pathogens (including weed propagules, nematodes, insects, fungi, and bacteria) is pre-plant soil disinfestation, using pesticides or other physical or biological methods. Methyl bromide (MB) is the most important soil fumigant chemical used by growers around the world. It is a broad spectrum pesticide with excellent activity against most potential soil pests. Apart from controlling major soil-borne pathogens and pests, soil fumigation with MB and other fumigants frequently provides increased crop growth and yield responses.

MB was identified as a risk to the stratospheric ozone layer in 1992 and targeted for world-wide phase-out in 1997 by means of the Montreal Protocol, an international treaty. Under the current terms of the agreement and of the federal *Clean Air Act*, pre-plant consumption of MB in the United States is scheduled to be gradually phased out by 2005 (USDA, 2000). The impending loss of MB as a soil fumigant has stimulated intensive efforts to develop and implement suitable replacement strategies. Among the potential alternative control methods being touted to replace methyl bromide is biofumigation that is among the most useful of the non-chemical disinfestation methods.

4.2. WHAT IS BIOFUMIGATION?

In recent years, biofumigation has emerged as an effective non-chemical alternative to manage nematode pests. As originally defined, the term "biofumigation" demonstrates the suppressive effects of Brassicaceae plant family on noxious soil-borne pathogens and is specifically attributed to the release of biocidal isothiocyanates (ITCs) due to the hydrolysis of glucosinolates (GSLs, thioglucosides) present in crop residues, catalysed by myrosinase (MYR, β-thioglucoside glucohydrolase) isoenzymes (Matthiessen and Kirkegaard, 2006). Stapleton *et al.* (1998) defined biofumigation as an agronomic practice that release volatile biotoxic compounds into the soil atmosphere during the decomposition of organic amendments. Cultivation and incorporation of cover/rotation crops, especially Brassicaceae plants, occasionally suppresses soil-borne diseases, including nematodes. Utilization of volatile toxic compounds, such as isothiocyanates generated from the glucosinolates in such crops, for soil-borne disease control is generally termed biofumigation (Kirkegaard *et al.*, 1993).

Biofumigation is a sustainable agronomic practice using naturally produced plant compounds to manage soil pests, including plant-parasitic nematodes (PPNs). This practice primarily relies on volatile organic compounds (VOCs) when they or their by-products are incorporated into soil. Many plants produce VOCs; however, the biofumigation practice is dominated by use of glucosinolates (GSLs) that are hydrolyzed into isothiocyanates (ITCs) capable of killing or driving away PPNs. Biofumigation efforts concentrate on the Brassicaceae family, with mustards and radish at the forefront of success (Daneel *et al.*, 2018). Plant material is grown as a cover crop and tilled into the soil at maturity, where it breaks down to release ITCs. Biofumigation plants or by-products may also be incorporated as seed meal or applied as extractions added to water (Estupiñan-López *et al.*, 2017).

The use of certain crops as biological fumigants ahead of crop production to manage soil-borne pests is receiving considerable interest in recent times. The crops that have shown the potential to serve as biological fumigants include plants in the mustard family (such as mustards, radishes, turnips, and rapeseed) and Sorghum species (Sudan grass, sorghum-Sudan grass hybrids). The crops from the mustard family show some promise to reduce soil-borne pests by releasing naturally occurring compounds called glucosinolates in plant tissue (roots and foliage). When chopped plant tissues are incorporated in soil, they are further broken down by enzymes (myrosinase) to form chemicals (glucosinolates) that behave like fumigants. Isothiocyanates are the breakdown products of glucosinolates, which are the same chemicals that are released from metam-sodium (Vapam) and metam-potassium (K-Pam), commonly used as chemical fumigants.

Biofumigation research focuses on the most economically damaging PPNs, including root-knot nematodes (RKNs) (*Meloidogyne* spp.), cyst nematodes (CNs) (*Heterodera* spp.), and lesion nematodes (LNs) (*Pratylenchus* spp.) (Jones *et al.*, 2013).

4.3. ADVANTAGES

Biofumigation has several potential benefits. The practice offers a range of nematotoxic chemicals that PPN's may not be adapted to resist, with a variety of species and byproducts continuously being expanded by the research community (Daneel *et al.*, 2018). Biofumigation may also lower costs for growers as well as environmental and human health.

» The biofumigation achieved by disking brassica residues into the soil can significantly suppress weeds, nematodes, and soil-borne plant pathogens.

» Brassicas, including numerous mustard species, provide biomass to the soil.

» Biofumigation also improves physical and biological soil characteristics.

» Other benefits of biofumigation include: improved soil texture, increased water holding capacity, and improved soil microbial community structure.

» Brassica cover crops are known to reduce runoff and preserve nitrogen.

» Green manuring with *Brassica juncea* produced a delayed, but remarkable increase of potentially mineralizable Nitrogen.

» Green manures can also increase nutrient availability through weathering of soil mineral components. This weathering may be caused by the production of acids by microorganisms during the decomposition of the green manure.

» Biofumigation is responsible for increased infiltration rate, reduced wind erosion and reduced soil compaction.

4.4. MODES OF UTILIZATION

The modes of application of brassica plants for biofumigation approach include (i) slashing and incorporation of aerial plant parts into the upper layers of moist soil as soil amendments (green manures and seed meal), (ii) poor host status of several species are exploited as cover/rotation crops, and (iii) utilizing processed plant products high in GSLs such as seed meal, or dried plant material treated to preserve isothiocyanates (ITC) activity. In addition, brassica plants have also been used as trap crops (that stimulate hatching/activity of PPN without allowing their reproduction) to manage sugar beet cyst nematode, *Heterodera schachtii* (Smith *et al.*, 2004).

4.4.1. Green manuring

Incorporated biofumigant green manures or plow-downs (Fig. 4.1) can potentially combine the beneficial elements of rotation crops with a more concentrated release of biocidal GSL-hydrolysis products at the time of incorporation. Mojtahedi *et al.* (1993) demonstrated significant suppression of root-knot nematodes and increase in yield (17-25%) on potato by rapeseed (*Brassica napus*) compared with wheat green manures and provided evidence for the role of GSLs.

Fig 4.1. Growing, incorporation and mixing of green manures in soil of plant material using tractor-drawn implements.

4.4.2. Crop rotation/intercropping

Biofumigation by rotation crops or intercrops, where above-ground material is harvested or left to mature above-ground, relies on root exudates of growing plants throughout the season, leaf washings or root and stubble residues. Both GSLs and ITCs have been detected in the rhizosphere of intact plants, and these have been implicated in the suppression of pests and pathogens in both natural and managed ecosystems.

4.4.3. Processed plant products

The seed meal or oil cake by-products which remain after pressing rapeseed or mustard seed for oil constitute a convenient, high GSL material suitable for soil amendment for high value horticultural crops. These products contain sufficient intact myrosinase to ensure effective hydrolysis of the GSLs upon wetting. Brassicaceous seed meals have demonstrated significant suppressive activity in a range of insects, nematodes, fungi and weeds. Both *Rhizoctonia solani* and *Pratylenchus penetrans* responsible for apple replant disease were suppressed by both high and low GSL rapeseed meal (Mazzola *et al.*, 2001).

Compared to green manuring, use of seed meals (residual material post extraction of oil from seed) for PPN management is advantageous because the latter can easily be spread and incorporated into the soil without any risk of frost damage and they do not serve as host to PPNs, which is the case for green manuring/cover crops (Zasada *et al.*, 2009). The defatted meals contain high level of GSL (primarily singrin) that leads to rapid production of corresponding allyl ITC in soil (Leoni *et al.*, 2004).

4.5. BIOFUMIGATION CROPS

Major *Brassica* spp. having biofumigation properties include *B. oleracea* (cole crops: broccoli, cabbage, cauliflower, Brussels sprouts, and kale), *B. napus* (rapeseed and canola), *B. rapa* (turnip), *Raphanus sativus* (radish), *B. campestris* (field mustard), *B. juncea* (Indian mustard), *Sinapis alba* (white/yellow mustard), *B. nigra* (black mustard), *B. carinata* (Ethiopian mustard), *Eruca sativa* (salad rocket) *etc.* (Fig. 4.2 and Table 4.1) (Fourie *et al.*, 2016).

Fig 4.2. Biofumigation crops.

Table 4.1. Brassica plant species used for biofumigation

Species	Selection	Sowing time	Biomass yield (tons/ha)	Contents in GLs (µmol/ g d m)	GLs yield (Moles/ ha)	Main GLs	% N content d m	Biocidal activity
Brassica juncea	ISCI 20	Autumn	93.8±1.1	12.9±0.4	201±10	Singrin	1.5	3
		Spring	46.5±10.6	25.8±0.4	246±32.5		2.0	
Brassica juncea	ISCI 61	Autumn	133.5±23	11±1.1	183.6±26	Singrin	1.4	2
		Spring	60±14.1	17.7±2.3	213±35		2.0	
Brassica juncea	ISCI 99	Autumn	109±16	16.9±0.5	300±49.9	Singrin	1.2	4
		Spring	54.5±9.5	33.6±1.6	333.8±41		1.6	
Rapistrum rugosum	ISCI 4	Autumn	84±33.2	27.6±3.2	562±158	Cheirolin	1.9	4
		Spring	65.3±21.4	24.7±1.0	160.7±45		2.8	
Eruca sativa	Nemat	Autumn	87±19.2	9.4±0.6	186±15	Erucin	1.5	1
		Spring	45.3±4.2	12.9±2.5	107.2±8.5		2.0	
Brassica nigra	ISCI 27	Autumn	102.8±25	17.1±0.3	255.5±8.1	Singrin	0.9	4
		Spring	46±15.5	20.7±0.2	204.5±61		1.5	

4.6. BIOFUMIGATION APPROACHES

4.6.1. Brassica biofumigation

The efficacies for soil-borne nematode disease control were generally variable and inconsistent, a trend which seemed more salient with nematodes than with soil-borne fungi (Matthiessen and Kirkegaard, 2006). *In-vitro* assays revealed that isothiocyanates from Brassicaceae plants were nematicidal to juveniles of *M. incognita* and *Tylenchulus semipenetrans* at concentrations as low as 10 *m*M (Zasada and Ferris, 2003; Lazzeri *et al.*, 2004a). However, several studies have shown that nematode control efficacy and nematode reproduction in treated soils are not correlated with glucosinolate contents in plants (Potter *et al.*, 1998; McLeod and Steele, 1999). Although glucosinolates are thought to play an important role in nematode suppression, non-glucosinolate compounds, such as other sulfur-containing compounds (Bending and Lincoln, 1999), as well as biological and physiological factors, may also be involved (Mazzola *et al.*, 2001). Although the effects of soil type on biofumigation efficacy have not been well studied, sandier soils with low organic matter content appear to allow better performance of the biofumigant.

Before selecting a plant for biofumigation purposes, the host range of the target nematode(s) needs to be checked, because most Brassicaceae used as cover crops are hosts of some important nematodes, such as *Meloidogyne* spp. (McLeod and Steele, 1999), and host Brassicaceae plants may increase nematode populations instead of reducing them. Biofumigant plants that are non-hosts or poor hosts, or are nematode-resistant, are strongly preferred. A study in Australia showed that *M. javanica* reproduced on *B. juncea* and *B. napus* at rates which increased the nematode population in the soil during their growth period in hot seasons; however, nematode proliferation could be prevented if these plants were grown in the winter under low temperatures (Stirling and Stirling, 2003). Use of dried pellets of glucosinolate- containing plants can solve the problem of nematode population buildup in the field, while adding a biofumigation effect (Lazzeri *et al.*, 2004b). Again, to obtain maximum control efficacy from biofumigation, several factors must be considered in addition to host status for nematodes, including: the optimum plant growth stage for soil incorporation (probably when the glucosinolate content reaches maximum levels), biomass of the plant, and adequate soil moisture for plant degradation and gas diffusion into the soil while preventing rapid gas escape from the soil to the atmosphere. Variations in glucosinolate profiles and concentrations exist among Brassicaceae crops and their developmental stages (Bellostas *et al.*, 2007).

An effective way of enhancing the nematode-control efficacy of biofumigation in small areas is to combine the incorporation of Brassicaceae plants with soil tarping using plastic film, which may prevent rapid emission of volatile nematicidal compounds from the soil to the atmosphere, and increase the soil temperature via soil-solarization effect if performed in hot seasons. Elevation in soil temperatures due to soil solarization has been shown to improve the control efficacy of amendments with Brassicaceae crops (Ploeg and Stapleton, 2001). Combinations of sub lethal soil temperatures (30–38.8 °C) and the biofumigation effect (toxic volatile compounds) may have a synergistic effect on nematode-control efficacy. Sub-lethal temperatures may render nematodes more sensitive to toxic compounds or to antagonistic microorganisms. Because nematode control has been obtained by soil solarization plus biofumigation with non-Brassicaceae plants, such as bell pepper plant residues (Piedra Buena *et al.*, 2007), this method does not appear to be specific to glucosinolate-containing plants. The nematode-control efficacy can also be enhanced by the combination of biofumigation with non-Brassicaceae plants (such as bell pepper plant residues).

Different mustards (e.g. *Brassica juncea var. integrifolia* or *Brassica juncea var. juncea*) should be used as intercrop on infested fields. As soon as mustards are flowering, they are mulched and incorporated into the soil. While incorporated plant parts are decomposing in a moist soil, nematicidal compounds of this decomposing process do kill nematodes. Two weeks after incorporating plant material into the soil a new crop can be planted or sown (it takes about two weeks for the plant material to decompose and stop releasing phytotoxic substances i.e., chemicals poisonous to plants).

It is recommended to alternate the use of agricultural residues with green manure, especially from Brassicaceae, using 5-8 kg/m^2 of green matter, although combinations of legumes and grass can be applied. In the case of the use of green manure cultivated in the same field, fast growing plants should be incorporated at least 30 days after having been planted, to avoid the increase of pathogen populations. Planting Brassicaceae after biofumigation can serve as bio-indicators of possible phytotoxicity, because the germination of these seeds is sensitive to phytotoxic substances. At the same time, they are very sensitive to nematodes and permit the detection of areas in the crop where biofumigation is not effective. They act like trap plants, and like biofumigants when incorporated into the soil.

In Spain, successful application of biofumigation was achieved in strawberries, peppers, cucurbits, tomato, Brassicaceae, cut flowers, citrus and banana. Biofumigation has also been recently applied to Swiss chard and carrot crops. The most utilized biofumigants have been goat, sheep and cow manure, and residues from rice, mushroom, Brassicaceae and gardens.

The effectiveness of biofumigation in controlling nematodes is nearly the same as with the use of conventional pesticides. Biofumigation may also regulate viral problems by controlling vector nematodes.

4.6.2. Non-Brassica biofumigation

4.6.2.1. Sorghum and Sudan grass: One important class of secondary metabolite (having biocidal property) produced by the plant is cyanogenic glucoside (Vetter, 2000). Dhurrin is the major cyanogenic glucoside in *Sorghum bicolor*, representing 30% of the dry weight of shoot tips in seedlings. Dhurrin is synthesized from the amino acid L-tyrosine catalysed by cell membrane-anchored enzymes cytochrome P450 and soluble UDP (uridine diphosphate)-glucosyltransferase (Laursen *et al.*, 2016). Similar to GSL-MYR system in cruciferous plants, in intact plant tissues, enzymes and substrate are located in separate cells – dhurrin in vacuole of the epidermal cell and catabolic enzymes in mesophyll cell. Upon disruption of cellular integrity due to biotic invasion, dhurrin is hydrolysed by the endogenous β–d–glucoside glucohydrolase (dhurrinase) to liberate glucose and produce an unstable p-hydroxymandelonitrile, which is quickly converted to toxin hydrogen cyanide (HCN) and p-hydroxybenzaldehyde by enzymatic action of α-hydroxynitrile lyase or at basic pH (De Nicola *et al.*, 2011). The concentration of dhurrin decreases as the plant ages (Bolarinwa *et al.*, 2016). Other crops such as Sudan grass (*Sorghum sudanense*), clover and flax are also known to produce nematotoxic cyanogenic glucosides (Widmer and Abawi, 2002). Cassava (produces linamarin as cyanogenic glucoside) roots were used conventionally to control nematodes in Brazil (Chitwood, 2002). Compared to sorghum, Sudan grass has some desirable characters such as dwarf nature, high palatability and it does not produce toxic compounds that may threaten livestock. Sorghum-Sudan grass hybrid (commonly called Sudex, derived from crosses of *S. bicolor* and *S. sudanense*, produce greater biomass than the either genotypes) is drought tolerant and contains reduced lignin which improves digestibility in animals and increases decomposition rate of plant material (Dover *et al.*, 2004).

Sorghum was used as cover/rotation crop (Nyczepir *et al.*, 1996) and green manure (Nyczepir and Rodriguez-Kabana, 2007) to manage the ring nematode, *Mesocriconema xenoplax* (predisposes to peach tree short life disease) in peach orchard. Considering that *M. xenoplax* is capable of producing cyanide detoxifying enzyme β–cyanoalanine synthase, the nematode suppressive effect of sorghum was embodied to the proliferation of antagonistic microflora *in situ* due to decomposition of sorghum organic matter. The effect of biodisinfestation (polyethylene tarps were used) and urea in suppressing nematodes was also evident in this study (Nyczepir and Rodriguez-Kabana, 2007). The successful biodisinfestation of *M. incognita* by soil amendment of sorghum-Sudan grass hybrid or Sudex in tomato plants was reported by Stapleton *et al.* (2010).

4.6.2.2. Marigold: α-Terthienyl (thiophene – polyacetylenic sulfur compound), a plant secondary metabolite is found in abundance in the roots of marigold (*Tagetes* spp., Asteraceae or Compositae family). Upon photoactivation of α-terthienyl with near ultraviolet light (325–400 nm), reactive oxygen species is produced which is known to be phytotoxic against insects and nematodes both *in vitro* and *in vivo* (Bakker *et al.*, 1979). However, in the rhizosphere, in the absence of light, irradiation of α-terthienyl (essential for nematicidal activity) may not take place. Since only the endoparasitic nematodes are affected due to the growing of *Tagetes* spp., it was assumed that the α-terthienyl is activated inside the living root system *via* other mechanisms than light (Bakker *et al.*, 1979). According to Hooks *et al.* (2010), root peroxidases produced due to nematode attack may activate α-terthienyl *in vivo* and hence *Tagetes* spp. is more effective as cover crop than as soil amendment. In addition to α-terthienyl, plants of Compositae family, including chrysanthemum, safflower *etc.* contain a number of biogenetically related allelopathic compounds such as terthienyl, dithienyl, dithioacetylene and acetylene; and together these compounds may exert the nematotoxic effect (Marahatta *et al.*, 2012).

4.6.2.3. Chinaberry: The extracts of chinaberry (*Melia azedarach*) plant have shown prominent nematicidal effect against *Meloidogyne* spp. in *in vitro*, greenhouse and field experiments in tomato (Ntalli *et al.*, 2010; 2018) and cucumber (Cavoski *et al.*, 2012). The suppression of nematodes by *Melia* fruit powder and water extracts were comparable to a commercial nematicide (oxamyl) treatment. In addition, enhanced plant growth due to triggering of plant defence mechanism and proliferation of soil microbes and microbivorous nematodes were documented (Ntalli *et al.*, 2018). The principal active compound of chinaberry is furfural, which exhibits fumigant nematicidal effect against PPNs (Ntalli *et al.*, 2018).

4.7. MODES OF ACTION

The secondary metabolite GSL (located in cell vacuole) is sequestered within the plant's tissues along with the hydrolysing enzyme myrosinase (MYR) (stored in cell wall or cytoplasm). In native form, sulfur-containing GSLs generally show weak biological activity, whilst due to the rupture or lysis of plant tissue GSL and MYR come together to produce a series of biologically active volatile compounds such as ITCs (in high concentration) (related to the active ingredient in the commercial fumigants Metham sodium and Dazomet, which release methyl isothiocyanate into the soil) (Fig. 4.3), nitriles, epithionitriles and thiocyanates (in low concentration) (Ploeg, 2008). However, exuded GSL could occasionally be hydrolysed by extracellular MYR liberated from *in situ* soil microbes (Ngala *et al.*, 2015).

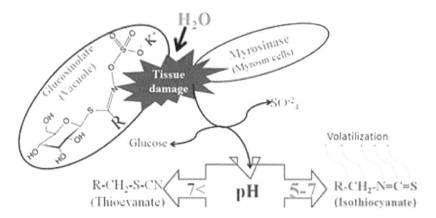

Fig. 4.3. Glucosinolate hydrolysis to release isothiocyanate.

In addition to the GSL hydrolysis products, decomposing brassica tissues produce other volatile sulfur-containing toxins, such as methyl sulfide, dimethyl sulfide, dimethyl disulfide, carbon disulfide, methanethiol *etc.*, which may play significant role in biofumigation process (Lord *et al.*, 2011). The active sites of ITC or other volatiles react with the biological nucleophiles of target nematodes, mainly thiol and amine groups of various enzymes which become irreversibly alkylated (Avato *et al.*, 2013). A mechanism of oxidative DNA damage may also be induced by ITC (Murata *et al.*, 2000). *In vitro* studies have suggested that the nematode motility is significantly inhibited due to the exposure to ITCs derived from brassica plants presumably because the host finding ability of nematodes is crippled. In an isolated study, it was suggested that the exudates of brassica species can potentially reduce the size of dorsal pharyngeal gland nucleus in potato cyst nematode, *Globodera rostochiensis*, ultimately affecting the nematode parasitism in potato (Dossey and Riga, 2010). It was demonstrated that the extracts of brassica leaves can cause significant reduction in the viability of encysted eggs of *G. pallida* (Murata *et al.*, 2000) indicating that the biofumigation effect can penetrate the hard cuticular layer of cysts. The ovicidal effect of brassica has also been demonstrated in root-knot nematode, *Meloidogyne incognita* (Oliveira *et al.*, 2011). The substantial quantities of ITC can be produced using black mustard (*B. nigra*) and Indian mustard (*B. juncea*) (Tollsten and Bergström, 1988) and could be utilized in a biofumigation cropping system.

Fig. 4.4 is a diagrammatic summary of the many ways in which an incorporated Brassica green manure can influence the yield and quality of a subsequent commercial crop. While the primary focus will be on the pathway shown for biofumigation as an isothiocyanate-based concept for suppression of soil-borne nematode diseases.

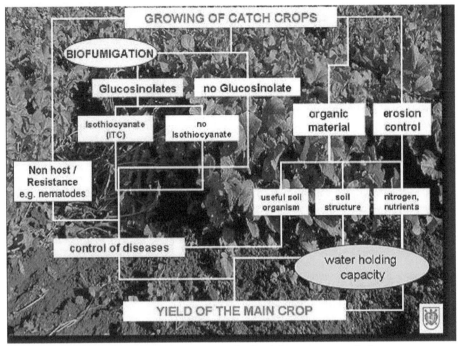

Fig. 4.4. Pathways for control of soil-borne nematode diseases

4.8. NEMATODE MANAGEMENT

Here are some examples of how other brassica crops are being used to manage nematodes:

» The use of oil radish as a green manure has dramatically reduced stubby root nematode (*Trichodorus*) and root lesion nematode (*Pratylenchus*) in Idaho potato fields (Anon, 2001).

» When oil radish is used as a trap crop for the sugar beet cyst nematode, its roots exude chemicals that stimulate hatching of nematode eggs. The larvae that emerge are unable to develop into reproductive females - reducing the population densities for the following crop (Hafez, 1998).

» Plantings of rape or mustard in rotation with strawberries have checked the increase of some nematodes (Brown and Matthew, 1997).

» Rapeseed and Sudan-grass green manures grown prior to potatoes at Prosser, Washington, provided 72 and 86% control of the root-knot nematode in potatoes (Stark, 1995). In the same study, on-farm research in western Idaho showed that rapeseed green manures decreased soil populations of root-lesion nematodes to a greater extent than did Sudan grass green manures. Fall Sudan-grass should be plowed down after it is stressed (i.e., the first frost, stopping irrigation). Winter rapeseed and canola should be incorporated in very early spring (Cardwell and Ingham, 1996).

4.8.1. Root-knot nematodes, *Meloidogyne* spp.

Mustards consistently perform as one of the best biofumigants against root-knot nematodes due to the high level of GSLs produced (Rudolph *et al.*, 2015). *Eruca sativa* and *Raphanus raphanistrum* s.

sp. *sativus* have been shown to be poor hosts. These poor hosts can serve as cover crops, trap crops, or a biofumigant crop. Effective trap crops include many members of the Brassicaceae family. However, not all Brassicaceae species deter RKNs. Many subspecies from the *Brassica juncea* and *B. rapa* species have been shown to serve as good hosts (Daneel *et al.*, 2018). Successful biofumigation of RKNs has been conducted using seed meal from cotton seed, arugula, land cress, and black mustard to impair hatching and reduce mobility (Estupiñan-López *et al.*, 2017).

4.8.1.1. Potato: Soil application of shredded mixed crucifers showed the highest percentage reduction in *M incognita* population of 83.6% in potato crop followed by radish (79.2%), broccoli (74.2%), and cabbage (68.4%). The use of mixed crucifers was recommended because it had the highest percentage reduction in nematode counts both on-station (83.57%) and on-farm (86.7%) and highest marketable yield (11.44 t/ha), and return on investment (ROI) of 192%.

Argula (*E. sativa*) green manure used in combination with reduced rates of 1, 3-D-Telone had significantly lowered PPN numbers such as *M. chitwoodi*, *M. hapla* and *Paratrichodorus allius* in potato (cv. Russet Burbank) rhizosphere compared to Telone at full rate. Additionally, this treatment did not reduce the beneficial free-living nematode populations or the non-pathogenic *Pseudomonas* populations (Riga, 2011)

4.8.1.2. Tomato: *Brassica juncea* Selection ISCI 99 acted mainly as a biofumigant while *Eruca sativa* cv. Nemat revealed an interesting trap crop effect on *M. incognita* infecting tomato. The use of pellets derived from Brassicaceae species, incorporated into the soil before the transplanting of tomato has shown a lower gall index in the roots. The oil extract of Brassicaceae species reduced the juvenile nematode population in the soil during the whole tomato cultivation cycle (Colombo *et al.*, 2008). All the treatments produced a tomato yield significantly higher than with untreated soil.

Kaskavalci *et al.* (2009) showed that grafting the susceptible tomato with a resistant rootstock in combination with broccoli-and *T. erecta*-mediated biofumigation enabled superior *M. incognita* management and increased tomato yield than any of the treatments alone.

4.8.1.3. Carrot: Soil population density of *M. incognita* was significantly lower in the carrot plots treated with green manure or hay of *B. juncea* or with green manure of *E. sativa*, either alone or combined with seed meal pellet, compared with untreated control. Nematicidal effects of biofumigation treatments did not differ from that of Fenamiphos. Moreover, commercial pellet and *B. juncea* green manure and hay resulted also in carrot yield almost double the control, whereas lower increases were provided by *E sativa* green manure, either alone or combined with the pellet.

4.8.1.4. Muskmelon: In muskmelon crop, *E. sativa* green manure resulted in a nematicidal effect (on *M. incognita*) statistically similar to Cadusaphos and higher than *B. juncea* amendment. Muskmelon yield recorded for chemical treatment and *B. juncea* were 79.2 and 70.8 t/ha, respectively, and were 3 times higher than untreated control (De Mastro *et al.*, 2008b).

4.8.1.5. Lettuce: Highest top weight and plant height of lettuce were noted in broccoli- and cabbage-amended soil. Root weight was highest in broccoli- and cauliflower- amended soil. Significantly lower number of galls was recorded in resistant cultivar Great Lakes as compared to that of susceptible cultivar Tyrol. No egg masses were produced in cultivar Great Lakes. Significant reductions in second stage juveniles (*M incognita*) in the soil were found in mustard and radish amendment.

Abawi and Vogel (Ploeg, 2000) achieved significant reduction of *M. hapla* galling and reproduction in lettuce by amending the soil with *T. patula* (cvs. Polynema and Nemagone) plant parts.

4.8.1.6. Aster: In the aster crop, the highest suppression of *M. incognita* population and the lowest gall formation on roots were achieved after *E. sativa* green manure and pellet, alone or in combination, which was also significantly lower than Fenamiphos. A consistent but lower reduction of nematode population was found after *B. juncea* green manure. Control plots showed significantly lower stem length and diameter, number of flowers per stem, and dry matter compared to soil treated with *E. sativa* and *B. juncea* green manures or pellets (De Mastro *et al.*, 2008a).

4.8.2. Cyst nematodes, *Heterodera* spp., *Globodera* spp.

The most promising avenue of biofumigation for the control of cyst nematodes has been achieved using several Brassicaceae species, including raw plant material, extracts, and seed meals (Ngala *et al.*, 2015).

4.8.2.1. Potato: Black mustard, Caliente mustard, marigolds, and Sudan grass have the potential to reduce potato cyst nematode (*Globodera rostochiensis*) by biofumigation. Chopped and incorporated into soil, the green manure releases chemicals that 'fumigate' the soil killing the cysts as well as fungal spores and weed seeds.

The high toxicity of *B. juncea* derived 2-propenyl GSL on encysted eggs of *Globodera pallida* leading to significant hatch inhibition was indicative of the fact that biofumigation effect can pass through the tough cuticular layer of cysts (Brolsma *et al.*, 2014). It was speculated that 2- propenyl GSL had stimulated the expression of heat shock proteins which involves high energy cost, and excessive energy used for heat shock protein expression invariably caused starvation and perturbed hatching in nematodes (Brolsma *et al.*, 2014). A strong positive correlation was observed between biofumigation-induced *G. pallida* mortality and microbial activity indicating the soil microbes playing some role in suppressing *G. pallida* populations during the biofumigation process (Ngala *et al.*, 2015a; b).

4.8.2.2. Sugar beet: The sustainable production of sugar beet in fields infested with the sugar beet cyst nematode (SBCN) (*Heterodera schactii*) was achieved by using Brassicaceous green manure cover crops which act as trap crops for nematodes (Matthiessen and Kirkegaard 2006). Similarly, growing green manure cover crops like fodder radish (*Raphanus sativus*) and white mustard (*Sinapis alba*) preceding sugar beet crop also gave effective management of the sugar beet cyst nematode (Lelivelt and Hoogendoorn 1993). Even though the larvae of *H. schactii* enter roots of sugar beet and develop, but their sexual differentiation is disrupted resulting in very low numbers of females in the following generation, causing a significant decline in the population and reducing infestation of subsequent sugar beet crops.

Following the major breakthrough in 1985 that oil radish variety RSO1841 significantly reduced SBCN population, field studies were conducted with Pagletta, Nemex, and R184 and more reduction in nematode population was achieved. Higher quality of sugar beet crop was generally produced on soils planted with oilseed radish or white mustard residues than with the regular management practices. In the greenhouse studies, it was found that two new oil radish varieties Defender and Comet significantly reduced the population of *H. schachtii* (95%). Studies in 2007 proved that among six varieties of green manure crops, maximum beet yield (t/ha) was from Defender (91), Colonel (89) and Arugula (90) planted pots. At present Defender is the most economical variety highly suitable for the SBCN management (Hafez and Sundararaj, 2004).

4.8.3. Citrus nematode, *Tylenchulus semipenetrans*

Biofumigation with yellow and oriental mustard cover crops reduced citrus nematode (*Tylenchulus semipenetrans*) by 90%. The beneficial fungi such as *Trichoderma* have been found at much higher levels in the mustard treated soils. This does not occur when the soil is fumigated with chemicals. Mustard cover crops are now common place in the Salinas Valley, California. Citrus nematode suppression was 92% greater after oriental and yellow mustard (except yellow mustard/plastic) than after cereal or legume, indicating mustard allelochemicals possibly suppressing nematodes.

In-vitro assays revealed that isothiocyanates from Brassicaceae plants were nematicidal to juveniles of *M. incognita* and *T. semipenetrans* at concentrations as low as 10 μm (Lazzeri *et al.*, 2004a, b).

4.8.4. Grapevine dagger nematode, *Xiphinema index*

Biofumigation may be an alternative for *X. index* control, which can also increase soil biodiversity. Biofumigation, when applied together with one year bare fallow, can be an effective alternative for controlling *X. index*. When nematode problems exist, biofumigation should be carried out immediately after uprooting the vineyard, to prevent their movement towards deep soil horizons (Bello *et al.*, 2004).

Biofumigation using a mixture of sheep and chicken manure, at a proportion of 7: 3, respectively and a dose of 10 kg/m², was effective for controlling *X. index* in loam/sandy soils between 0-60 cm in depth, where the roots normally develop in the beginning of replant. The efficacy of biofumigation in combination with one year bare fallow in controlling nematodes is confirmed for loam/clay/sandy soils (Bello *et al.*, 2004).

4.8.5. Banana nematodes, *Radopholus similis, Helicotylenchus multicinctus, Rotylenchulus reniformis* and *Hoplolaimus indicus*

Considering that multiple applications of nematicide is not economically feasible in perennial crop such as banana, intercropping with *T. erecta* had reduced multiple PPNs including *R. similis, H. multicinctus, R. reniformis* and *H. indicus* (Ploeg, 2000). Since marigold crop residues are often left in the field the interpretation of inter/cover/rotation crop effect and green manure effect has been difficult with marigold.

4.8.6. Lesion nematode, *Pratylenchus penetrans*

4.8.6.1. Strawberry: Marigold as cover/rotation crop provide comparable or better efficacy than chemical fumigants in reducing PPN numbers in soil. For example, greater control of *P. penetrans* was achieved in strawberry (with better fruit yield) while rotating with *Tagetes patula* (cv. Single Gold) than fumigation with metam sodium (Evenhuis *et al.*, 2004). It was suggested that Tagetes may control nematode populations to greater depths than the soil fumigant. Additionally, the possibility of long-lasting toxic effect of crop debris remaining in the soil after Tagetes rotation cannot be ruled out. It is to be noted that the half-life of commonly used soil fumigants is mere 8–14 days depending on the temperature, moisture and presence of microorganisms in soil (Noling, 2003).

4.8.6.2. Tobacco: Marigolds (*T. patula* and *T. erecta*) as rotation crop reduced *P. penetrans* populations below the economic threshold level (compared to chemical fumigation) in a cash crop, tobacco with increased yield (Reynolds *et al.*, 2000).

Brassica and non-brassica plants used for biofumigation to manage PPNs in field are presented in Table 4.2.

Table 4.2. Biofumigation to manage PPNs in field using brassica and non-brassica plants.

Biofumigant crop	Target host crop/ pest	Type of study	Management status	Reference
1. Brassica plants				
S. alba	*Potato- Globodera pallida, G. rostochiensis*	As trap/cover crop	Hatch inhibition of juveniles from cysts	Scholte & Vos, 2000
Broccoli	Tomato - *Meloidogyne incognita*	Fresh residues were added with chicken manure in field	Root galling considerably reduced at high temp. (>25 °C)	Lopez-Perez *et al.*, 2005
B. juncea, B. rapa, Kale *B. napus, R. sativus*	Zucchini squash, Cantaloupe, Tomato - *M. incognita*	Cover crop amendment before planting of host crop	Considerable reduction in nematode population during planting, improved marketable yield	Monfort *et al.*, 2007
B. rapa, B. juncea, B. napus, E. sativa	Tomato – *M. incognita*	Green manure in nursery bed	Reduction in gall number and increase in seedling height in tomato	Randhawa & Sharma, 2007
B. juncea	Carrot - *Heterodera carotae*	Seed meal	Reduction in viable egg nos. in cysts during harvest	Grevsen, 2010
Sinapis alba	*Potato- G. rostochiensis*	Incorporation of Chopped plant parts	Decreased abundance of PPN, increased beneficial nematode community	Valdes *et al.*, 2012
Brassica juncea, S. alba, Eruca sativa	Tomato – *M. javanica*	Incorporation of chopped plant parts prior planting	Significant suppression of *M. javanica*	Kruger *et al.*, 2015
B. carinata	Capsicum – *M. incognita*	Seed meal and sheep manure	Lower root gall index with greater fruit yield	Guerrero-Díaz *et al.*, 2013
Cabbage	Potato, tomato, bell pepper -*Pratylenchus, Meloidogyne, Heterodera*	Incorporation of chopped plant parts prior planting	Significant reduction in juvenile population of several PPNs	Kago *et al.*, 2013
B. juncea, Raphanus sativus	*Potato – G. pallida*	Soil incorporation of chopped biomass in field	Significant increase in mortality of encysted eggs	Ngala *et al.*, 2015
Broccoli	Bell pepper – *M. incognita*	Cover crop in field soil	Considerable reduction in nematode numbers	Rudolph *et al.*, 2015
S. alba, E. sativa	French bean – *M. incognita*	Green manuring	67% reduction in nematode numbers in field	Muiru *et al.*, 2017

B. napus, B. rapa, Kale, B. juncea, R. sativus, S. alba	Grapevine – M. javanica	Chopped leaves incorporated into soil	All green manures substantially lowered nematode numbers	McLeod & Steel, 1999
B. campestris, B. juncea	Orange – Tylenchulus semipenetrans	Soils amended with green manure	Suppressive effect on nematode populations	Walker & Morey, 1999
B. juncea, B. napus, S. alba	Apple – Pratylenchus penetrans	Seed meal amendment	Suppressive effect on nematode populations	Mazzola et al., 2009
B. juncea	Grapevine – M. javanica	Soil incorporation of seed meal & slashed shoots	Many fold reduction in nematode population densities	Rahman & Somers, 2005
2. Non-brassica plants				
Sorghum, Tagetes patula, T. minuta	Tomato – Meloidogyne spp.	Cover/rotation crop	Nematode damage was significantly reduced	Otipa et al., 2009
T. patula	Carrot – M. javanica	Grown in rotation	Reduced nematode number & increased fruit yield	Huang, 1984
T. erecta	Tomato- M. javanica	Intercrop	Reduced nematode number & increased fruit yield	Abid & Maqbool, 1990
T. erecta, Sorghum- Sudan grass	Taro – M. javanica	Cover crop incorporated in soil	Reduced nematode density	Sipes & Arakaki, 1997
T. patula, Sudan grass	Lettuce – M. hapla	Green manure in micro-plots	Reduced root-galling severity & reproduction with increased yield	Abawi & Vogel, 2000
Sudan grass, T. minuta	French bean – M. incognita	Green manuring	Sign. higher crop yield, least nematode infection	Muiru et al., 2017
Sorghum	Peach- Mesocriconema xenoplax	Incorporation of chopped biomass with/without urea & polyethylene tarp	Comparable with MB fumigation in suppressing nematodes in early stage	Nyczepir & Rodriguez- Kabana, 2007
T. patula	Mulberry – M. incognita	Intercrop	Reduced root galls	Govindaiah et al., 1991
T. patula	Pineapple – Rotylenchulus reniformis	Grown in rotation	Nematode control was similar to clean fallow	Ko & Schmitt, 1996
T. patula	Strawberry – Pratylenchus penetrans	Grown in rotation	Reduced nematode density & increased crop yield	Evenhuis et al., 2004

4.9. MAXIMIZING BIOFUMIGATION POTENTIAL

4.9.1. Enhancing GSL profiles

Having established that suppression by incorporated *Brassica* tissues is associated with GSL hydrolysis products, and indications of which hydrolysis products are most toxic, there are significant opportunities to enhance the biofumigation potential of Brassicaceous green manures. Firstly, by selecting brassicas which produce the greatest amount of the GSL-precursors most toxic to the target organisms, and secondly by managing the incorporation process to maximize the exposure of the organisms to the toxic compounds at the most vulnerable stage. Strategies to increase the production of GSL hydrolysis products by brassicas have been summarized elsewhere (Kirkegaard and Sarwar 1998), and rely upon; (i) significant variation in the type and concentration in individual GSLs among different species, cultivars and plant parts, (ii) independent variation in biomass production, and (iii) differential toxicity of different hydrolysis products to particular organisms. Together these represent opportunities to select or develop biofumigant types which may provide up to 10^5-fold increase in biofumigation potential over varieties selected at random.

4.9.2. Improving efficacy in field

The potential efficacy of ITCs present in *Brassica* green manures can be considered by comparison with amounts of the synthetic soil fumigant methyl ITC (MITC) which are applied (517-1294 nmol/g). Assuming a maximum biomass of 15 t/ha, a tissue GSL concentrations of 100 µmole/g and an incorporation depth of 10 cm (soil bulk density 1.4), the potential ITC production (assuming 100% conversion) would be equivalent to 1070 nmol/g, which is in the range of commercial MITC application. Although the efficiency of conversion from incorporated tissues can be as low as 15% (Borek *et al.*, 1995), several ITCs have been shown to be up to 10 times more toxic than MITC. It is difficult however, to predict the impact of the lower concentrations of ITC released over an extended time period relative to commercial fumigant applications. More information is required on the fate, persistence and efficacy of the biocidal compounds released from incorporated tissues.

The impact of the biocidal compounds released from the tissue can also be increased by matching the timing of incorporation and release of biocides to the most vulnerable stages of the pest organism's life-cycle. While sufficient time must be provided for the organic material to decompose to avoid both physical and potential allelopathic interferences in the following crop, delaying for too long may allow some pathogens to recover.

Further studies are in progress to improve our understanding of the accumulation of GSLs in *Brassica* plants, the fate and activity of the biocidal hydrolysis compounds released from incorporated tissue in the soil, and the most effective ways of incorporating biofumigant crops into integrated pest management strategies.

4.9.3. Increasing ITC production using plant stress

Brassicas increase production of ITC in response to plant stress (e.g. nutrient stress, mechanical damage, disease, *etc.*).

Biofumigation is based on the use of glucosinolate-containing plants for the control of soil-borne nematode diseases. Upon tissue damage, glucosinolates are hydrolyzed by endogenous enzymes (myrosinase) and a range of biologically active compounds are formed. Isothiocyanates (ITCs) are the quantitatively dominating products formed at neutral pH. Most of these compounds are volatile and only sparingly soluble in aqueous systems, and depending on the R-group structure and the presence of nucleophiles, further transformation of ITCs occurs. At lower pH and in the presence of certain molecules able to deliver two redox equivalents, the proportion of nitriles increases at the expense of ITC.

The effect of ascorbic acid and glutathione on the production of nitriles at pH 5 was investigated by micellar electro-kinetic capillary chromatography (MECC). The presence of 0.25 μmol ascorbic acid increased the production of nitriles although at higher concentrations the proportion of nitriles decreased. Increasing amounts of GSH favored the production of nitriles (40% of the total degradation products were nitriles in the presence of 2 μmol GSH). The oxidation of GSH gives the redox equivalents needed for the liberation of the sulfur from the unstable intermediate of the glucosinolate hydrolysis leading to the formation of the nitrile.

The following cultural practices/guidelines are recommended in order to maximize the efficacy of biofumigation technique in minimizing the PPNs below the economic threshold level in any cropping system:

>> Brassica crops should be grown in low rainfall areas to ensure the optimal growth of plants until flowering (at this stage plant contains maximum GSL).

>> The plants should be harvested during flowering stage and aerial plant parts should be finely chopped and evenly mixed in the soil profile where maximum number of PPNs occurs.

>> Green manure incorporated soil should be watered to field capacity and covered with mulches to increase temperature and trap VOCs.

>> Knowledge about the GSL content of various commercially available brassica genotypes is necessary before adopting them for biofumigation.

>> Knowledge regarding the target nematode pest and their host status on candidate biofumigant crop is necessary before using brassica plants as cover/rotation crop.

>> The soil incorporation of sorghum/Sudan grass young biomass (1–2 months old) is more effective than older tissue (3–4 months old) because dhurrin content decreases with the maturity and height of the plant.

>> For green manuring, sorghum/Sudan grass must be grown during summer/autumn and chopped before winter because dhurrin content gradually decreases in standing crop due to frost during winter.

>> Marigold should be grown for at least two months prior planting of susceptible crop. Cover/rotation crop is a desirable option than green manuring. For perennial crops, intercropping is a better option.

>> Sealing with water or plastic after incorporation will improve the efficacy (as with all fumigants). Soil conditions should not be overly dry or excessively wet.

4.10. CULTURAL PRACTICES THAT IMPACT THE EFFICACY OF BIOFUMIGATION

Successful biofumigation is complex; the release and breakdown of chemicals must be timed and managed correctly. Management and methods of incorporation depends largely on the specific biofumigant crop and target organism (Akhtar and Malik, 2000). Conditions that may have an impact on efficacy include soil characteristics such as temperature, pH, water content, organic matter, and nutrient availability (Westphal *et al.*, 2017).

4.10.1. Soil characteristics

Soil with a neutral pH, relatively warm temperature, and sufficient moisture will support chemical reactions that result in the highest levels of ITCs (Gimsing and Kirkegaard, 2009; Collange *et al.*, 2011; Omirou *et al.*, 2013). Higher levels of soil water allow for GSLs to spread through soil quickly and facilitate hydrolysis (Omirou *et al.*, 2013). Irrigation of Brassicaceae crops can help them to reach their full potential under low water conditions. Ploeg and Stapleton (2001) found that when adding broccoli (*B. oleracea*) amendments to soil, the level of nematode suppression was very low at 20°C but increased with temperatures up to 35°C. An effective method to raise the temperature of small areas of soil is solarization, i.e., cover the area with plastic to trap solar heat and to confine nematotoxic chemicals to the area (Blok *et al.*, 2000; Oka, 2010).

4.10.2. Method, rate, and timing

The best way to ensure that the maximum levels of ITCs are released into soil is through effective breakdown of plant cells. Plant tissues that are incorporated into soil without first being broken down will release lower levels of ITCs, sometimes less than 1% (Gimsing and Kirkegaard, 2009). Cells must be broken open to release GSLs and begin the chemical process required for biofumigation. This breakdown may be facilitated through mechanical means, or even freezing (Morra and Kirkegaard, 2002). Often, plant tissue is macerated with a slasher or flail chopper and then immediately incorporated into the soil (Kruger *et al.*, 2013).

Speed is an important consideration when incorporating biofumigant materials into soil; fast incorporation traps volatizing compounds in the soil. The highest levels of in-soil GSLs have been recorded within the first 30 min of their addition, despite some levels of GSLs or ITCs remaining present for up to a week (Gimsing and Kirkegaard, 2006; 2009). Zasada and Ferris (2004) observed that biomass providing high enough levels of ITCs reduced nematode population by 65–100%, and that survival was reduced with biomass applications of greater than 2.9% (w/w).

Biofumigant plants contain higher levels of GSLs during specific life stages (Bellostas *et al.*, 2004), while nematode hatch can be dependent upon soil temperature (Kaczmarek *et al.*, 2014). Endoparasitic nematodes spend most of their lives within their host where they are protected and are most susceptible to control measures during their rare stages of infectious movement within the soil. The method and timing of biofumigant crop termination and incorporation impact the amount and quality of secondary metabolites released into the soil (Kruger *et al.*, 2013). For maximum biofumigation effect, crop planting, termination, and incorporation into soil should coincide with the most vulnerable life stage of the target PPN.

Bellostas *et al.* (2004) investigated the levels of GSLs in the cruciferous family during different stages of development and in several sections of the plant. They found that the level of GSLs produced was a trade off with plant biomass. The tissues likely to contain the highest levels of GSLs were reproductive and root tissues, with the highest levels of GSL variation observed in roots (Kruger *et al.*, 2013). This characteristic suggests that the maximum nematotoxic benefits may be achieved through incorporation of mature, whole plant tissue. This would ensure that the most concentrated GSLs are incorporated and the most biomass is utilized.

4.11. INTEGRATED NEMATODE MANAGEMENT

Although a plethora of research have been conducted on using these crops for PPN management, agricultural stakeholders remain to be convinced to integrate biofumigation into their INM practices.

Indeed, it is unlikely that biofumigation as a standalone technique will eliminate target PPNs in soil, but this technique can easily be integrated with other strategies such as soil solarization, minimal use of nematicides, use of resistant varieties *etc.* in a cost-effective manner to provide acceptable levels of nematode control over multiple seasons. For instance, Argula (*E. sativa*) green manure used in combination with reduced rates of 1,3-D-Telone had significantly lowered PPN numbers such as *M. chitwoodi, M. hapla* and *Paratrichodorus allius* in potato (cv. Russet Burbank) rhizosphere compared to Telone at full rate. Additionally, this treatment did not reduce the beneficial free-living nematode populations or the non-pathogenic *Pseudomonas* populations (Riga, 2011). Kaskavalci *et al.* (2009) showed that grafting the susceptible tomato with a resistant rootstock in combination with broccoli- and *T. erecta*-mediated biofumigation enabled superior *M. incognita* management and increased tomato yield than any of the treatments alone. Dazitol (a natural product developed from mustard seeds and contains as high as 4.37% ITC; (Cao *et al.*, 2007) can be considered as an alternative to commercial fumigants to be integrated in INM module. The role of solarization in enhancing biofumigation efficacy has already been discussed in earlier sections.

4.12. CONCLUSION

Biofumigation has the potential to effectively control PPNs, and is a sustainable practice using naturally forming plant chemicals to kill or drive away PPNs. Biofumigation may work as a stand-alone treatment or in combination with other strategies such as sanitation, organic amendments, or solarization (Wang *et al.*, 2006). In particular, several mustards and cole crops from the Brassicaceae family, in addition to other species such as marigold, are successful producers of nematotoxic GSLs and ITCs (Zasada and Ferris, 2004; Daneel *et al.*, 2018). This method may require increased hands-on management for scouting and identification of nematode pests, as well as careful selection of suitable biofumigants and application strategies. However, particularly for systems already employing cover cropping, there is a massive opportunity for a biofumigant cover crop to provide both the inherent benefits of cover cropping and additionally offering pest control (Matthiessen and Kirkegaard, 2006).

For farmers who already apply organic matter or who grow cover crops, the switch to biofumigation to control nematodes may be a sensible one. With careful planning and the tools to manually break-down and incorporate a biofumigant, the method has the potential for high success in managing PPNs.

Although a plethora of research have been conducted on using the biofumigant crops for PPN management, agricultural stakeholders remain to be convinced to integrate biofumigation into their INM practices. Indeed, it is unlikely that biofumigation as a standalone technique will eliminate target PPNs in soil, but this technique can easily be integrated with other strategies such as soil solarization, minimal use of nematicides, use of resistant varieties *etc.* in a cost-effective manner to provide acceptable levels of nematode control over multiple seasons.

Chapter - 5

Precision Agriculture-The-State-of-Art Nematode Management

5.1. INTRODUCTION

Nematodes are not uniformly distributed within fields, and there may be substantial acreages in most fields where nematodes are either not present, or are not an economic concern. Nematode distribution varies significantly throughout a field. Farmers usually apply one rate of a nematicide across a field to control nematodes. Field-wide application results in nematicides being applied to areas without nematodes and the application of sub-effective levels in areas with high nematode densities. Applying a nematicide uniformly over a field can be costly and environmentally questionable. The site-specific nematicide placement is a means of limiting nematicide applications to where they are needed within a field. This system can reduce nematicide usage more than 75% compared to uniform-rate applications while still enhancing crop yields.

The spatial distribution of nematodes is characterized by the presence of circular or oval-shaped infected spots, with marked spatial and temporal variability. The existence of these spots is related to the low mobility capacity of nematodes, which is a few meters in a year, unless they are transported by some disseminating agent, such as agricultural machines. Recent advances in precision agriculture technologies and spatial statistics allow realistic, site-specific estimation of nematode damage to crop plants and provide a platform for the site-specific delivery of nematicides within individual fields. Nematode infestations tend to be spatially clustered within agricultural fields and result in crop yield penalties in some areas but not in others (Evans *et al.*, 2002; Wyse-Pester *et al.*, 2002; Monfort *et al.*, 2007). The relatively high cost of fumigant nematicides, the difficulty in their application, and environmental concerns with using fumigants has encouraged many growers to consider ways to target specific zones within fields for nematicide application to minimize waste and expense (Mueller *et al.*, 2011). The advent of "precision agriculture" technologies provides the possibility of site-specific nematode management rather than the "whole-field" approach that has historically been used. In order for a site-specific nematicide placement strategy to be utilized at the farm level, a clear indication of the potential for profitability as well as the efficacy of this approach must be perceived. Logically, estimation of profitability of site specific nematode management and the development of application recommendations for individual fields must be based on the estimation of yield potential (penalty) where nematodes are yield-limiting. Site specific crop yield data, as with most other agronomic data obtained at high resolutions within fields, are expected to be spatially structured, and failure to properly account for spatial structure of the yield data may result in inefficient parameter estimates for yield response functions.

The role of homogeneous vegetation in pest outbreaks, which can also be brought about by the uniform application of pesticides over large areas, has long been recognized by applied crop protection scientists. The simplification of pest communities, their impoverishment of many natural enemies and reduction of their clustered distribution are also brought about by pesticide applications (Johnson and Tabashnik, 1999). A judicious application of pesticides (insecticides, fungicides, nematicides, and herbicides) at a scale smaller than a field was not possible until recently, due to mechanized large scale crop production systems. The implementation of precision agriculture, also known as precision farming, prescription farming or site-specific farming, has made it possible for fine-tuning of inputs in the field, through the new technologies that have become available in the last decade or so. The site specific pest management systems combined with an understanding of pest population in time and space is suggested as one of the strategies to reduce the negative impact of pesticide use and to avoid the traditional applications based on average pest density (Park *et al.*, 2007).

The success of site-specific nematode management depends on an affordable map of the nematode distribution within a field as the basis for making management decisions. The cost of sampling and making the map must be less than the cost reduction of site-specific management. The cost of sampling may be reduced if nematode density is consistently correlated with a field characteristic that is less expensive to sample. In that case, observations of the characteristic may be substituted for the more expensive observations of nematode density or a map of the correlated characteristic may be used to target nematode sampling to where there is the most uncertainty about the management decision (Dieleman and Mortensen, 1999; Heisel *et al.*, 1999). Field characteristics that have a documented influence on nematode densities and distributions vertically and horizontally over time in a soil profile include soil texture (Georgis and Poinar, 1983; Koenning *et al.*, 1996; Queneherve, 1988), soil nutrients (Francl, 1993; Marshela *et al.*, 1992), and soil moisture (Gorres *et al.*, 1998).

A set of crop protection management practices that vary inputs at the appropriate spatial and temporal scales within a field based on predicted economic and ecological outcomes is called precision crop protection. Unlike most definitions, ecological as well as economic criteria and the temporal scale so important for pest management are included in this definition. The recognition that a great deal of pest variability exists within agricultural fields which results in large yield differences across the field, is the major conceptual novelty of this cultural practice. Hence, application of variable crop protection inputs in different parts of the field in order to optimize the crop response (i.e. yield) is the main goal of precision crop protection.

Optimization of farm profits and minimization of agriculture's impact on the environment is a knowledge-based technical management system that is achieved through site-specific pest management, which has been in the market for more than 30 years with new applications each year. Precision crop protection involves obtaining pest information about a field and updating it continuously to fine tune management strategies. It uses the data to correlate responses from specific pest management decisions for an economic response under those conditions. Precision pest management determines the exact amounts of crop protection inputs (such as pesticides) required for taking management decisions in order to get higher yields and maximum profits.

Based on the spatial variability in nematode occurrence and their habitat, how the efficacy of nematode management measures could be improved and adverse environmental impacts could be minimized is discussed in this chapter. This agro-ecologically based strategy uses new technologies that allow farmers to spatially vary plant protection inputs in the field and that are being adopted in precision nematode management based on the ecological effects of plant diversity and nematicide application on the population dynamics and density of nematodes. Geographic information systems (GIS), global positioning systems (GPS), remote sensing (RS), variable rate technology (VRT), and yield monitors (YM) are some of the new technologies used in precision agriculture for the management of nematode pests.

5.2. PRECISION NEMATODE MANAGEMENT

Uniform application of nematicides across the field and on an area-wide basis brings in habitat homogeneity which is congenial for nematode outbreaks. Similarly, eradication of all non-crop plants in the field and its vicinity by herbicide applications may make it easier for nematodes to locate crop plants and deprives beneficial predators and parasitoids of important food sources (Shelton and Edwards, 1983). Hence, there is a need to improve the efficacy of nematode-control strategies and to minimize adverse environmental impact by responding to the spatial variability in nematode occurrence and habitats. Crop diversity through the use of multi-line cultivars, varietal mixtures and intercropping enhances natural enemy populations that reduce the density of nematode populations through predation, parasitism and competitive interactions. The nematode population outbreaks can be retarded by fine-tuning of nematicide applications that are made only at 'hot-spots' where nematode densities reach action thresholds, which would create a mosaic of communities that differ in species composition and inter-specific interactions. The release of toxic chemicals into the environment would be greatly reduced by agro-ecologically based nematode management strategy i.e., precision nematode management (Weisz et al., 1996; Brenner et al., 1998).

Application of broad-spectrum nematicides is responsible for killing of natural enemies of the target nematode which can also lead to the resurgence of nematode populations (Hardin et al., 1995). Site-specific applications through precision farming would minimize the exposure of predators and parasitoids to nematicides, and favor biological control which subsequently reduces nematode damage. Site-specific nematode management is responsible for the buffering effect of inter-patch dispersal and would make nematode population outbreaks less likely because it links sink and source populations. Sampling procedures are considerably influenced by the alteration of nematode distribution patterns in the field (approaches random or even distribution) requiring less sampling effort.

The environmental nematicide load is significantly reduced by precision farming leading to the reduction of material costs, as the necessary nematicide amount is 8-10% lower (calculated in active ingredient) than in case of traditional treatment. Takács-György et al. (2014) estimated that the amount of pesticides saved on the level of EU-25 countries is 31.7-84.5 thousand tons in case 15% of farms apply precision farming, 63.4-169.1 thousand tons in case 25% of them introduce it, while in the most favorable case (40%) it is 126.8-338.1 thousand tons.

The control with the use of chemical nematicides in the seeding furrow can be done locally and also at variable rates. Localized application will be done only in areas where the nematode population exceeds a certain limit. Variable rate application is based on the fact that the protection time provided by the nematicides is proportional to the application rate, and the tolerance of the plant to the nematodes increases as the size of the root system increases, by a dilution effect of the nematodes. Thus, higher doses are used in sites with larger populations of nematodes to protect plants for a longer period of time until they are able to support nematodes that are not controlled by nematicides. For this type of application, the necessary equipment is readily available in most farms.

The information needed to make soil and crop management decisions that fit the specific conditions found within each field is provided by precision nematode management that combines the best available technologies. It enables to take more informed management decisions by using GPS, GIS, and RS to revolutionize the way data are collected (at resolutions of 1 to 5 m) and analyzed. Precision nematode management has the potential to have detailed records covering every phase of the crop production process, thus enhancing sound nematode management decisions.

5.3. PRECISION TOOLS

The main tools that have recently become available to make precision nematode management a realistic farming practice today include global positioning system (GPS), geographic information system (GIS), remote sensing (RS), variable rate technologies (VRT), and yield monitors (YM) (Fig. 5.1).

5.3.1. Global positioning system (GPS)

GPS is a locating system based on a satellite that identifies an earth-based position using latitude, longitude, and in some cases, elevation (Fig. 5.2). GPS provides continuous position information in real time, while in motion. Having precise location information at any time allows soil and crop measurements to be mapped. GPS receivers, either carried to the field or mounted on implements allow users to return to specific locations to sample or treat those areas. For applications in agriculture, it is used for machine guidance and control (variable-rate-input applications) to provide line-of-sight signals and 24-hour coverage by orbiting 20,200 km above the Earth. The satellites complete orbits in slightly less than 12 hours by travelling in one of six orbital planes.

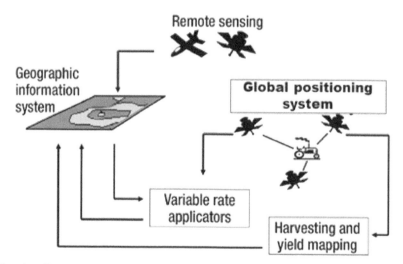

Fig. 5.1. Components of map-based precision nematode management systems.

Fig. 5.2. Global Positioning System (GPS)

5.3.2. Geographic information system (GIS)

GIS is a specifically designed data-management system to store spatial data in order to create variable-intensity maps (Fig. 5.3). GISs are computer hardware and software that use feature attributes and location data to produce maps. An important function of an agricultural GIS is to store layers of information, such as yields, soil survey maps, remotely sensed data, crop scouting reports and soil nutrient levels. For crop production purposes, it collects information pertaining to field history, input operations, GPS-based yield maps and soil surveys, aerial photography, satellite imagery, and nematode scouting data from various sources. The information about the current crop, assess treatments, and to potentially generate projected harvest maps, by collecting input data on weather, nematode densities, seed varieties, and planting populations.

Fig. 5.3. Geographic Information System (GIS)

5.3.3. Remote sensing (RS)

The acts of detection and/or identification of electromagnetic energy (ultra-violet, near infrared, or thermal infrared) phenomena, such as light and heat without having the sensor in direct contact with the object are called remote sensing (RS) (Fig. 5.4) (Frazier *et al.*, 1997). It is the collection of data from a distance. Data sensors can simply be hand-held devices, mounted on aircraft or satellite-based. Remotely-sensed data provide a tool for evaluating crop health. Plant stress related to nematode diseases and other plant health concerns are often easily detected in overhead images. Remote sensing can reveal in-season variability that affects crop yield, and can be timely enough to make management decisions that improve profitability for the current crop. A wide range of potential applications including the detection of crop stress; monitoring variability in crops, soils, nematodes, weeds, insects, and plant disease is provided by remote sensing.

5.3.4. Variable rate technologies (VRT)

The variable rate applicator has three components: control computer, locator and actuator. The application map is loaded into a computer mounted on a variable-rate applicator. The computer uses the application map and a GPS receiver to direct a product-delivery controller that changes the amount and/or kind of product, according to the application map.

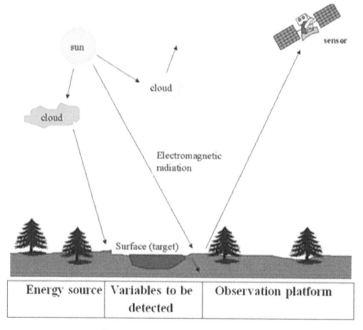

Fig. 5.4. Remote Sensing (RS)

VRT are application equipment which is mounted on fertilizer applicators and sprayers in order to control their delivery rates in different parts of the field based on a decision support system and (or) management plan (Fig. 5.5). All the data collected from various sources like GPS referenced data, RS images, and GIS generated maps, are used to produce a site-specific application plan based on sound agronomic principles.

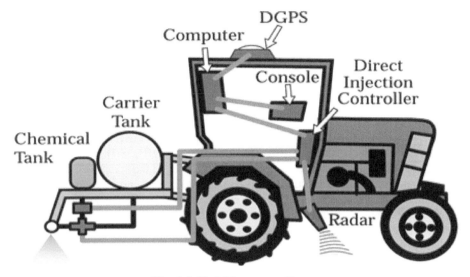

Fig. 5.5. Variable rate applicator

5.3.5. Yield monitors (YM)

Yield monitors continuously measure and record the flow of grain in the clean-grain elevator of a combine harvester. When linked with a GPS receiver, yield monitors can provide data necessary for yield maps. YM sensors are used for monitoring yield during the harvest to quantify yields across the field. The data on crop performance (grain flow, grain moisture, area covered, and location) for a particular year are collected from a yield monitor (an electronic tool), coupled with Global Positioning System (GPS) technology (Fig. 5.6). Yield monitors for commodities like cotton, forage silage, peanuts, and sugar beets are readily available in the market.

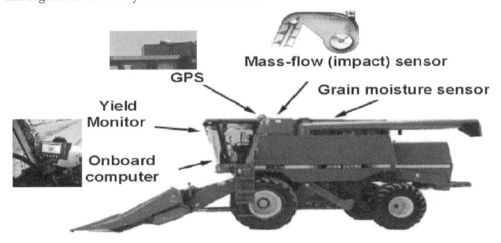

Fig. 5.6. Combine harvester equipped to monitor yield and produce yield maps

5.4. DECISION-SUPPORT SYSTEM

These are map-based methods for the implementation of precision nematode management based on data obtained by grid sampling used to generate a site-specific map, which is then coupled with a variable-rate applicator in the field. The aerial or satellite images [remote sensing (RS)] can also be used to generate such site-specific maps. It is possible to correct deficiencies (nematodes) in specific parts of the field by using various filters and imaging techniques (e.g. infrared photography), in order to detect variations in the health and stand of crop plants.

Data collection, data analysis/ interpretation, and application/variable rate technology are the three management components in precision nematode management (Fig. 5.7). These three components should be combined with a decision-support system (Fig. 5.8) to facilitate the delivery of variable levels of a specific management practice to various parts of the field and in a single field operation. Implementation of management response is based on decision-support system [an interactive software-based system that integrates and organizes all types of useful information (GPS referenced data, RS images, and GIS generated maps), to facilitate variable application rates to various parts of the field in a single operation].

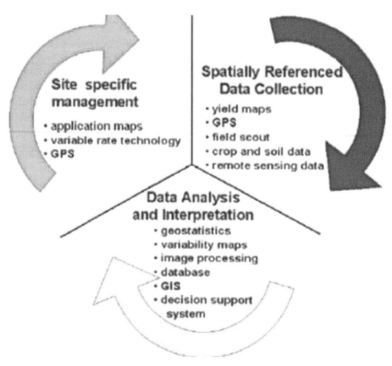

Fig. 5.7. Data processing and management cycles for site-specific management.

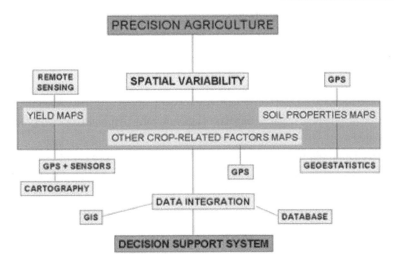

Fig. 5.8. Precision agriculture shown as a set of data maps, geotechnologies and a decision support system.

5.5. BENEFITS OF PRECISION FARMING

The numerous benefits provided by precision pest management are as follows:

» Better scheduling, sequencing of equipment, planning of field operations, equipment movement, *etc.* are possible due to improved equipment efficiency.

» Increased documentation of food safety.

» Monitoring and supervision, including better records of field operations, location of equipment, production output, and employee performance are improved.

» More accurate and precise application of nematicides to reduce the potential for leaching and runoff helps to enhance environmental stewardship. This is more important because environmental stewardship is incorporated throughout the precision nematode management decision support system.

» Production processes, crop conditions, and required inputs are facilitated by improved records.

» Reduced variability in growing conditions, improved varietal choices, crop rotation, *etc.* are responsible for risk reduction.

» The ability to identify, diagnose and communicate crop and field nematode problems is greatly improved.

» The concept of "doing the right thing in the right place at the right time" has a strong intuitive appeal which gives farmers the ability to use all operations and crop inputs more effectively.

» More effective use of inputs results in greater crop yield and/or quality, without polluting the environment.

» Precision agriculture can address both economic and environmental issues that surround production agriculture today.

5.6. NEMATODE MANAGEMENT

The primary objective of precision nematode management is to produce a healthier crop by adjusting needed crop protection inputs within the field rather than at the field level through spot-treatment of only those areas of the field needing nematode control, resulting in the reduction of nematicide costs and environmental degradation. Site-specific management of nematodes is an area that is fast developing. Crops can be managed better, with fewer trips across the field, resulting in more economic returns and reducing potential negative impacts of agricultural activities on the environment by linking soil, crop, nematode, and environmental features into one program.

Integrated nematode management (INM) has been the approach of choice for nematode management for the last three decades, and several principles of INM are highly compatible with the ideology behind precision farming which includes:

» Off-the-farm inputs like nematicide applications to be reduced.

» Use of action thresholds, i.e. taking corrective measures based on economic and ecological criteria.

» Use of cultural and biological control measures, and resistant varieties, to enhance sustainability.

The optimization of inputs such as nematicides, and minimization of economic and environmental damage provided in precision agriculture, is similar to INM. However, the spatial component so central to precision agriculture, is lacking in INM.

The most challenging step toward the use of precision agricultural technologies for nematode management include the creation of management maps and control VR applicators by collecting reliable data on nematode density across the field (Fleischer *et al.*, 1999). In view of technological limitations arising from the cryptic and dynamic nature of nematode pathogens, the application of precision agriculture to nematode management has been a slow process. However, the clustered distribution of nematode pests makes them suitable for management through precision farming. The use of precision-farming methods for nematode management is certainly justified in light of the severe adverse effects of nematicides on the environment, and the potential for drastically reducing the release of toxic chemicals.

The spatial distribution of nematodes is characterized by the presence of circular or oval-shaped infected patches, with marked spatial and temporal variability (Fig. 5.9). The existence of these patches is related to the low mobility capacity of nematodes, which is a few meters in a year, unless they are transported by some disseminating agent, such as agricultural machines. However, temporal variability occurs because plants submitted to a high initial population may have reduced root growth, reducing the supply of food to the nematodes in that place and thus reducing the final population. Places with an intermediate population allow better establishment of the crop, greater supply of food and with that greater nematode population. This makes the spots with larger populations vary from one place to another in each season.

There is no need for application of nematicide across the whole area when only 5 or 10% of the field actually has nematode problems. Recent advances in precision agriculture technologies and spatial statistics allow realistic, site-specific estimation of nematode damage to field crops and provide a platform for the site-specific delivery of nematicides within individual fields. Precision agricultural technologies have some potential in nematode management, but strategies to overcome the high cost and other problems inherent in nematode population assessments are needed. Still, site-specific management technologies e.g., global positioning systems (GPS) and geographical information systems (GIS), may facilitate a reduction of nematode management costs in some intensive cropping systems (**Nutter** *et al.*, 2002, Wyse-Pester *et al.*, 2002).

Fig. 5.9. Cotton field infected with root-knot nematodes showing patchy appearance

For a few nematodes such *Heterodera glycines* on soybean that affect foliage color and density, a combination of remote sensing and GIS technologies could become effective tools for nematode population assessments and their effects on crop yield (Nutter *et al.*, 2002). For site-specific nematode management, however, Evans *et al.* (2002) showed that large-scale sampling for potato cyst nematodes can involve serious problems regarding sampling and nematode inoculum potential that would likely result in misleading field maps. In contrast, Schomaker and Been (1999) developed the very precise Flevoland simulation model, based on intensive sampling (a large soil sample for each meter) in selected portions of potato-cyst infested fields that facilitated the identification of scattered infestation foci. Application of the above model and sampling could decrease nematicide usage by 86% in fields with highly contagious dispersal patterns of cyst nematodes in the Netherlands.

Based on early results, this management tool should allow specially prescribed nematode control in high-intensive crop production such as *Radopholus similis* on banana and root-knot nematodes on potato.

5.6.1. Cotton root-knot nematode, *Meloidogyne incognita*

Soil within field varies from sandy (high nematode population) to more densely packed silt loams and clays (low/nil nematode population). EC readings were highly negatively correlated with sand content. Positive correlation exists between nematode population densities and sand content (Fig. 5.10). Soil structure maps were generated by electrical conductivity sensors.

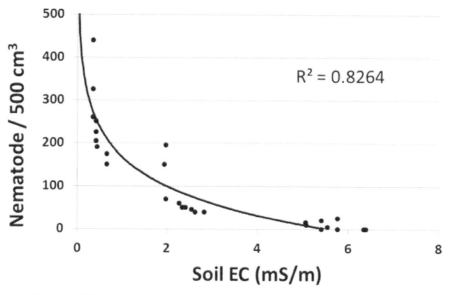

Fig. 5.10. EC readings were highly negatively correlated with sand content

Greatest nematicide response was noticed in areas with lowest EC values indicating highest content of sand. Seed treatment gave good nematode protection in silt loams/clay soils with higher EC values (Fig. 5.11). Nematicide treatment by soil zone resulted in 36%-42% reduction in amount of nematicide applied relative to whole-field application. Cotton lint yields were increased by 84 kg/ha.

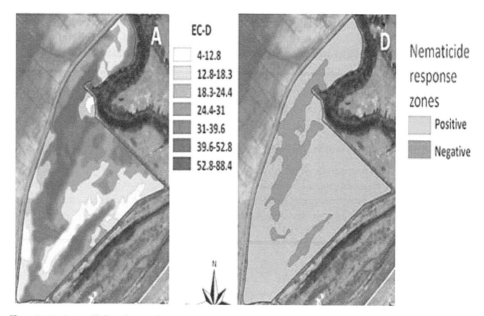

Fig. 5.11. *Left* – **EC values of cotton soil.** *Right* – **Nematicide response zones of cotton soil**

Nematode control in cotton is primarily dependent on the application of nematicides because of a lack of effective resistant cultivars (Koenning et al., 2004). The cost of a nematicide application in the seeding furrow for cotton crops is around US$150 per hectare, which represents around 6% of the production cost of the crop. With variable rate application, reductions of 40% in the volume of nematicide applied have been observed, mainly in the areas of a field where the nematodes (M. incognita) are not yet present and the product does not need to be applied. The yield of cotton in the areas where this technology has been adopted has been equivalent or greater than in areas with flat rate application (Table 5.1).

Table 5.1. Effect of variable nematicide application rates on saving of nematicide quantity and increase in cotton lint yield

Treatment	Application	Dose	Lint yield (lbs/A)
Aldicarb (Temik)	Whole-field	6.0 lbs./A	650
Aldicarb (Temik)	Variable rate	4.0 lbs./A	687
D-D (Telone)	Whole-field	3.0 gals/A	663
D-D (Telone)	Variable rate	0.6 gals/A	696
Control	---	---	566

Considering the typical case, the investment of US$10 per hectare for the mapping of the spatial distribution of the nematodes can provide US$60 per hectare of savings in the amount of nematicide used, obtaining the same cotton yield. In addition to the US$50 benefit per hectare, the use of the site-specific application brings an environmental gain not yet quantified, since the nematicides used are highly toxic and its application in places where it is not necessary can contribute to the elimination of other organisms and cause biological imbalances.

Variable rate nematicide (Temik and Telone) application system resulted in 5% higher yield and 34% (Temik) and 78% (Telone) lower nematicide usage compared to the single rate.

Site-specific and uniform application with 1, 3-D yielded similarly, and significantly better than the untreated strips. The application of 1, 3-D in uniform application had lint averaging 1,090 and 1,205 kg/ha for the Ashley and Mississippi Counties fields compared with the untreated with 941 and 1,093 kg/ha, respectively. When 1,3-D was applied site-specific, yields were still significantly higher than the untreated and averaged 1,025 and 1,183 kg/ha lint for the Ashley and Mississippi Counties fields, respectively (Table 5.2, Fig. 5.12) (Overstreet *et al.*, 2014).

Table 5.2. Cotton yield in fields infested with *Meloidogyne incognita* and treated with 1,3 – Dichloropropene in Ashley and Mississippi Counties, Arkansas, during 2008.

Treatment	Lint yield (kg/ha)	
	Ashley County	Mississippi County
Uniform application	1090	1205
Site-specific application	1025	1183
Untreated	941	1093

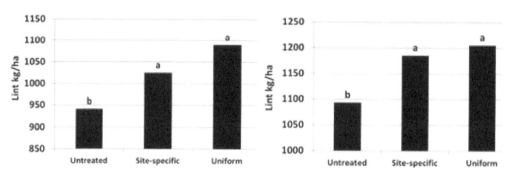

Fig. 5.12. Cotton yield in a fields infested with *Meloidogyne incognita* **in Ashley County (***Left***) and in Mississippi County (***Right***), Arkansas, during 2008.**

Site-specific treatment resulted in a 36% reduction in fumigant applied relative to uniform application for the Ashley County field and 42% reduction for the Mississippi County field.

5.6.2. Cotton reniform nematode, *Rotylenchuls reniformis*

Hyperspectral reflectance of cotton plants affected by the reniform nematode in fields is determined by GPS, RS and GIS. Nematode contour maps are developed to demonstrate spatial distribution of nematode and to establish zones of variable nematicide application rates. Cotton yields were higher in site-specific application treatments compared with conventional single rate applications.

5.6.3. Potato root-knot nematode

Site-specific application of DD fumigant was undertaken in eastern Idaho based on root-knot nematode population density in over 1200 ha of potato, which resulted in 30% reduction in chemical usage and production cost savings of US$ 180 ha⁻¹.

5.7. CONCLUSION

Deployment of new technologies for nematode management in precision agriculture through several approaches has been discussed. Information on nematode biology (e.g. typical within-field distribution patterns) and parameters correlated with nematode infestation (e.g. soil/plant nitrogen levels, climatic factors) are essential for site-specific control of nematodes. The management tools to reduce risk to the farmers are provided by GIS, GPS, and RS. The more informed management decisions can be made by producers, since they have simultaneous access to the numerous types of data needed. A more intimate knowledge of the system at hand is needed to understand how biotic and abiotic factors affect nematode populations, yield and the fate of nematicides in the environment for attaining greater precision in nematode management programs. Few equipment have already been developed such as nematicide applicators equipped with sensors to detect nematode patches and control delivery rate, and a soil sampler that will characterize not only nutrient content but also the potato cyst nematode levels in the soil, which are readily available in the market (Legg and Stafford, 1998). Further, there is a need for new tools for the monitoring of nematodes on a small spatial scale (Fleischer *et al.*, 1999), for geo-statistical analyses to detect spatial relations between variables in the environment (Liebhold *et al.*, 1993), and for the delivery of variable rates of nematode-control measures other than nematicides, such as mass release of biological control agents. The management decisions can be applied in a more precise manner by using VRT techniques based on information collected from GIS in combination with GPS and RS. The agricultural crop production costs and crop and environmental damage can be potentially reduced by following precision nematode management.

Chapter - 6

Nanobiotechnology-Based Nematode Management

6.1. INTRODUCTION

The world's population will grow to an estimated 9.5 billion by 2050, and it is widely recognized that global agricultural productivity must increase to feed a rapidly growing world population. In order to increase the agricultural productivity, it is necessary to use the modern technologies such as nanotechnology and nanobiotechnology in agricultural and food sciences. Nanotechnology has a tremendous potential to revolutionize agriculture and allied fields, including aquaculture and fisheries. Nanoagriculture focuses currently on target farming that involves the use of nanosized particles with unique properties to boost crop and livestock productivity (Scott and Chen, 2002). Nanomaterials play an important role in promoting sustainable agriculture and provide better foods globally (Gruère, 2012). In developing countries, nanotechnology has got important applications for enhancing agricultural productivity, along with other emerging technologies such as biotechnology including genetics, plant breeding, pest management, fertilizer technology, micro-irrigation, precision agriculture, and other allied fields (Jha *et al.*, 2011).

Nanoscale refers to size dimensions typically between approximately 1–100 nm (and more appropriately, 0.2–100.0 nm) because it is at this scale that the materials differ with respect to their physical, chemical, and biological properties from those at a larger scale. A single nanometer (nm) is 1 billionth of a meter. Nanotechnology refers to the understanding and control of matter at nanoscale, where a unique phenomenon enables novel applications (Committee on Technology, 2014). Nanosensors and other field sensing devices can be used in detection and management of insects, pathogens, nematodes, and weeds. Nanosensors can also provide information about timely application of agrochemicals like pesticides, nematicides, and herbicides.

At present, several management strategies are applied such as biological, chemical, organic, cultural, nanobiotechnology to control pathogenic nematodes. Use of nematicides of chemical origin are although effective, on another hand they cause environmental perturbations. The emerging of two novel techniques, nanotechnology and biotechnology has resolved many concerns that prevail with the traditional strategies of nematode management in plants and environment. Nanotechnology based agricultural systems have developed with a worthy scope to manage phytonematodes using drug-carrier and a controllable drug targeting and releasing system as it can enhance the quality of life and world's economy. Through advancement in nanotechnology, there are a number of state of-the-art techniques available including applications of several types of nanoparticles as protectants and carriers in the form of 'nanonematicide'. Several pathogenic phytonematodes are very effectively managed with the means of nanotechnology.

6.2. DIMENSIONS OF NANOMATERIALS

This classification is based on the number of dimensions of a material, which are outside the nanoscale (<100 nm) range.

> » In *zero-dimensional* (0D) nanomaterials all the dimensions are measured within the nanoscale (no dimensions are larger than 100 nm). Most commonly, 0D nanomaterials are nanoparticles.
> » In *one-dimensional* nanomaterials (1D), one dimension is outside the nanoscale. This class includes nanotubes, nanorods, and nanowires.
> » In *two-dimensional* nanomaterials (2D), two dimensions are outside the nanoscale. This class exhibits plate-like shapes and includes graphene, nanofilms, nanolayers, and nanocoatings.
> » *Three-dimensional* nanomaterials (3D) are materials that are not confined to the nanoscale in any dimension. This class can contain bulk powders, dispersions of nanoparticles, bundles of nanowires, and nanotubes as well as multi-nanolayers.

6.3. NANOTECHNOLOGY APPLICATIONS

Nanotechnology is science, engineering, and technology conducted at the nanoscale, which is about 1 to 100 nanometers. Nanoscience and nanotechnology are the study and application of extremely small things and can be used across all the other scientific fields including agriculture and life sciences. Nanotechnology has the potential to change the entire scenario of Plant Nematology with the help of new tools developed for the treatment of nematode diseases, rapid detection of nematode pathogens using nanosensor-based kits, improving the ability of plants to absorb nanonematicides. Nanoformulations are safer and have eco-friendly outlook for nematode management. Therefore, now-a-days, nanotechnological approach has been implicated against plant parasitic nematodes as they have mutualistic mode of action against nematodes and no phytotoxicity.

AgNP has shown evidence being a potentially effective nematicide (Roh *et al.*, 2009), and its toxicity is associated with induction of oxidative stress in the cells of targeted nematodes (Lim *et al.*, 2012). Moreover, it has been reported that chronic exposure of Al2O3 nanoparticles shows toxicity against the nematodes with end points of lethality, growth, reproduction, stress response and intestinal auto fluorescence (Wang *et al.*, 2009; Wu *et al.*, 2011). However, a limited research on nanoparticles application against controlling of nematode has been reported till date. Therefore, approaches are needed to offer new capabilities for preventing or treating pathogenic nematodes by using nanoparticles, which would result in the more effective monitoring in the way currently not possible.

Nanomaterials are being developed that offer the opportunity for more efficient and safe administration of nematicides by controlling precisely when and where they are released (Kuzma and VerHage, 2006). Use of nanotechnology in Plant Nematology can revolutionize the sector with new tools for nematode detection, targeted treatment, enhancing the ability of plants to fight nematodes and withstand environmental pressures. Smart sensors and smart delivery systems will help the agricultural industry combat nematode pests. In the near future, nanostructured catalysts will be available which will increase the efficiency of nematicides, allowing lower doses to be used.

Plants are one of the important sources of nanoparticles based nematicides, and it contains different types of nanoparticles as a nematicide component in agriculture applications (Fig. 6.1). Researchers have focused their attention on understanding the mechanism of synthesis of botanical-based nanoparticles for controlling nematodes, to prevent diseases caused by them, and to protect our environment instead of chemical nematicides.

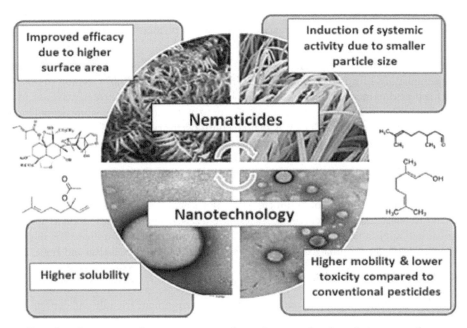

Fig. 6.1. Benefits of association between nematicides and nanotechnology. It improves the properties of the bioactive compounds and increases their effectiveness in nematode control.

6.4. BENEFITS

» Nematode management through the formulations of nanomaterials-based nematicides.
» Enhancement of agricultural productivity using bio-conjugated nanoparticles (encapsulation) for slow release of nematicides.
» Nanoparticle-mediated gene or DNA transfer in plants for the development of nematode disease-resistant varieties.
» Use of nanomaterials for preparation of different kinds of biosensors, which would be useful in remote sensing devices required for precision nematode management.
» Early detection of nematode pests and pollutants including nematicide residues.
» Improved efficacy due to higher surface area.
» Induction of systemic activity due to smaller particle size.
» Higher solubility.
» Higher mobility and lower toxicity in comparison to conventional pesticides.

6.5. NANONEMATICIDES

Nanonematicides cover a wide variety of products, some of which are already on the market. They cannot be considered as a single entity; rather such nanoformulations combine several surfactants, polymers (organic), and metal nanoparticles (inorganic) in the nanometer size range. The lack of water solubility is one of the limiting factors in the development of crop-protection agents. Microencapsulation has been used as a versatile tool for hydrophobic nematicides, enhancing their dispersion in aqueous media and allowing a controlled release of the active compound. Polymers often used in the nanoparticle production have been reported (Perlatti *et al.*, 2013).

In the research and development stage, nano-sized nematicides or nano- nematicides are mostly nano-reformulations of existing nematicides (Green and Beestman, 2007). Nanoformulations are generally expected to increase the apparent solubility of poorly soluble active ingredients, to release the active ingredient in a slow/targeted manner, and/or to protect against premature degradation (Kah *et al.*, 2013). Nanonematicides offer a way to both control delivery of nematicide and achieve greater effects with lower chemical dose. Agrochemical companies are reducing the particle size of existing chemical emulsions to the nanoscale, or are encapsulating active ingredients in nanocapsules designed to split open, for example, in response to sunlight, heat, or the alkaline conditions in nematode's stomach. Viral capsids can be altered by mutagenesis to achieve different configurations and deliver specific nucleic acids, enzymes, or antimicrobial peptides acting against the nematodes (Pérez-de-Luque and Rubiales, 2009). Silver nanoparticles have received significant attention as a nematicide for agricultural applications (Afrasiabi *et al.*, 2012). The potential of nanomaterials in nematode management as modern approaches of nanotechnology has been reported (Rai and Ingle, 2012).

6.6. GREEN SYNTHESIS METHODS

The synthesis of stable metal-based nanoparticles with environment friendly and convenient green route by using biopolymers, plant extracts and microorganisms has attracted the researchers to adopt green nanotechnology. Because of its nontoxicity, cost effective, eco-friendly synthesis at large-scale and can be used very effectively against nematodes. It acts as a viable alternative for traditional chemical synthesis procedures (Fig. 6.2).

The green synthesis of silver nanoparticles from the plant extracts and microbes is a boon for advancement of research in nanotechnology and has a potential impact in safe delivery of nematicides for crop protection. Synthesis of extracellular nanoparticles with variety of locally available biological agents is a novel and economical concept for bioprospecting. It provides new avenues to exploit wide variety of biological species for product development (Sharma *et al.*, 2010). Green synthesis practice helps in reducing generation of hazardous waste by using environmentally safe solvents and nontoxic chemicals. The effective use of nanoparticle synthesized by green synthesis is very important.

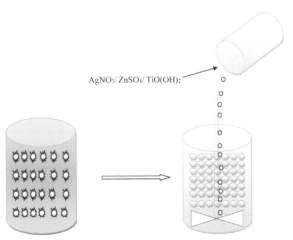

AgNO₃/ ZnSO₄/ TiO(OH)₂

Plant leaf extract or supernatant of bacteria or fungi   Ag/ZnO/TiO₂ nanoparticles

Fig. 6.2. Biological synthesis of silver/zinc oxide/titanium dioxide nanoparticles.

6.6.1. Plant extracts

Various types of plant extracts can be used for the synthesis of metal nanoparticles (Fig. 6.3 and Table 6.1). Singh *et al.* (2010) demonstrated the cost effective and environment friendly technique for green synthesis of silver nanoparticles (30 nm) by conversion of silver nitrate solution into silver nanoparticles with weed leaf extract of *Argemone maxicana* as reducing and capping agent. The antimicrobial activity of synthesized nanoparticles was commendable. They also found that, reduction of silver ions into silver nanoparticles was due to the precipitation of leaf proteins and metabolites. Stable silver nanoparticles can also be synthesized with the leaf extract of eucalyptus hybrid and *Cycas* (Jha and Prasad, 2010). Mubarak Ali *et al.* (2011) reported the plant extract mediated synthesis of silver and gold nanoparticles.

Neem (*Azadirachta indica*)   Aloe vera   Tea (*Camellia sinensis*);   *Catharanthus roseus*

Lemongrass (*Cymbopogon sp.*)   *Cinnamomum camphora*   *Datura metel*   Geranium leaf

Fig. 6.3. Various types of plants used for the synthesis of metal nanoparticles

Table 6.1. Green synthesis of metallic nanomaterials by plants

Plant species/Active ingredient	Plant material/ solvent	Type of metallic nanoparticle	Size range (nm)	Reference
Neem/Azadirachtin	Kernel, water	Au, Ag	50–100	Shankar *et al.*, 2004
Tea/ Catechins, theaflavins and thearubigins	Leaves, water	Ag, Au	30–40	Vilchis-Nestor *et al.*, 2008
Jatropha curcas/ Curcacycline A & Curcacycline B	Latex, water	Pb	10–12.5	Joglekar *et al.*, 2011
Geranium/Terpenoids	Leaves, water	Au	16–40	Shankar *et al.*, 2003
Lemon grass/Sugar derivative molecules	Leaves, water	Au	200–500	Shankar *et al.*, 2005
Oat	Stems, water	Au	5–85	Armendariz *et al.*, 2004
Aloe vera	Leaves, water	Au, Ag	50-350	Chandran *et al.*, 2006
Cinnamomum camphora/ Polyol components and heterocyclic components	Leaves (dried biomass)	Ag, Au	55-80	Huang *et al.*, 2007
Garlic	Cloves	Ag	4-22	Ahamed *et al.*, 2011
Calotropis procera	Leaves	Ag	150–1000	Babu and Prabu, 2011
Papaya	Fruits	Ag	25-50	Jain *et al.*, 2009
Catharanthus roseus	Leaves	Ag	48-67	Kannan *et al.*, 2011
Chenopodium album	Leaves	Ag, Au	10–30	Dwivedi & Gopal, 2010
Sweet orange	Peels	Ag	35±2 (25°C) 10±1 (60°C)	Kaviya *et al.*, 2011b
Turmeric		Ag	---	Sathishkumar *et al.*, 2010
Datura metel		Ag	16-40	Kesharwani *et al.*, 2009
Amla	Fruits	Ag, Au	Ag 10-20 Au 15–25	Ankamwar *et al.*, 2005
Eucalyptus hybrid		Ag	50-150	Dubey *et al.*, 2009
Euphorbiaceae/ Flavonoids and phenolic acids	Latex	Cu/Ag	Ag 18 nm Cu 10.5 nm	Patil *et al.*, 2012
Moringa oleifera		Ag	57	Prasad and Elumalai, 2011
Mucuna pruriens		Au	6–17.7 nm	Arulkumar & Sabesan, 2010
Banana	Peels	Ag	20	Bankar *et al.*, 2010
Parthenium	Leaves	Au	50	Parashar *et al.*, 2009b
Guava		Au	25-30	Raghunandan *et al.*, 2009
Cumin/ Eugenol	Seeds	Ag	29–92	Banerjee, 2011
Ocimum	Leaves	Ag	9.5	Mallikarjuna *et al.*, 2011
Mango	Leaves	Au	17-20	Gan *et al.*, 2012

6.6.2. Microorganisms

Apart from plant extracts, microorganisms (bacteria and fungi) can also be effectively used in synthesis of silver nanoparticles in a cost effective, eco-friendly, economic and efficient way (Mandal *et al.*, 2006). Pathogenic organisms such as *Cryphonectria* and *Phytophthora infestans* were exploited for the synthesis of silver nanoparticles (Dar *et al.*, 2013). Several microorganisms can be used for the synthesis of metal nanoparticles (Table 6.2).

Table 6.2. Green synthesis of metallic nanomaterials by microorganisms

Source	Type of metallic nanoparticle	Location	Size range (nm)	Reference
Bacteria				
Pseudomonas stutzeri	Ag	Intracellular	< 200	Klaus *et al.*, 1999
Morganella sp.	Ag	Extracellular	20-30	Parikh *et al.*, 2008
Lactobacillus strains	Ag, Au	Intracellular	---	Nair & Pradeep, 2002
Plectonema boryanum (Cyanobacteria)	Ag	Intracellular	1-10, 1-100	Lengke *et al.*, 2004
Escheichia coli DH 5a	Au	Intracellular	25-33	Liangwei *et al.*, 2007
Clostridium thermoaceticum	CdS	Intracellular & extracellular	---	Cunningham *et al.*, 1993
Actinobacter spp.	Magnetite	Extracellular	10-40	Bharde *et al.*, 2005
Shewanella algae	Au	Intracellular, pH=7 Extracellular pH=1	10-20 50-500	Konishi *et al.*, 2007
Rhodopseudomonas sp.	Ag	Intracellular, Extracellular	6-10	Manisha *et al.*, 2014
Thermomonospora sp.	Au	Extracellular	8	Ahmad *et al.*, 2003b
Rhodococcus sp.	Au	Intracellular	5-15	Ahmad *et al.*, 2003a
Klebsiella pneumonia	Ag	Extracellular	5-32	Shahverdi *et al.*, 2007
Shewanella oneidensis	Uranium	Extracellular	---	Marshall *et al.*, 2007
Yeast				
MKY3	Ag	Extracellular	2-5	Kowshik *et al.*, 2003
Candida glabrata, Schizosaccharomyces pombe	CdS	Intracellular	200	Dameron *et al.*, 1989
Fungi				
Phoma glomerata	Ag	Extracellular	60-80	Birla *et al.*, 2009
Fusarium oxysporum	Ag	Extracellular	20-40	Duran *et al.*, 2005
Verticillium	Ag	Intracellular	25±12	Mukherjee *et al.*, 2001
Fusarium oxysporum, Verticillium sp.	Magnetite	Extracellular	20-50	Bharde *et al.*, 2006
Aspergillus fumigatus	Ag	Extracellular	5-25	Bhainsa & D'souza, 2006
Aspergillus niger	Ag	Extracellular	3-30	Jaidev & Narasimha, 2010

| *Trichoderma asperellum* | Ag | Extracellular | 13-18 | Mukherjee *et al.*, 2008 |
| *Phaenerochaete chrysosporium* | Au | Extracellular | 10-100 | Sanghi *et al.*, 2011 |

6.7. DETECTION OF NEMATODES AND NEMATICIDE RESIDUES

6.7.1. Detection of nematodes

Detection and utilization of biomarkers that accurately indicate nematode disease stages is a new area of research. Measuring differential protein production in both healthy and diseased states leads to the identification of the development of several proteins during the infection cycle. These nano-based diagnostic kits not only increase the speed of detection but also increase the power of the detection (Prasanna, 2007). In the future, nanoscale devices with novel properties could be used to make agricultural systems "smart". For example, devices could be used to identify plant health issues before these become visible to the farmer. Such devices may be capable of responding to different situations by taking appropriate remedial action. If not, they will alert the farmer to the problem. In this way, smart devices will act as both a preventive and an early warning system. Such devices could be used to deliver chemicals in a controlled and targeted manner.

Bio-sensors, consisting of an organic-based detection mechanism, such as enzymes, are able to detect these specific threats (Perumal and Hashim, 2014). Due to their size-related properties, nanobiosensors show an increase in accuracy, detection limits, sensitivity, selectivity, temporal response and reproducibility, compared to conventional biosensors (Huang *et al.*, 2011). Therefore, nanobiosensors provide a very precise tool that can be used to prevent nematode outbreaks and monitor soil quality, which enhances quality and quantity of yields (Ram *et al.*, 2014).

6.7.2. Detection of nematicide residues

Nanoparticles (NP) based nanosensors can be used to detect nematicide residues. Nanosensors for nematicide residue detection offers high sensitivity, low concentration detection limits, super selectivity, fast responses and small sizes (Liu *et al.*, 2008). The nanosensors aimed to detect the nematicide residues of methyl parathion (Kang *et al.*, 2010) and parathion (Wang and Li, 2008), have proved to give high degree of accuracy and handiness. Additionally, Van Dyk and Pletschke (2011) have also emphasized the higher sensitivity of enzyme-based NP biosensors in the detection of residues of organochlorines, organophosphates and carbamates. In some of these biosensors, nano C, Au, hybrid Ti (titanium), Au-Pt (platinum) and nanostructured lead dioxide PbO_2, TiO_2 and Ti were used to immobilize the enzymes on sensor substrate and to increase the sensor sensitivity.

6.8. SMART DELIVERY OF NEMATICIDES

Up to 70% of nematicides do not reach their target because they are unstable in the environment and difficult to be taken up (Solanki, 2015). Nano-based smart delivery systems have the ability to provide more efficient and targeted delivery to specific plant cells due to their size-related properties (Solanki, 2015). They also show enhanced stability in the environment, which improves the availability of nematicides to crops (Solanki, 2015; Kah *et al.*, 2013). Smart delivery systems further enhance the delivery of nematicides through their ability of slow or controlled release (Solanki, 2015; Kah *et al.*, 2013). This is known to extend the effectiveness of nematicides from three to over thirty days (Adak *et al.*, 2012). In addition, the effect of nematicides was found to be twice as strong with half the dose applied (Xiang *et al.*, 2013). Enhanced delivery of nematicides improves the resistance of crops towards threats like droughts, pests and pollution (Kah and Hofmann, 2014). Therefore it improves the quality and quantity of yields (Solanki, 2015). Nano-biosensors can enhance this process even further by enabling smart delivery systems to precisely release nematicides in response to environmental triggers and biological demands (Solanki, 2015). This provides opportunities for real-time monitoring and control.

The aims of nanoformulations are generally common to other nematicide formulations and consist in:

» Increasing the apparent solubility of poorly soluble active ingredient.

» Releasing the active ingredient in a slow/targeted manner and/or protecting the active ingredient against premature degradation.

6.8.1. Nanoencapsulation

Encapsulation is defined as process in which the given object is surrounded by a coating or embedded in homogeneous or heterogeneous matrix, thus this process result in capsules with many useful properties (Fig. 6.4) (Rodríguez *et al.*, 2016). The benefits of encapsulation methods are for protection of nematicides from adverse environments, for controlled release, and for precision targeting (Ozdemir and Kemerli, 2016).

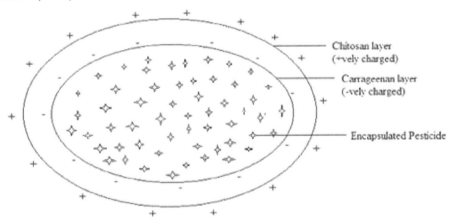

Fig, 6.4. Encapsulation of nematicides

Nanoscale systems like encapsulation and entrapment of agrochemicals such as nematicides by using polymers, dendrimers, surface ionic attachments and other mechanisms may be used in controlled and slow release of nematicides, which allow the slow uptake of active ingredients and in turn reducing the amount of nematicide application by minimizing the input and waste. Importance of nanoscale delivery system in agriculture is because of its improved solubility and stability to degradation due to the environmental factors. The nanoscale delivery vehicles increase effectiveness by binding firmly to the plant surface and reduce the amount of nematicides by preventing run-off into the environment (Fig. 6.5) (Chen and Yada, 2011).

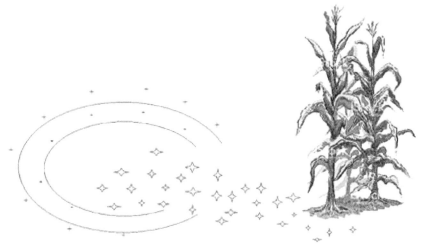

Fig. 6.5. Controlled release of nematicides from nanocapsules

Agrochemical companies are reducing the particle size of existing chemical emulsions to the nanoscale, or are encapsulating active ingredients in nanocapsules designed to split open, for example, in response to sunlight, heat, or the alkaline conditions in nematode's stomach. Nanocapsules can enable effective penetration of nematicides through cuticles and tissues, allowing slow and constant release of the active substances.

6.9. NEMATODE MANAGEMENT

Nanonematicides are formulations of active ingredient of a nematicide in nanoform that have slow degradation, targeted delivery, and controlled release of active ingredient for longer period that make them environmentally safe and less toxic in comparison with conventional chemical nematicides. Several studies have reported an enhancement in the efficacy of certain biological substances on nematodes and a reduction of losses due to physical degradation through encapsulation of these substances in nanoparticulate systems.

Use of nanoparticles in nematode management is a novel and fancy approach that may prove very effective in future with the progress of application aspect of nanotechnology (Ladner *et al.*, 2008). The effect of applications of nanoparticles in nematode management can be divided into two perspectives; direct effect of nanoparticles on nematodes and use of nanoparticles in formulating nematicides (Table 6.3). Due to their ultra-sub microscopic size, nanoparticles gain the high degree of reactivity and sensitivity and thus have potential to prove very useful in control of plant parasitic nematodes (PPNs), in addition, nematicidal residue analysis. As chemical and physical properties of nanoparticles vary greatly as compared to larger form, thus it has become imperative to evaluate the effect of nanoparticles on nematodes to harness the beneficial effects of this technology in plant protection, especially against PPNs. Ultra-small size and very high reactivity will affect the activity of PPNs (Gatoo *et al.*, 2014).

Table 6.3. Nanoparticle formulated nematicides along with dose and application techniques for management of different nematodes.

Crop	Nematode pest	Nanoparticle formulated nematicide application rate	Application techniques
Potato	*Globodera* spp.	Aldicarb at 2.24-3.36	Incorporated in row
		Oxamyl at 4.0-5.5	
		Carbofuran at 4.0-5.5	
Tomato, cucurbits	*Meloidogyne* spp.	Aldicarb at 3.36	Incorporated in 30 cm bands
		Ethoprophos at 0.9-2.9	Incorporated in bands
		Oxamyl at 0.6-1.2	
		Fenamiphos at 1.6-3.3	
		Dazomet at 30-50 g/m^2	Incorporated in bands and irrigated time interval before planting
Citrus	*Tylenchulus semipenetrans*	Fenamiphos at 10.8-21.6	Annual treatment applied along drip-line
		Aldicarb at 5.5-11.0	
Grape	*Xiphinema index*	Aldicarb at 10 ml	In band for nursery use
	Meloidogyne spp.	Fenamiphos (EC formulation) at 10 ml	
Banana	*Radopholus similis*	Carbofuran at 2-4 g a.i./plant	Applied around plant 2-3 times per year
	Helicotylenchus multicinctus	Ethoprophos at 2-4 g a.i./plant	
	Pratylenchus spp.	Fenamiphos at 2-4 g a.i./plant	
	Meloidogyne spp.	Isazofos at 2-4 g a.i./plant	
		Ebufos at 2-4 g a.i./plant	

An indigenous bacterial strain isolated from indigenous gold mines identified as *Bacillus* sp. GPI-2 was found to convert AuCl$_4$ into AuNP of size 20 nm with the spherical shape. These bio gold AuNPs were evaluated for their ability to kill nematodes (Siddiqi and Husen, 2017). Gold and silver NPs possess nematicidal activity that may provide an alternative to high-risk synthetic nematicides or inconsistent biological control agents. High frequency (biweekly) and high application doses (90.4 mg/m^2) of gold and silver NPs may be required to achieve effective field efficacy for root-knot nematode (Kalishwaralal *et al.*, 2008). Combining AgNP with an irrigation system such as fertigation or tank-mixture with compatible chemicals that supplement the AgNP nematicidal effect may increase applicability of AgNP. Further understanding of the mechanism in the nematicidal action of Silver NP also warrants improvement of gold and silver NPs efficacy (Ganesh Babu and Gunasekaran, 2009).

Silver nanoparticle (AgNP) has shown evidence of being a potentially effective nematicide (Roh *et al.*, 2009), and its toxicity is associated with induction of oxidative stress in the cells of targeted nematodes (Lim *et al.*, 2012). Ag-nanoparticles of *Urtica urens* extracts concomitant with rugby were effective in the management of *M. incognita*, since it increased nematicidal activity 11-fold more than the least toxic extract against eggs (Nassar, 2016). The toxicity of three nanoparticles, silver, silicon oxide and titanium oxide, to the root-knot nematode, *M. incognita*, was recorded in laboratory and pot experiments (Ardakani, 2013).

Green silver nanoparticles (GSN) (12.75 mg100 mL^{-1}) were effective in controlling the root-knot nematode (both *T. turbinata* and *U. lactuca* algae), similar to chemical control in eggplants; resulted in significant reduction in number of nematode juveniles (J2) in soil, number of galls, number of female nematodes, and number of egg-masses per root system compared with those of control (infected plants). GSN (17 mg mL^{-1}) obtained from *U. lactuca* was more effective in reducing second-stage juveniles (J2s) of *M. javanica* (69.44%) population in soil. All treatments improved eggplants growth parameters. Change in DNA profile using of both RAPD and EST markers was noted.

In a pot experiment, all treatments of AgNP and 0.02% TiO$_2$NP completely controlled *M. incognita*. Treatments of 0.02, 0.01 and 0.005% of AgNP as well as 0.02% of TiO$_2$NP were toxic to tomato plants and significantly reduced tomato root and stem length and fresh weights in comparison to control.

Application of AgNPs improved tomato plant growth and reduced root-knot nematode (*M. incognita*) infection in comparison to silver nitrate and control treatments. The highest increment of fresh weight as well as the lowest numbers of galls and egg-masses was obtained when tomato plants was treated with AgNP produced by ginger extract at 1 mM (Table 6.4) (El-Deen and El-Deeb, 2018; Ahmed and Bahig, 2018).

Table 6.4. Effect of soil treatment with AgNPs on galls and egg-masses of tomato (index of 0 to 9) caused by *M. incognita* under greenhouse conditions

Treatments	Conc. (mM)	Fresh plant wt. (g)	No. of galls	RGI	No. of egg-masses	Egg-mass index
Biological AgNP	0.25	48.2 c	4.7 cd	2 ef	0 f	1 d
	0.50	52.0 b	2.0 d	2.0 ef	0 f	1 d
	1.00	55.4 a	1.3 d	1.3 f	0 f	1 d
Chemical AgNP	0.25	41.1 d	6.0 cd	2.7 de	1 ef	1.7 cd
	0.50	37.9 de	10.3 c	3.3 d	3.7 de	2 c
	1.00	23.4 g	20.7 b	4.3 c	5.3 d	2.3 c
AgNO$_3$	0.25	37.5 e	25.0 b	5 c	12 c	4 b
	0.50	33.6 f	28.0 b	5 c	14.7 bc	4 b
	1.00	20.9 g	48.0 a	6 b	6 b	4 b
Control (Check)	---	16.3 h	55.0 a	7 a	24	5 a
LSD at 0.05	---	3.2	8.01	0.85	2.9	0.72

Note. Each value represents the mean of four replicates. Within each concentration, values with the same letter do not differ significantly according to LSD (P≤ 0.05).

Jo *et al.* (2013) reported that application of silver nanoparticles significantly reduced the nematode population and improved the turf quality.

Mohammad *et al.* (1981) reported that leaf extracts of *U. urens* effectively control the citrus nematode, *Tylenchulus semipenetrans*. Decker (1982) listed *U. urens* (stinging nettle) among plants that control root-knot nematode and Nasiri *et al.* (2014) found that exudates of *U. dioica* effectively controlled the *Pratylenchus, Aphelenchoides* and *Helicotylenchus* nematodes.

Tomato seedlings treated with Abamectin loaded plant virus nanoparticles (PVN[Abm]) had healthier root growth and a reduction in root galling (*Meloidogyne hapla*) demonstrating the success of this delivery system for the increased efficacy of Abm to control nematode damage in crops. Encapsulating Abm within the *Red Clover Necrotic Mosaic Virus* (RCNMV) produced a plant virus nanoparticle (PVN) delivery system for Abm.

6.10. PHYTOSYNTHESIZED NANOFORMULATIONS FOR NEMATODE MANAGEMENT

6.10.1. Laboratory studies

6.10.1.1. Inhibition of egg hatching of *M. incognita*: The inhibition in egg hatching was highest (91.49 %) at the highest concentration of 0.0005 per cent of AgNPs (The AgNPs synthesized using *Tridax procumbens*) at 7 days after inoculation followed by 87.35% in 0.0001, and 82.64 in 0.00005% (Fig. 6.6) (Surega, 2015).

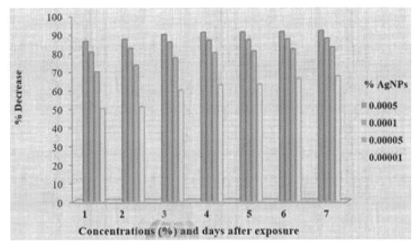

Fig. 6.6. Effect of phytosynthesized AgNPs on egg hatchability of *M. incognita*

6.10.1.2. Morphological changes caused by the phytosynthesized AgNPs on root-knot nematode egg masses: SEM studies on the ultrastructure of root knot nematode eggs on exposure to AgNPs revealed that the eggs were completely deformed. Healthy eggs were intact whereas the treated eggs had lost their structural integrity (Fig. 6.7) (Surega, 2015).

6.10.1.3. Juvenile immobility and mortality: Cromwell *et al.* (2014) observed that more than 99 per cent of J_2 of *M. incognita* became inactive on exposure to AgNPs in water at 30 to 150 µg/ml under laboratory conditions.

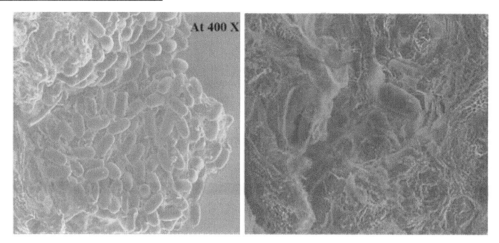

Fig. 6.7. Effect of phytosynthesized AgNPs on morphological changes in *M. incognita* egg mass at different magnifications by SEM (Left-Untreated, Right – Treated).

Tridax procumbens mediated synthesized AgNPs, SiO_2 and TiO_2 gave cent per cent immobility and mortality of J_2 of *M. incognita in vitro* at 800, 400 and 200 mg ml/lit except SiO_2 (Ardakani, 2013).

There was cent per cent mortality of juveniles of *M. incognita* at higher concentrations of 0.0005 and 0.0001 per cent of the phytosynthesized AgNPs within the shortest period of exposure of 24 hr and the trend was continued at all the period of exposure. Similarly the cent per cent mortality of juveniles of *M. incognita* was also caused by the lower concentration of 0.00005 but after sixth day of exposure (Fig. 6.8) (Surega, 2015).

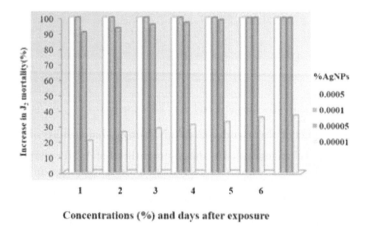

Fig. 6.8. Effect of green synthesized AgNPs *in vitro* **on mortality of juveniles of** *M. incognita*

6.10.1.4. Morphological changes caused by the green synthesized AgNPs on root-knot nematode juveniles: The dead juveniles of root-knot nematode on exposure to the most effective concentration of 0.0005 per cent of AgNPs appeared to be deformed with rupture of cell wall (Fig. 6.9) (Surega, 2015).

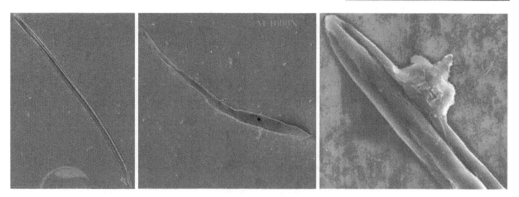

Fig. 6.9. Morphological changes caused by phytosynthesized AgNPs on juveniles of *M. incognita* (*Left* – **Untreated at 800x,** *Middle* **at 1000x and** *Right* **at 6000x - Treated).**

6.10.2. Glasshouse studies

6.10.2.1. Influence of phytosynthesized AgNPs on attraction of *M. incognita* to tomato roots

The highest reduction (94.67 %) in the attraction of J2 of M. *incognita* was recorded 24 hr after inoculation by 0.0005 per cent concentration and it was ranging from 59.15 to 91.14 per cent in other concentrations of 0.00001 to 0.0001 of the AgNPs irrespective of period of exposure from 1 to 7 days (Fig. 6.10) (Surega, 2015).

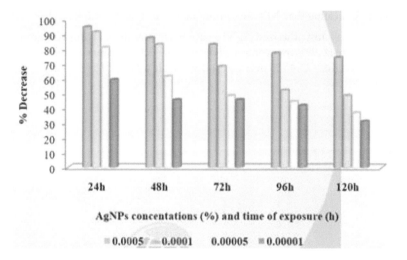

Fig. 6.10. Effect of phytosynthesized AgNPs on attraction of *M. incognita* **juveniles of towards roots of tomato** *in vitro*

6.10.2.2. Influence of AgNPs on tomato root penetration of *M. incognita*: The highest reduction (79.76 to 90.16 %) of root penetration of *M. incognita* (J2) was recorded by highest concentration of 0.0005 per cent of phytosynthesized AgNPs one day after inoculation (Fig. 6.11) (Surega, 2015).

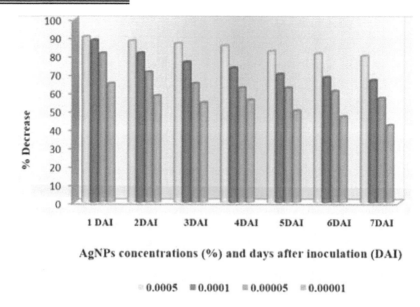

AgNPs concentrations (%) and days after inoculation (DAI)

0.0005   0.0001   0.00005   0.00001

Fig. 6.11. Effect of phytosynthesized AgNPs on tomato root penetration of *M. incognita* **juveniles**

6.10.2.3. Influence of AgNPs on Plant growth parameters of tomato: Under pot experimentations, AgNPs and TiO_2 were found to reduce the shoot height and root length of tomato compared to untreated control indicating that NPs are causing toxic effect on tomato (Ardakani, 2013).

Application of phytosynthesized AgNPs gave maximum increase in plant growth parameters (shoot height -13.11 to 26.99% over control, Shoot weight - 13.92% over control, root length - 26.57% over control, root weight - 23.45 % over control, and fruit yield - 20.54% over control) (Figs. 6. 12 to 6.14) (Surega, 2015).

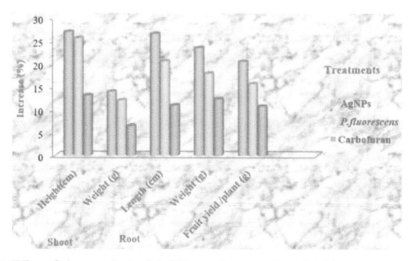

Fig. 6.12. Effect of phytosynthesized AgNPs on plant growth and yield parameters on tomato infected with *M. incognita*

Fig. 6.13. Effect of phytosynthesized AgNPs on plant growth of tomato infected with *Meloidogyne incognita* **(T1: AgNPs at 0.0005%, T2:** *Pseudomonas fluorescens* **@ 2 .5 kg/ha, T3: Carbofuran 3G @ 1kg a.i. /ha, T4: Untreated control).**

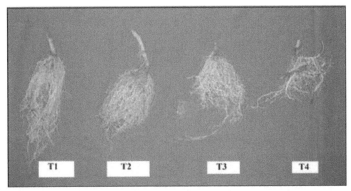

Fig. 6.14. Effect of phytosynthesized AgNPs on root biomass of tomato infected with *Meloidogyne incognita* **(T1: AgNPs at 0.0005%, T2:** *Pseudomonas fluorescens* **@ 2 .5 kg/ha, T3: Carbofuran 3G @ 1kg a.i/ha, T4: Untreated control).**

6.10.2.4. Influence of AgNPs on *M. incognita* **population on tomato:** The phytosynthesized AgNPs at 0.0005 per cent was highly effective to reduce the *M. incognita* population in soil by 88.16 %, in roots by 92.79 %, number of egg masses/g root by 79.80 %, eggs/egg mass by 65.14 %, and gall index (0.2 as compared to 3.0 in carbofuran) (Fig. 6.15) (Surega, 2015).

6.10.3. Field studies

In turf grass, AgNPs reduced *M. graminis* population in soil by 92 and 82 per cent after 4 and 2 days of exposure, respectively under field conditions. Similarly the biweekly application of 90.4 mg/m² of AgNPs caused reduction in gall formation in root and improved quality of turf grass for two years without phytotoxicity (Cromwell *et al.*, 2014).

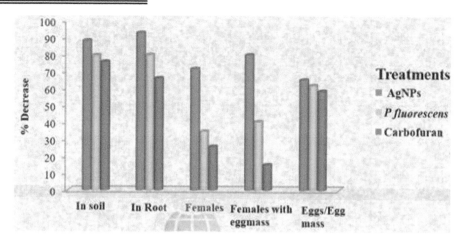

Fig. 6.15. Effect of phytosynthesized AgNPs on management of *M. incognita* on tomato.

6.11. MECHANISM OF ACTION OF AgNPs

The information available on the mechanisms of AgNPs in the suppression of nematode population is very meagre. The available information also implied the effect of nanoparticles at gene level. Therefore in-depth study needs to be documented on the mechanisms of nanoparticles in the management of nematodes.

6.11.1. Suppression in nematode reproduction

Roh *et al.* (2009) investigated the ecotoxicity of AgNPs on *Caenorhabditis elegans*. In this study, the authors reported that AgNPs at 0.1 and 0.5 mg/lit exerted considerable toxicity on *C. elegans* as the nanoparticles decreased reproductive potential of the nematode and increased expression of the superoxide dismutases-3 *(sod-3)* and abnormal dauer formation protein *(daf-12)* genes. Functional genomic studies using mutant analyses also proved the expressions of *sod-3* and *daf-12* gene and it has been related to failure in the reproductive potential of *C. elegans*. The oxidative stress was also attributed as an important mechanism for the toxicity of AgNPs.

Lim *et al.* (2012) elaborately studied the mechanism of AgNPs on *C. elegans* by focusing the involvement of oxidative stress for the failure of reproductive capacity of nematodes. Initially the AgNPs tested as potential oxidative stress inducers increased the formation of reactive oxygen species (ROS) in *C. elegans*. Subsequently the potential upstream activated signaling pathway was studied in response to AgNPs exposure by paying special attention to the *C. elegans* PMK-1 p38 mitogen activated protein kinase (MAPK). The results indicated that AgNPs exposure led to increased ROS formation, increased expression of PMK-1 p38 MAPK and hypoxia inducible factor (HIF-1), glutathione S-transferase (GST) enzyme activity decreased reproductive potential in *C. elegans*. Therefore, the results suggested that oxidative stress is an important mechanism of AgNPs induced reproduction toxicity on *C. elegans* and that PMK-1 p38 MAPK played an important role in this mechanism. Similarly Yang *et al.* (2012) reported that AgNPs which were less soluble due to size or coating with polyvinylpyrrolidone and gum arabic also acted via oxidative stress to affect target organism of *C. elegans*.

6.11.2. Suppression in nematode development

Meyer *et al.* (2010) studied the physicochemical, behavior, uptake, toxicity and mechanism of AgNPs with *C. elegans* being used as biological model and concluded that AgNPs inhibited the growth and development of the nematode.

6.12. CONCLUSION

There is a significant vision to utilize nanotechnology in vast, prospective ways relevant specially to plant nematode identification and management. It has been revealed that the application of various types of nanoparticles and nanocarriers and considering other relevant aspects on this technique play a vital role in managing the damage caused by nematodes, with special reference to root-knot, cyst, and other disease-causing nematodes. Various methods are employed to synthesize nanomaterials in order to attain very effective results. The direct application of nanoparticles expressively reduced the damage caused by phytonematodes in several economically important crops. Furthermore, other nanomaterials (nanocapsules, nanotubes, *etc.*) are also being used for transferring and control releasing of highly active components of biopesticides, organic pesticides, host-resistance-inducing chemicals, and inhibitors to manage pathogenic nematodes in soil and plants. Besides disease control, metal-based [Ag, Cu, Ti, Se, Au, cobalt (Co)] nanosensors or enzyme-based biosensors are extensively used in nematode disease diagnosis and in residue analysis of nematicides.

Nanotechnology holds the promise of controlled delivery of nematicides, to improve nematode management, plant growth enhancement and nutrient utilization. Nanoencapsulation shows the benefit of more efficient and targeted use of nematicides in environment friendly greener way. With the advancement of nanotechnology, application of green chemistry in synthesis of nanomaterials by using plant extracts and microorganisms has reduced the use of toxic solvents and guarantees eco-protection. Nanotechnology in conjunction with biotechnology has significantly extended the applicability of nanomaterials in crop protection. Even though the toxicity of nanomaterials has not yet clearly understood, it plays a significant role in crop protection because of its unique physical and chemical properties. The application of nanomaterials for nematode management is relatively new in the field of agriculture and it needs further research investigations. Barring the miniscule limitations, nanomaterials have a tremendous potential in making crop protection methodologies cost effective and environmentally friendly. Productivity enhancement through nanotechnology-driven precision farming and maximization of output and minimization of inputs through better monitoring and targeted action is desirable. Nanotechnology enables plants to use nematicides more efficiently.

Chapter - 7

Role of Enriched Vermicompost in Nematode Management

7.1. INTRODUCTION

There is an urgent need to improve crop production and waste disposal mechanism due to decrease in land availability for cultivation and exponential increase in human population. Organic waste composting is a technique which converts organic wastes into useful composts, which could be used as biofertilizer for sustainable agriculture growth. Conventional composting through microbes is a thermophilic process, in which many microbes are lost due to excess temperature emitted during the composting process. While vermicomposting is a mesophilic process, which conserves all microbes and earthworm associated with it to provide associated beneficial effect for degradation of organic matter by preserving the diverse community of all beneficial microflora.

The process of composting of organic material using earthworms is known as vermicomposting. Vermicompost is a product prepared by composting which is an accelerated biodegradation process of organic materials using earthworms and various organisms through non-thermophilic decomposition. Earthworms directly influences the microbial community of soil and it maintains normal chemical and physical properties of soil, due to which it is popularly called the "farmer's friend".

Vermicompost provides more biologically active and nutritive biofertilizers in soil as earthworms transform different organic waste material into useful vermicompost material by grinding, churning and digesting these substances in association with microbes which is essential in biogeochemical processes (Maboeta and Rensburg, 2003). Earthworms enhance the beneficial microbes and suppresses harmful microbes to convert different infectious hospital wastes into risk-free materials (Mathur et al., 2006).

Vermicompost contains higher concentration of nutrients and is basically used as organic fertilizer. They are rich in many nutrients including calcium, nitrates, phosphorus and soluble potassium, which are essentially required for plant growth (Edwards and Burrows, 1988). Different plant growth hormones like gibberellins, auxins and cytokinins are also present in vermicompost, which has microbial origin.

Earthworm causes increase in availability of soil organic matter through degradation of dead matters by microbes, leaf litter and porosity of soil. Vermicompost is a non-thermophilic biodegradation process of waste organic material through the action of microorganism with earthworm. The product produced through vermicompost is highly fertile, very fine soil particles with marked porosity, adequate aeration, low C: N ratios and high water-holding capacity (Domínguez and Edwards, 2010). The termed "drilosphere" is coined for microflora and microfauna in soil influenced by earthworms (Lavelle and Pashanasi, 1989).

Vermicompost-treated plants show accelerated growth which might be due to the presence of growth hormones, micronutrients such as carotenoids, flavones, and phenolic compounds in vermicompost (Kumar *et al.*, 2011). It was seen that vermicompost not only exhibited more potentiality as nematicide but probably also acted as better growth promoter. Vermicompost can also exert some beneficial effects in the management of plant parasitic nematodes.

7.2. NUTRITIONAL COMPOSITION

Vermicomposting enhances levels of different material in casted soil than available mineral concentration due to microbial activity in its gut (Edwards and Bohlen, 1996). Hand *et al.* (1988) reported that the earthworms enhance nitrogen mineralization in the soil, consequently resulting in more availability of nitrate in the soil. The vermicompost is also involved in reduction of organic carbon and carbon nitrogen ratio than in the normal composts. The combined earthworm and microorganism action causes lowered loss of different organic matter from the soil substrates as CO_2 introduces 20–43% of total organic carbon material in soil after the completion of vermicomposting period. Vermicompost also contains all essential nutrients including nitrates, phosphate, exchangeable calcium and soluble potassium which are quickly absorbed by plants (Edwards, 1998; Atiyeh *et al.*, 2002). Also observed more micro and macro nutrients in the vermicompost which are rich in the earthworm casts.

7.2.1. C/N ratio

The carbon and nitrogen (C/N) ratio is most important parameter during composting process which clearly indicates about the decomposition rate. Plants are able to take mineral nitrogen in the form of nitrates, only when carbon and nitrogen ratio falls below 20 (Dash and Senapati, 1985). The proper ratio of carbon and nitrogen is therefore required for the proper plant growth. Earthworms cause reduction in carbon level thereby increasing the nitrogen content in fresh organic matter.

7.2.2. Nitrogen

Nitrogen is very essential constituent of all amino acids and protein. Deficiency of nitrogen directly decreases the growth of plants leading to chlorosis, stunted and slow growth. According to Hand *et al.* (1988) mineralization with nitrogen was highly facilitated in earthworm presence and it leads to deposition of nitrate in the soil.

7.2.3. Phosphorus

Earthworms activity causes increase in total phosphorus concentration in soil in comparison to the food source available in soil. This clearly indicates that the vermicomposting causes increase in phosphorus level through the mineralization of phosphoric organic compounds (Hartenstein, 1983; Mitchell and Edwards, 1997).

7.2.4. Iron (Fe)

Iron (Fe) is also an important element required for growth and productivity of all plants. Only very trace amount of iron is required in comparison to other minerals by plant like carbon, oxygen, hydrogen, nitrogen, phosphorus, sulfur and potassium for proper plant growth. The iron functions like a cofactor, as it has a catalytic site for many essential enzymes activity which are even required for chlorophyll synthesis.

7.2.5. Magnesium (Mg)

It is an important component used in formation of chlorophyll, which play vital role in photosynthesis. It is also required for carbohydrate metabolism and acts as enzyme activator in nucleic acid synthesis. Magnesium serves as a carrier of phosphate compound in plants and also supports uptake of many essential elements into plant. It enhances production of oils and fats through the translocation of carbohydrates.

7.2.6. Manganese (Mn)

Manganese (Mn) plays vital role in nitrogen assimilation as enzyme activator. It is very important constituent of chlorophyll. Low plant manganese usually causes leaves to turn yellow due to reduced chlorophyll content. Organic soils usually contain intermediate amounts of manganese.

7.2.7. Zinc (Zn)

Low presence of zinc leads to high yield of crops. Zinc efficiency has been reported in many enzymatic activities of plants (Rengel, 2001). Zinc utilization mechanism in plant tissue is most important mechanism of zinc in plant tissues. Heavy metal bioaccumulation study showed that increased duration of vermicompost concentration of Zn and Cu decreases heavy metal bioaccumulation in soil (Hobbelen *et al.*, 2006).

It is observed that the worm castings contain higher percentage (nearly two-fold) of both macro and micronutrients than the garden compost (Table 7.1) (Nagavallemma *et al.*, 2004).

Table 7.1. Nutrient composition of vermicompost and garden compost are given.

Nutrient element	Vermicompost (%)	Garden compost (%)
Organic carbon	9.8 - 13.4	12.2
Nitrogen	0.51 - 1.61	0.8
Phosphorus	0.19 - 1.02	0.35
Potassium	0.15 - 0.73	0.48
Calcium	1.18 - 7.61	2.27
Magnesium	0.093 - 0.568	0.57
Sodium	0.058 - 0.158	<0.01
Zinc	0.0042 - 0.110	0.0012
Copper	0.0026 - 0.0048	0.0017
Iron	0.2050 - 1.3313	1.1690
Manganese	0.0105 - 0.2038	0.0414

7.3. ROLE OF VERMICOMPOST IN PLANT GROWTH PROMOTION

Use of vermicomposts as biofertilizers has been increasing recently due to its extraordinary nutrient status, and enhanced microbial and antagonistic activity. Vermicompost produced from different parent material such as food waste, cattle manure, pig manure, *etc.*, when used as a media supplement, enhanced seedling growth and development, and increased productivity of a wide variety of crops (Edwards and Burrows, 1988; Wilson and Carlile, 1989; Buckerfield and Webster, 1998; Edwards, 1998; Subler *et al.*, 1998; Atiyeh *et al.*, 2000c). Vermicompost addition to soil-less bedding plant media enhanced germination, growth, flowering and fruiting of a wide range of greenhouse vegetables and ornamentals (Atiyeh *et al.*, 2000a, b, c), marigolds (Atiyeh *et al.*, 2001), pepper (Arancon *et al.*, 2003a), strawberries (Arancon *et al.*, 2004b) and petunias (Chamani *et al.*, 2008). Vermicompost application in the ratio of 20: 1 resulted in a significant and consistent increase in plant growth in both field and greenhouse conditions (Edwards *et al.*, 2004), thus providing a substantial evidence that biological growth promoting factors play a key role in seed germination and plant growth (Edwards and Burrows, 1988; Edwards, 1998). Investigations revealed that plant hormones and plant-growth regulating substances (PGRs) such as auxins, gibberellins, cytokinins, ethylene and abscisic acid are produced by microorganisms (Barea *et al.*, 1976; Arshad and Frankenberger, 1993).

Wide variety of plant species grows effectively in vermicompost rich soil, including many horticultural crops like tomato, cauliflower *etc.* (Gutiérrez-Miceli *et al.*, 2007), aubergine (Gajalakshmi and Abbasi, 2004), garlic, pepper (Arancon *et al.*, 2005), strawberry, green gram and sweet corn (Lazcano *et al.*, 2011). Vermicompost is also very much effective in enhancing production of many medicinal plants rich in aromatic compounds (Prabha *et al.*, 2007), cereals such as rice and sorghum (Bhattacharjee *et al.*, 2001), fruit crops such as papaya and banana, and ornamentals like geranium (Chand *et al.* 2007), petunia, marigolds and poinsettia. Effect of vermicompost was also observed in forest trees including eucalyptus, acacia and pine tree (Donald and Visser, 1989). Vermicompost are very beneficial and used as a partial or total substitute for chemical fertilizer in agriculture and artificial greenhouse potting media. Likewise, few studies show that water-extracts obtained from vermicompost, vermiwash were used as foliar sprays, which enhances growth of tomato plants (Tejada *et al.*, 2008), strawberries and sorghum. Vermicompost also stimulates seed germination in green gram and other plant species (Karmegam *et al.*, 1999), tomato plants (Zaller, 2007), pine trees and petunia. Vermicomposts are used effectively for vegetative growth of leaf, stimulating growth of root and shoot (Edwards *et al.*, 2006). These effects cause increase in root branching and leaf area and alterations in morphology of seedling plant (Lazcano *et al.*, 2009). Vermicompost stimulates flowering in plants, increasing flowers produced (Arancon *et al.*, 2008), and increase in fruit yield (Singh *et al.*, 2008).

7.4. BACTERIAL DIVERSITY ASSOCIATED WITH EARTHWORMS

A variety of bacterial species have been reported associated with earthworms/ vermicompost though the bacterial species varied with its isolation site including soil, intestine, and excrements. Almost 43 bacterial species were isolated from earthworm intestines and 25 obtained from fresh excrements of which, 9 were common. Among 40 bacterial species isolated from soil and intestine, 13 were shared species; 9 were gram-positive, and 6 *Bacillus* species were spore forming. Comparison of soil and excrements bacteria revealed similarity of only 6 isolated species, of which three species were gram-positive and three species were gram-negative. *Brevundimonas diminuta* (α-Proteobacteria), *Kocuria palustris* (Actinobacteria) and *D. acidovorans* (β-Proteobacteria), were isolated from all three substrates. Comparison of bacteria isolated from the intestine of *Aporrectodea caliginosa*, *Lumbricus terrestris*, and *Eisenia fetida* earthworms revealed that the highest number of 43 bacterial taxa was isolated from *A.*

caliginosa digestive tract; while from *L. terrestris* and *E. fetida*, 22 and 21 taxa were isolated, respectively. Few members of bacteria were isolated from all earthworm species, which includes Bacteroidetes (classes Flavobacteria and Sphingobacteria), Actinobacteria, Proteobacteria (classes α-, β-, γ-) and Firmicutes (class Bacilli). Five bacterial species isolated from earthworm exhibited relatively low similarity between the sequenced 16S rRNA gene fragments (approx. 1490 nucleotides) and the genes of known bacterial taxa (93–97%), which includes *Ochrobactrum* sp. 341-2 (α-Proteobacteria), *Sphingobacterium* sp. 611-2 (Bacteroidetes), *Massilia* sp. 557-1 (β-Proteobacteria), *Leifsonia* sp. 555-1, and a Microbacteriaceae, isolate 521-1 (Actinobacteria). Micromycetes were observed in digestive tracts of fasted earthworm species. The incubation temperature had no effect on the number of fungal CFU isolated from the intestines. Fungi isolated from the earthworms after 20 days of starvation, are *Bjerkandera adusta* and *Syspastospora parasitica* identified by light-colored sterile mycelia, as well as *Geotrichum candidum*, *Alternaria alternata*, *Acremonium murorum* (*A. murorum* var. *felina*), *A. versicolor*, *Aspergillus candidus*, *Rhizomucor racemosus*, *Mucor hiemalis*, *Cladosporium cladosporioides*, Fusarium (*F. oxysporum, Fusarium* sp.), and *Penicillium* spp.. The density of fungal colony in the air-dry intestine was 103–104 CFU; this value is very close to the fungal populations density in soil mineral horizons. These fungi are most resistant to the conditions within earthworm digestive tract.

7.5. ENRICHMENT WITH BENEFICIAL MICROORGANISMS

7.5.1. Enrichment with bacteria

Earthworm's gut microflora has high ability to increase plant nutrient availability. They highly influence the soil dynamics and chemical processes, by adding its litter and affecting the soil micro-flora activity (Edwards and Bohlen, 1996). Earthworms and microorganisms interaction seem to be very complex. They excrete plant growth-promoting substances and making soil fertile. *Pseudomonas oxalaticus* an oxalate-degrading bacterium was isolated from intestine of different species of earthworm and *Streptomyces lipmanii* from actinomycetes group was identified in the gut of *Eisenia lucens*. Scanning electron micrographs showed presence of endogenous microflora in guts of earthworms, *L. terrestris* and *Octolasion cyaneum*. Gut of *E. foetida* contained various anaerobic N_2-fixing bacteria such as C. *Beijerinckii, Clostridium butyricum* and *C. paraputrificum*.

7.5.2. Enrichment with fungi

A total of 194 fungal entities comprising 117 mitosporic fungi, 45 ascomycetes, 15 zygomycetes, 14 SM morphotypes and three basidiomycete morphotypes were reported from the vermicompost. Mitosporic fungi including the ascomycetes in their anamorphic state are the most dominant. The thermotolerant fungus, Scedosporium state of *Pseudallescheria boydii* also display a significantly high load in vermicompost, However *Penicillium* and *Aspergillus* showed highest load in vermicompost.

7.5.3. Method for enrichment of vermicompost with bioagents

» One ton of vermicompost has to be enriched by mixing with 2 kg of each of *Pseudomonas fluorescens* + *Trichoderma harzianum* + *Purpureocillium lilacinum*. It has to be covered with mulch and optimum moisture of 25 - 30% has to be maintained for a period of 15 days.

» The inoculated vermicompost should be thoroughly mixed once a week for maximum multiplication and homogenous spread of the microorganisms in the entire lot of vermicompost (Fig. 7.1).

Fig. 7.1. Mixing of bioagents with vermicompost.

7.6. NEMATODE MANAGEMENT

It has been well documented that addition of organic amendments decreases the populations of plant parasitic nematodes (Addabdo, 1995; Sipes *et al.*, 1999; Akhtar and Malik, 2000). Vermicompost amendments appreciably suppress plant parasitic nematodes under field conditions (Arancon *et al.*, 2003). Vermicomposts also suppressed the attack of *Meloidogyne incognita* on tobacco, pepper, strawberry and tomato (Swathi *et al.*, 1998; Edwards *et al.*, 2007; Arancon *et al.*, 2002; Morra *et al.*, 1998) and decreased the numbers of galls and egg masses of *Meloidogyne javanica* (Ribeiro *et al.*, 1998).

The application of vermicompost resulting in reduction of free-living nematodes populations owing to its adverse effects on these nematodes. It was seen that the introduction of vermicompost at 1 kg/m^2 considerably impaired the reproduction of *Meloidogyne incognita* in tobacco plants (Swathi *et al.*, 1998). Morra *et al.* (1998) reported that application of solid vermicompost effectively suppressed the population as well as attack of *Meloidogyne incognita*. Arancon *et al.* (2002, 2003) conducted a series of field experiments on suppression of plant parasitic nematodes by application of solid vermicompost in tomato, pepper, strawberries, and grapes. They revealed that application of solid vermicompost ranging from 2 to 8 kg per hectare can significantly reduce the population of plant parasitic nematodes. They also noticed an increased population of fungivorous and bacteriovorous nematodes in solid vermicompost-treated plots (Figs. 7.2 to 7.4). Application of vermicompost tea reduced penetration and hatching of nematode but not reproduction over a period of time. Effectiveness of vermicompost for suppression of plant parasitic nematodes such as root-knot nematode (*Meloidogyne* spp.) in particular has been significantly observed (Ribeiro *et al.*, 1998; D'Addabbo *et al.*, 2011; Mahalik and Sahu, 2018).

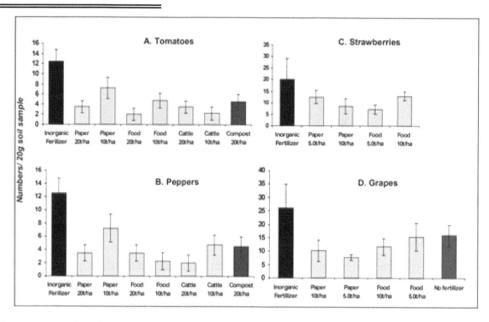

Fig. 7.2. Numbers (Means ± SE) of plant parasitic nematodes in inorganic fertilizer-treated , vermicompost-treated, compost-treated and unfertilized soils planted with tomatoes (A), peppers (B), strawberries (C), and grapes (D).

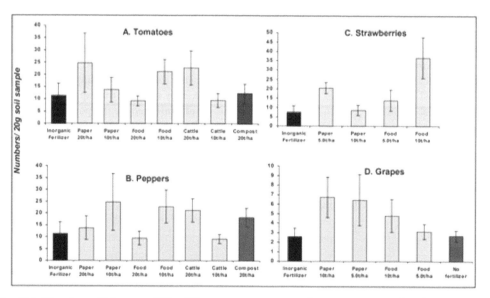

Fig. 7.3. Numbers (Means ± SE) of fungivorous nematodes in inorganic fertilizer treated, vermicompost-treated, compost-treated and unfertilized soils planted with tomatoes (A), peppers (B), strawberries (C), and grapes (D).

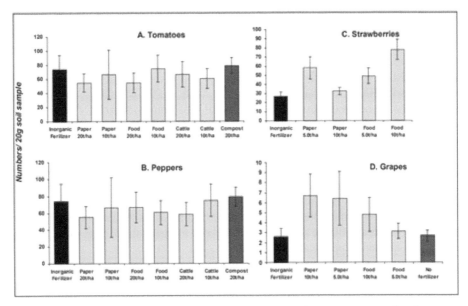

Fig. 7.4. Numbers (Means ± SE) of bacterivorous nematodes in inorganic fertilizer-treated, vermicompost-treated, compost-treated , and unfertilized soils planted with tomatoes (A), peppers (B), strawberries (C), and grapes (D).

Gabour *et al.* (2015) observed inhibitory effect of vermicompost application on the populations of the plant-parasitic nematode *Rotylenchulus reniformis*. In addition to vermicompost, recent studies have shown that the application of vermicompost tea has the potential to control plant-parasitic nematodes. In this sense, Edwards *et al.* (2007) studied a significant suppression in the number of galls caused by *Meloidogyne hapla* in tomato when the plants were subjected to aerated vermicompost tea. The effects of vermicompost are likely on nematodes due to the mortality of nematodes by the release of nematicidal substances such as hydrogen sulfate, ammonia, and nitrite produced (Rodríguez-Kábana, 1986). Promotion of the growth of nematode predatory fungi that attack their cysts (Kerry, 1998); favoring of rhizobacteria that produce toxic enzymes and toxins (Siddiqui and Mahmood, 1999); or indirectly by favoring populations of nematophagous microorganisms, bacteria, and fungi, which serve as food for predatory or omnivorous nematodes, or arthropods such as mites, which are selectively opposed to plant-parasitic nematodes (Bilgrami, 1997).

Commercial vermicomposts, produced from cattle manure, food and recycled paper wastes, were applied at rates of 5 t/ha, 10 t/ha and 20 t/ha, to field plots planted with tomatoes (*Solanum lycopersicum*), bell peppers (*Capsicum anuum grossum*), strawberries (*Fragaria ananasa*) or grapes (*Vitis vinifera*). Control plots were treated with inorganic fertilizers only, and all vermicompost treated plots were supplemented with inorganic fertilizers, to equalize levels of available N in all plots. Populations of plant-parasitic nematodes were depressed significantly by the three vermicomposts in all four field experiments compared with those in plots treated with inorganic fertilizer. Conversely, populations of fungivorous and bacterivorous nematodes tended to increase consistently compared with those in the inorganic fertilizer-treated plots.

Water soluble fraction (WSF) enriched with *Trichoderma* spp. stimulates plants such as tomato (Radin and Warman, 2011; Zandonadi *et al.*, 2016), *Callophyllum brasiliense* (Busato *et al.*, 2016), lettuce

(Pane *et al.*, 2014), bell pepper (Zaccardelli *et al.*, 2018) and maize (Zandonadi *et al.* 2019; Vujinovi´c *et al.* 2020). It can modulate the development of the root, the primary and secondary metabolisms, and the balance in hormones and reactive oxygen, resulting in enhanced nutritional efficiency and root-knot nematode (RKN) suppression (Zanin *et al.*, 2018; Pane *et al.*, 2014; Edwards *et al.*, 2010; Arancon *et al.*, 2020). WSF enriched with *Trichoderma* boosts the yield of tomato and bell pepper plants infected with RKN (dos Santos Pereira *et al.*, 2020). A 10-fold increase in yield of tomato plants was observed when water-extractable fraction of vermicomposts enriched with *T. virens* was applied when compared to the control. Regarding biocontrol of RKN, the same treatment was able to reduce the number of J2 and eggs by 62% on bell pepper and 46% on tomato plants, respectively, when compared to the control. This study examined the potential physiological mechanisms related to the stress alleviation of *M. incognita* in infected tomato and bell pepper plants using WSF extracted from vermicompost of cow manure (WSFv), vermicompost enriched with *Trichoderma asperellum* (WSFta) or with *T. virens* (WSFtv). The present study tested two hypotheses: (i) the effect of the application of the water-extractable fractions from vermicompost is shown by the enhancement of nutrient concentration; and (ii) the addition of *Trichoderma* mitigates the damage of the plant growth infected-tomato and bell pepper plants by nematodes.

Soil application of vermicompost @ 2.5 tons/ha enriched with *Purpureocillium lilacinum* @ 2.5 kg + vermicompost @ 2.5 tons of /ha enriched with *Pseudomonas fluorescens* @ 2.5 kg/ha proved effective to manage root-knot nematode, *Meloidogyne* spp. and hence increased bitter gourd fruit yield (5.210 t/ha as compared to 2.090 t/ha in control) (Singh *et al.*, 2019).

Tuberose corms treatment with *Purpureocillium lilacinum* at 20 ml/ kg corms along with soil application of 2 tons/ha of vermicompost enriched with 5 l of *P. lilacinum* gave maximum increase in the plant growth parameters like spike length (71.25 cm), spike weight (44.34 g), root length (15.38 cm), root weight (7.18 g), reduced nematode gall index (1.5 as compared to 5.0 in control) and increased yield (26.4 t/ha as compared to 22.2 t/ha in control) (Grace *et al.*, 2019).

However, Mishra *et al.* (2017) observed that suppression of root-knot nematode (*Meloidogyne* spp.) by vermicompost tea has been inconsistent. In some reports, vermicompost does not show a positive result in reducing the population of plant parasitic nematodes. For example, Szczech *et al.* (1993) observed that vermicompost did not reduce *Heterodera schachtii*. Vermicompost exhibited no suppressive effects on the number of *M. hapla* galls infesting cabbage and tomato roots (Kimpinski *et al.*, 2003).

Vermicompost enriched with *Purpureocillium lilacinum* or *Streptomyces avermitilis* provided significantly effective control of *M. incognita* infecting eggplant. Vermicompost derived from municipal wastes showed a low C/N ratio (1: 14) with an excess of nitrogen that exhibited nematicidal activity against *M. incognita*. Enriched vermicompost with bio-agents showed a synergistic effect upon nematode population, root galling and number of females. The potential of vermicompost enriched with *P. lilacinum* and *S. avermitilis* as safe alternatives to control *M. incognita* infecting eggplant through an integrated management program and bring sustainability to agriculture (Khairy *et al.*, 2021).

7.7. MANAGEMENT OF DISEASE COMPLEXES

Root knot nematode (*Meloidogyne incognita*) and soft rot bacterium *Pectobacterium carotovorum* sub sp. *carotovorum* are responsible for root rot disease complex in carrot (*Daucus carota*) throughout the world, reducing both quality and quantity of the marketable yield. Seed treatment and soil application of 2 tons ha⁻¹ of vermicompost enriched with 5 L of *Bacillus subtilis* IIHR BS-2 recorded maximum increase in plant growth parameters and root yield (28.8%), maximum reduction in *M. incognita* population (69.33%) and soft rot disease incidence (70.20%) (Rao *et al.*, 2017).

The root-knot nematode, *Meloidogyne incognita* interacts with *Fusarium oxysporum* f. sp. *melonis* in causing a wilt disease complex in gherkin. Under field conditions, seed treatment (20 ml kg⁻¹) and subsequent soil application of vermicompost (2 t ha⁻¹) enriched with BCA consortia (5 L) (*Trichoderma viride* IIHR TV-2 and *Bacillus subtilis* IIHR BS-21) before planting, followed by monthly application at 1 t ha⁻¹ showed maximum increase in the plant growth parameters and reduction in nematode population in soil (66.37%), roots (62.45%) and per cent wilt incidence (42.56%) compared to untreated plants. Marketable gherkin yields also increased by 34.42% over the control. The efficacy of vermicompost enriched with microbial consortia in managing pathogen complexes and can be included as promising component in integrated nematode and disease management for gherkin and other horticultural crops (Kamalnath *et al.*, 2019).

7.8. INTEGRATED NEMATODE MANAGEMENT

Seed treatment with *Purpureocillium lilacinum* @ 5 ml/kg + soil application of vermicompost @ 2.5 ton/ha enriched with *P. lilacinum* (@ 10 ml/kg) recorded highest increase of 51.28 %, 87.0%, 55.53%, and 67.62% in plant height, root length, shoot dry weight, and root dry weight over untreated check, respectively with reducing final nematode population in soil (171.0 J2/200 cc soil) and in roots of okra (41.25/ 5g. root) with the lowest root knot index (2.0) followed by seed treatment with *Pochonia chlamydosporia* @ 5 ml/ kg + soil application of vermicompost @ 2.5 ton/ha enriched with *P. chlamydosporia* @ 10 ml/kg. Moreover, seed treatment with *P. lilacinum* @ 5 ml/kg + soil application of vermicompost @ 2.5 ton/ha enriched with *P. lilacinum* @ 10 ml/kg gave the highest fruit yield (7.19 tons/ha) which was 36.6% higher than untreated plot (5.26 tons/ha) and performed as the most economical treatment for root-knot nematode management in okra with highest incremental benefit: cost ratio of 2.75 (Mahalik and Sahu, 2018).

7.9. MODES OF ACTION

The mechanisms by which vermicomposts suppress plant diseases caused by the plant parasitic nematodes are still speculative but it may be due to increased competition from fungivorous and bacterivorous nematodes resulting from increased availability of food sources after vermicompost application. There is good evidence (Edwards, 1998) that earthworms greatly increase overall microbial activity in organic wastes greatly by providing fragmented organic materials for microbial growth of soil bacteria and fungi. Soils that were treated with inorganic fertilizers only had much less organic matter available for microbial growth compared to those in the vermicompost-treated soils.

Various mechanisms might be implicated in the suppression of phytoparasitic nematodes such as decomposition of the compost into the soil and ammonia production, stimulation of soil microbial biomass, and release of biocidal substances having nematicidal activity (Oka and Yermiyahu, 2002).

There are several feasible mechanisms that attribute to the suppression of plant parasitic nematodes by vermicompost application and it involves both biotic and abiotic factors. Organic matter addition to the soil stimulates the population of bacterial and fungal antagonists of nematodes (*e.g.*, *Pasteuria penetrans*, *Pseudomonas* spp. and chitinolytic bacteria, *Trichoderma* spp.), and other typical nematode predators including nematophagous mites viz., *Hypoaspis calcuttaensis* (Bilgrami, 1997), Collembola and other arthropods which selectively feeds on plant parasitic nematodes (Thoden *et al.*, 2011). Vermicompost amendment promoted fungi capable of trapping nematode and destroying nematode cysts (Kerry, 1988) and increased the population of plant growth-promoting rhizobacteria which produce enzymes toxic to plant parasitic nematodes (Siddiqui and Mahmood, 1999). Vermicompost addition to soils planted with tomatoes, peppers, strawberry and grapes showed a significant reduction of plant parasitic nematodes and increased the population of fungivorous and bacterivorous nematodes compared to inorganic fertilizer treated plots (Arancon *et al.*, 2002). In addition, few abiotic factors viz., nematicidal compounds such as hydrogen sulphide, ammonia, nitrates, and organic acids released during vermicomposting, as well as low C/N ratios of the compost cause direct adverse effects while changes in soil physico-chemical characterises *viz.*, bulk density, porosity, water holding capacity, pH, EC, CEC and nutrition possess indirect adverse effects on plant parasitic nematodes (Rodriguez-Kabana, 1986; Thoden *et al.*, 2011).

The effects of applications of vermicomposts to soils were much greater on fungivorous nematode populations than on bacterivorous nematode populations. Earthworms depend upon fungi as a main source of food and tend to increase fungal activity in their casts by excreting fungal spores (Edwards and Fletcher, 1988) which may also explain why there were greater increases in populations of fungivorous nematodes than in those of bacterivorous nematodes. Moreover, plant parasitic nematodes are attacked by cyst fungi and nematode trapping fungi populations of which could have increased in response to vermicompost applications (Kerry, 1988). The greater availability of microorganisms as a source of energy could increase the competitive ability of both bacterivorous and fungivorous nematodes as compared to plant parasitic nematodes.

7.10. CONCLUSION

The enhancement of nutrients and beneficial microbial population in the vermicompost is yet another important evolving trend where the vermicompost is value added with nutrients and or microorganisms resulting in improved growth, crop protection and yield of crop plants. Vermicompost supports a diverse microbiome, introduces and enhances populations of antagonistic microorganisms, releases nematicidal compounds, increases the tolerance and resistance of plants, and encourages the establishment of a "soil environment" that is unsuitable for PPNs. The successful combination of functional microorganisms with organic amendments gave effective management of soil-borne pathogens. This technology of enhancing the organic material such as vermicompost with biocontrol agents has several advantages in terms of socio-economic feasibility as it can reduce the cost incurred on biopesticides and be easily multiplied on a large scale by farmers themselves in their own farm. There is a vast scope for utilizing this microbe as a biopesticide for nematode disease management in open fields, polyhouses and organic farms across several crops. Vermicompost enhances soil biodiversity by promoting the beneficial microbes which in turn enhances plant growth directly by production of plant growth-regulating hormones and enzymes and indirectly by controlling plant pathogens, nematodes and other pests, thereby enhancing plant health and minimizing the yield loss. Hence, these biopesticides are beyond doubt to bring about a revolution in the field of pesticide industry as a safe alternative to several chemical fungicides and nematicides.

Chapter - 8

Seed Bio-Priming for Management of Nematodes

8.1. INTRODUCTION

Efficient seed germination is important for agriculture. Successful establishment of early seedling indeed requires a rapid and uniform emergence and root growth. Germination of orthodox seeds commonly implies three distinct phases (Fig. 8.1) consisting of Phase I: seed hydration process related to passive imbibition of dry tissues associated with water movement first occurring in the apoplastic spaces; Phase II: activation phase associated with the re-establishment of metabolic activities and repairing processes at the cell level; and Phase III: initiation of growing processes associated to cell elongation and leading to radicle protrusion. Phases I and III both involve an increase in the water content while hydration remains stable during Phase II. It is commonly considered that before the end of Phase II, germination remains a reversible process: the seeds may be dried again and remain alive during storage and able to subsequently re-initiate germination under favorable conditions.

Biopriming is a novel method of seed treatment that combines biological (inoculation of seed with beneficial organisms to protect seeds) and physiological aspects of disease management. It is recently used as an alternative method for controlling many seed and soil-borne pathogens using selected bacterial and fungal antagonists. Biological seed treatments provide an alternative to chemical control with added advantages of induced disease resistance, eco-friendly nature, and sustainable disease management.

Biopriming involves seed imbibition together with antagonistic bacterial/fungal inoculation of seed (Callan *et al.*, 1990). As other priming method, this treatment increases rate and uniformity of germination, but additionally protects seeds against the seed and soil -borne pathogens. Hydration of seeds infected with pathogens during priming can result in a stronger microbial growth and consequently impairment of plant health. However, applying antagonistic microorganisms during priming is an ecological approach to overcome this problem (Parvatha Reddy, 2013). Moreover, some bacterial/fungal antagonists used as biocontrol agents are able to colonize rhizosphere and support plant in both direct and indirect way after germination stage (Callan *et al.*, 1997). It was found that biopriming is a much more effective approach to nematode disease management than other techniques such as pelleting and film coating (Müller and Berg, 2008). Nowadays, the use of biopriming with plant growth-promoting bacteria (PGPB) as an integral component of agricultural practice shows great promise (Glick, 2012; Timmusk *et al.*, 2014).

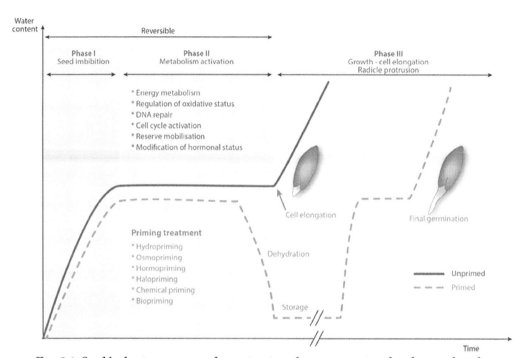

Fig. 8.1. Seed hydration curves and germinating phases in unprimed and primed seeds.

8.2. WHAT IS BIO-PRIMING OR BIOLOGICAL SEED TREATMENT?

Biological seed treatments for control of seed and seedling diseases offer the grower an alternative to chemical fungicides. While biological seed treatments can be highly effective, it must be recognized that they differ from chemical seed treatments by their utilization of living microorganisms. Storage and application conditions are more critical than with chemical seed protectants and differential reaction to hosts and environmental conditions may cause biological seed treatments to have a narrower spectrum of use than chemicals. Conversely, some biocontrol agents applied as seed dressers are capable of colonizing the rhizosphere, potentially providing benefits to the plant beyond the seedling emergence stage (Nancy *et al.* 1997).

Seed treatment with bio-control agents along with priming agents may serve as an important means of managing many of the soil and seed-borne diseases, the process often known as 'bio-priming'. The bio-priming seed treatment combines microbial inoculation with pre-plant seed hydration. Bio-priming involves coating seed with a bacterial biocontrol agent such as *Pseudomonas aureofaciens* AB254 and hydrating for 20 hr. under warm (23 °C) conditions in moist vermiculite or on moist germination blotters in a self-sealing plastic bag. The seeds are removed before radical emergence. The bacterial biocontrol agent may multiply substantially on seed during bio-priming (Callan *et al.* 1990).

PGPR application through seed bio-priming involves soaking the seeds for pre-measured time in liquid bacterial suspension, which starts the physiological processes inside the seed while radicle and plumule emergence is prevented (Anitha *et al.* 2013) until the seed is sown. The start of physiological process inside the seed enhances the abundance of PGPR in the spermosphere (Taylor and Harman, 1990). This proliferation of antagonist PGPR inside the seeds is 10-fold than attacking pathogens which enables the plant to survive those pathogens (Callan *et al.* 1990) increasing the use of biopriming for biocontrol too.

Bio-priming process had potential advantages over simply coating seed with *P. aureofaciens* AB254. Seed priming often results in more rapid and uniform seedling emergence and may be useful under adverse soil conditions. Sweet corn seedling emergence in pathogen infested soil was increased by *P. aureofaciens* AB254 at a range of soil temperatures, but emergence at 10 °C was slightly higher from bio-primed seeds than from seeds coated with the bacterium (Mathre *et al.* 1994).

Induction of defense-related enzymes by biocontrol agents and chemicals, osmotic priming of seeds using poly ethylene glycol (PEG) solution is known to improve the rate and uniformity of seed germination in several vegetable crops (Smith and Cobb, 1991). The observed improvements were attributed to priming induced quantitative changes in biochemical content of the sweet corn seeds (Sung and Chang, 1993).

A successful antagonist should colonize rhizosphere during seed germination (Weller, 1983). Priming with PGPR increase germination and improve seedling establishment. It initiates the physiological process of germination, but prevents the emergence of plumule and radical. Initiation of physiological process helps in the establishment and proliferation of PGPR on the spermosphere (Taylor and Harman, 1990). Bio-priming of seeds with bacterial antagonists increase the population load of antagonist to a tune of 10 fold on the seeds thus protected rhizosphere from the ingress of nematode pathogens (Callan *et al.*, 1990).

8.3. BIOAGENTS USED

18.3.1. Bacteria

» *Bacillus pumilus, B. subtilis, B. megaterium*

» *Pseudomonas aureofaciens, P. fluorescens*

» *Azotobacter chroococcum*

» *Azospirillum lipoferum*

» *Pastueria usage*

8.3.2. Fungi

» *Trichoderma harzianum, T. viride, T. hamatum*

» *Purpureocillium lilacinum*

» *Pochonia chlymydosporia*

» *Myrothecium verrucaria*

8.3.3. Arbuscular mycorrhizal fungi

» *Glomus mosseae*

8.4. PROCEDURE

» Pre-soak the seeds in water for 12 hours.

» Mix the formulated product of bioagent (*Trichoderma harzianum* and/or *Pseudomonas fluorescens*) with the pre-soaked seeds at the rate of 10 g per kg seed.

» Put the treated seeds as a heap.

» Cover the heap with a moist jute sack to maintain high humidity.

» Incubate the seeds under high humidity for about 48 hr. at approximately 25 to 32 °C.

» Bioagent adhered to the seed grows on the seed surface under moist condition to form a protective layer all around the seed coat (Fig 8.2).

» Sow the seeds in nursery bed.

» The seeds thus bioprimed with the bioagent provide protection against seed and soil borne plant pathogens, improved germination and seedling growth (Fig 8.3).

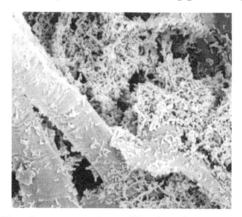

Fig 8.2. Multiplication of *Pseudomonas aureofaciens* **AB 254 on bioprimed tomato seed hairs surface.**

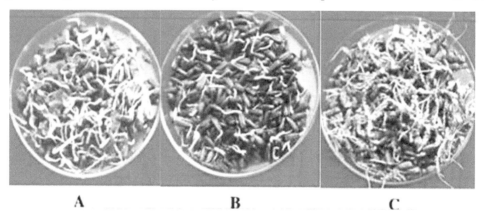

Fig 8.3. Effect of biopriming on seed quality parameters of sunflower. A. Bio-primed seeds with *Pseudomonas fluorescens* **in jelly, B. Control, C. Bio-primed seeds with** *Pseudomonas fluorescens* **in vermiculite**

Rice seed bio-priming with *Trichoderma harzianum/T viride/T virens/ Pseudomonas fluorescens or m*ixed formulation of *T harzianum and P fluorescens at* 5 - 10 g/kg of seed (Fig 8.4) gave effective control of various seed, soil and seedling nematode diseases. Biopriming also increases seed germination in tomato (Fig. 8.5).

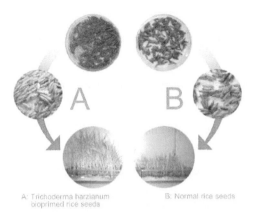

Fig 8.4. Rice seed bio-priming with *Trichoderma harzianum* **strain PBAT-43.**

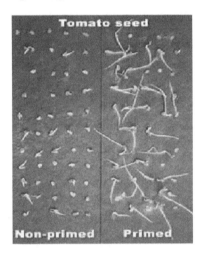

Fig 8.5. Increase in germination of tomato seeds by bio-priming.

8.5. NEMATODE MANAGEMENT

Management of seed and soil-borne nematodes using seed bio-priming is presented in Table 8.1.

Table 8.1. Management of seed-borne nematodes using bio-priming of seeds

Crop/s	Nematode/s
Wheat, Rye	*Anguina tritici*
Rice	*Aphelenchoides besseyi* *Ditylenchus angustus*
Groundnut	*Pratylenchus brachyurus* *Ditylenchus destructor*
Aster	*Aphelenchoides ritzema-bosi*

8.5.1. *Bacillus spp.*

Nematicidal microorganisms have been used as seed bio-priming treatments and one of the few examples of a commercially available biological seed treatment is the nematicidal bacterium *Bacillus firmus* - the active ingredient in the product is Poncho/Votivo (Bayer Crop Science, 2016). This product is used for control of plant parasitic nematodes on a range of crops including corn, cotton, sorghum, soybean, and sugar beet (Table 8.2) (Wilson and Jackson, 2013). As a spore-former, *B. firmus* is well suited to withstand the stresses associated with commercial seed treatment processes with the company claiming 2 years product stability under cool dry conditions. A range of other nematicidal microorganisms are used commercially.

Table 8.2. Approximate recommended application rates for *Bacillus firmus* seed treatments

Crop	Nematode	*Bacillus firmus*/seed	*Bacillus firmus*/ha
Maize[a]	*Heterodera* spp.	2×10^6	1.5×10^{11}
Cotton[b]	*Hoplolaimus* spp.	2×10^6	2×10^{11}
Sorghum[c]	*Criconema* spp.	2×10^5	3.7×10^{10}
Soybean[d]	*Pratylenchus* spp.	5.2×10^8	1.24×10^{11}
Sugar beet[e]	*Meloidogyne* spp.	2.4×10^6	2×10^{10}

[a] Assumes 74,000 plants/ha

[b] Assumes 1,00,000 plants/ha (irrigated rate)

[c] Assumes 1,000 grain weight = 25 g, sowing rate – 1,85,000/ha (irrigated rate)

[d] Assumes sowing rate 2,40,000 plants/ha

[e] Assumes sowing rate 85,000 plants/ha

Bio-priming of Soybean seeds with *Bacillus firmus* provides up to 60 days protection from soybean cyst nematode, *Heterodera glycines* and increased yield by 438-656 kg/ha over control. Bio-priming of Maize seeds with *Bacillus firmus* controlled needle (*Longidorus* sp.) and sting (*Belonolaimus* sp.) nematodes and increased yield by 525-700 kg/ha over insecticide (Poncho 250) (Table 8.3).

Table 8.3. Seed bio-priming using *Bacillus firmus* to manage soil-borne nematodes

Crop	Target nematodes	Increase in yield kg/ha	Reference
Soybean	Cyst nematode, *Heterodera glycines*	438-656 over check	Bayer Crop Science (2016)
Maize	Needle and sting nematodes	525-700 over pesticide	
Corn, cotton, sorghum, sugar beet, soybean	*Heterodera* spp., *Meloidogyne* spp. *Pratylenchus* spp.	---	

Treatment of tomato seeds with several strains of *B. subtilis* (Sb4-23, Mc5-Re2, and Mc2-Re2) significantly reduced the numbers of galls and egg masses of *M. incognita* compared with the untreated control (Adam *et al.*, 2014).

8.5.2. *Pasteuria* spp.

Pasteuria spp. are also well recognized as endospore-forming bacterial endoparasites of plant parasitic nematodes (Fig. 8.6). While difficult to mass produce, *Pasteuria* spp. have demonstrated potential as seed treatments for control of reniform nematodes: population control was comparable to a seed-applied nematicide (thiodicarb) at a seed coating application rate of 1.0×10^8 spores/seed (Schmidt *et al.*, 2010).

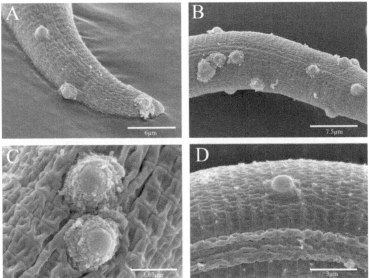

Fig. 8.6. Attachment and infection of *Pasteuria* spp. Pr3 on reniform nematodes.

Pasteuria spp. were observed to infect and complete their life-cycle in juvenile, male and female reniform nematodes. *Pasteuria* spp. germination occurred rapidly after attachment on day 1, with and without cotton plants. Mycelial structures and thalli were present by day 3 in soil with cotton plants and by day 7 in soil alone. Mature *Pasteuria* endospores were formed by day 15 in soil with cotton plants and by day 23 in soil alone.

The results of the seed treatment trial with 1.0×10^8 *Pasteuria* Pr3 endospores/seed or thiodicarb/imidacloprid showed 58.6% and 64.3% decrease in total reniform nematodes, respectively, as compared to the untreated control (Figs. 8.7). Equivalent reductions were observed in both the soil-borne and root-borne reniform nematode populations. The maximum soil spore loading rate of 1.2×10^8 spores/500 cm^3 reduced reniform nematode females by 82.3% as compared to the untreated control.

Pasteuria Pr3 endospores produced by *in vitro* fermentation demonstrated efficacy as a commercial bionematicide to control *R. reniformis* on cotton in pot tests, when applied as a seed bio-priming treatment. Population control was comparable to a seed-applied nematicide/insecticide (thiodicarb/imidacloprid) at a seed coating application rate of 1.0×10^8 spores/seed.

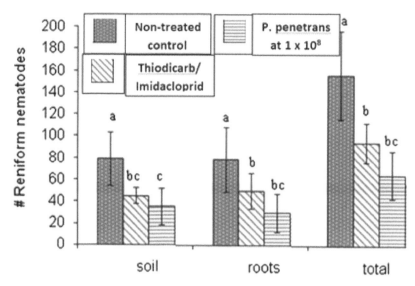

Fig. 8.7. Effect of seed bio-priming with *Pasteuria* **Pr3 endospores or the nematicide/insecticide; thiodicarb/imidacloprid on number of** *Rotylenchulus reniformis* **on cotton at 4 weeks post treatment. Different letters indicate statistical significance at P ≤ 0.05.**

Pasteuria Pr3 endospores produced by *in vitro* fermentation demonstrated efficacy as a commercial bionematicide to control *R. reniformis* on cotton in pot tests, when applied as a seed bio-priming treatment. Population control was comparable to a seed-applied nematicide/insecticide (thiodicarb/imidacloprid) at a seed coating application rate of 1.0 x 10^8 spores/seed.

8.5.3. *Pseudomonas fluorescens*

Biopriming tomato seed with plant growth promoting rhizobacteria (*Pseudomonas fluorescens, Bacillus* spp. and *Burkholderia* spp.) showed higher root length, shoot length, vigor index and reduced gall/knot formation after 21 days of seed germination. Among different PGPR isolates tested, *P. fluorescens* gave maximum increase in plant growth parameters and least root galling, followed by *Burkholderia* sp. and *Bacillus* sp. (Table 8.4, Fig. 8.8) (Gupta *et al.*, 2021).

Table 8.4. Effect of selected PGPR isolates on germination, vigor and root galling on tomato plants infected with *Meloidogyne incognita*.

PGPR isolates	Germination %	Seedling vigor	No. of galls/plant
Pseudomonas fluorescens	100 ± 0.0[d]	3880.66 ± 9.13[b]	0±0.0[a]
Enterobacter sp.	57.77 ± 8.8[b]	837.5 ± 52[c]	63 ± 6.5[ab]
Bacillus sp.	80 ± 0.0[cd]	1836 ± 19[cd]	0 ± 0.0[a]
Burkholderia sp.	93.33 ± 6.6[d]	2956 ± 28.23[d]	0 ± 0.0[a]
Control	42.22 ± 13.5[a]	335.1 ± 28[a]	51 ± 2.6[b]

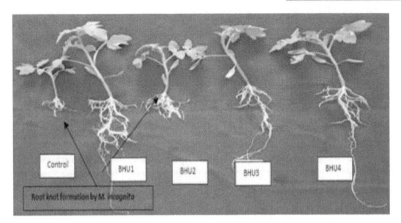

Fig. 8.8. Effect of selected PGPR isolates in tomato plants under nematode infection (BHU 1- *Pseudomonas fluorescens,* **BHU 2 -** *Enterobacter* **sp., BHU -** *Bacillus* **sp., BHU 4 -** *Burkholderia* **sp.).**

Seed biopriming with the bacterial strain, *Pseudomonas fluorescens* has a very good potential to augment plant growth activity and potential biological control managing root-knot nematodes in tomato. In future, it may lead to incorporation of other compatible microorganisms along with *P. fluorescens* in enhancing their positive attributes to replace chemical application with promising biocontrol agent.

8.5.4. *Glomus mosseae*

Black gram seed biopriming with *Glomus mosseae* suppressed *M. incognita* soil population by up to 14–49% under pot culture and 35–46% reduction over nematode alone treatment (Sankaranarayanan and Sundarababu, 2010). This reduction in the severity of disease caused by *M. incognita* in mycorrhizal plants might be due to the altered biochemical constituents in the host plant (Sikora and Schonbeck, 1975; Siddiqui *et al.,* 1999) or improved plant nutrition, especially phosphorus (Hussey and Roncadori, 1982) or alteration of compounds of root exudates or alteration of the physiological components of AMF root due to increased lignin levels in the exodermis of mycorrhizal plants (Dehne and Schonbeck, 1975). The presence of increased quantities of sugars, amino acids, like phenylalanine and serine and phosphorus may each or collectively play a role in suppressing the development of *M. incognita* in mycorrhizal plants (Krishnaprasad, 1971).

8.5.5. *Purpureocillium lilacinum*

Okra seed biopriming with *P. lilacinum* gave better response in the root-knot index and fruit yield than the untreated control. The elective effect of *P. lilacinum* was attributed to the toxic and enzymatic principles such as acetic acid, proteolytic acid, and chitinolytic enzymes released during fungal interaction with the nematode, which might cause killing or inhibiting *M. incognita* directly or indirectly promoting plant growth and ultimately enhancing yield component of okra (Mahapatra and Sahani, 2007).

Among seed biopriming treatments with *Bacillus subtilis, Pseudomonas fluorescens* and *Purpureocillium lilacinum; P. lilacinum* was found the most effective treatment on both galls and egg masses, achieving 88.23 and 76.94% reduction, respectively (Khalil *et al.* 2012). *P. lilacinum* strain UP1 was effective against *M. incognita* infecting tomato under pot experiments and significantly reduced the number of galls, nematodes and egg masses compared with Nemacur (Oclarit *et al.* 2009). Also, Kavitha *et al.* (2007)

indicated that *P. fluorescens* decreased nematode penetration and galling by 54 and 70%, respectively (Sharma *et al.*, 2008). *P. lilacinum* caused substantial egg deformation in *M. incognita*. These de formed eggs never matured or hatched (Jatala *et al.*, 1985). In addition to killing juveniles and females of *M. incognita* and *Globodera pallida* in the laboratory test, this fungus infects eggs of *M. incognita*. It destroys the embryos within five days because of simple penetration of the egg cuticle by individual hypha aided by mechanical and/or enzymatic activities (Jatala, 1986). Also, *P. lilacinum* suppressed root-knot infections, which resulted in fewer galls developing in the root system (Linderman 1992; Siddiqui *et al.* 2001; Prakob *et al.* 2007). The serine protease produced by *P. lilacinum* might play a role in penetration of the fungus through the eggshell of the nematodes (Bonants *et al.*, 1995). On the other hand, early developed eggs were more susceptible than the eggs containing fully developed juveniles. As observed by transmission electron microscopy, fungal hypha penetrated the *M. javanica* female cuticle directly (Khan *et al.*, 2006).

Seed treatment with *P. lilacinum* at the rate of 20g/kg seeds, nursery bed treatment @ of 50g *P. lilacinum*/m^2 and application of FYM (5 tons) enriched with 5kg of *P. lilacinum*/ha proved to be significantly effective in the management of *M. incognita* and significantly increased tomato yield (Rao *et al.*, 2012).

8.5.6. *Trichoderma harzianum*

The pot house studies revealed that seed biprimig with fungal and bacterial bio-agents, *viz. Trichoderma harzianum, T viride* and *Pseudomonas fluorescens* @ 5 g, 7.5 g and 10 g/kg seed against root-knot nematode. They found that the maximum plant growth parameters and a minimum number of galls/ plant, egg masses/plant, eggs/egg mass, nematode population/100cc, and total nematode population were recorded with *T. harzianum* @ 10 g/kg seeds followed by *T. viride* @ 10 g/kg seeds, *P. fluorescens* @ 10 g/kg seeds and *T. harzianum* @ 5 g/kg seeds over control (Nama and Sharma, 2017).

8.5.7. Several bioagents

Different fungal and bacterial antagonists (*T. harzianum, T. viride, P. lilacinum, P. chlamydosporia* and *P. fluorescens*) used as seed biopriming treatments gave effective control of root-knot nematode, *M. incognita* and increased the plant growth of okra plants sixty days after sowing in all treatments compared to the untreated control (Kumar *et al.*, 2012).

The combination of two fungal bioagents (*Aspergillus* spp. - toxic and *Purpureocillium* spp. - egg parasitic) was found more effective than a single bioagent against *M. incognita* infecting cowpea, resulting in better plant growth (Verma *et al.*, 2009). Both types of fungal bioagents (*Aspergillus terreus* - toxic and *P. lilacinum* - egg parasitic), used in a talc-based formulation, successfully managed root-knot nematode infecting tomato under field conditions (Goswami and Sharma, 1999; Kumar *et al.*, 2012). Therefore, this formulation could be proposed as an ideal component of an integrated pest management package (Mittal, 2006; Verma, 2009).

8.5.8. Consortium of *Pseudomonas fluorescens* (Pfbv22) and *Bacillus subtilis* (Bbv57)

Consortium application of *P. fluorescens* (Pfbv22) and *B. subtilis* (Bbv57) as seed treatment each @ 5 g kg^{-1} seed significantly reduced the root-knot (*M. incognita*) and reniform nematode (*R. reniformis*) infestation and enhanced the plant growth and fruit yield of tomato (Jonathan *et al.*, 2009).

8.6. MANAGEMENT OF DISEASE COMPLEXES

Seed biopriming with *T. harzianum* reduced the number of galls, root galling, egg masses and severity of root rot disease complex caused by the root-knot nematode *M. javanica* and the fungus *Rhizoctonia solani* disease complex on soybean plants under greenhouse conditions (Mahdy *et al.*, 2006). Similarly, significant increase in the plant growth and decrease in the final nematode population was obtained in soybean plants treated with *T. harzianum* and *P. fluorescens* over streptocycline (Barua and Bora, 2008). The experiment's findings on the management of disease complex caused by *M. incognita* and *R. solani* with seed treatment of *T. harzianum* and *P. fluorescens* on okra indicated that both the bio-agents to be significantly effective in reducing the damage and increasing the plant growth parameters of soybean (Bhagawati *et al.*, 2009).

8.7. MECHANISMS OF ACTION

Bio-priming is directly involved in the enrichment of plant development by the excretion of compounds and mineral solubilization (Sukanya *et al.*, 2018). Phosphorus solubilizing microorganism like *Bacillus, Beijerinckia, Enterobacter, Microbacterium, Pseudomonas* and *Serratia* release rock crystal dissolving compounds like organic anions, protons, hydroxyl ions, carbon dioxide or liberation of extracellular enzymes namely phosphatase leading to solubilization of the phosphorus and make it available to the plant (Zaidi *et al.*, 2009; Glick, 2012). Some PGPR used in bio-priming have the capacity to mobilize potassium from potassium-bearing minerals (Mica, Illite and Orthoclase) by emission of organic acids (citric acid, tartaric acid and oxalic acid) which directly dissolves the rock potassium or chelate the silicon ion (Sheng and He, 2006). Siderophore production by biocontrol agents in primed seeds has inhibited the nematode disease and improved the plant growth (Keswani *et al.*, 2014; Jain *et al.*, 2012). Bio-priming with PGPR in various crops having the capacity to produce growth hormones such as auxin, cytokinin, gibberellin *etc.* which recorded high germination rate, higher shoot and root growth and homogenous crop stands (Glick, 2012; Noel *et al.*, 1996; Verma *et al.*, 2001). Bio-priming is the best solution for better growth and expansion of micro-propagated plants which have abridged photosynthetic activity, poorly working stomata and undersized root and shoot system (Kavino *et al.*, 2010). Bio-priming with PGPR reducing the period required for lignification of micro-propagated plants and accelerates production process (Ramamoorthy *et al.*, 2002). Bio-priming leads to biochemical changes viz., enhanced production of proteins, hormones, phenol and flavonoid compounds contribute to enhanced plant growth and improved performance. Growth responses in herbaceous plants are calculated by nitrogen reserve compounds like nitrates, amino acids and proteins (Volenec *et al.*, 1996). Soluble protein fractions in bio-primed seeds and seedlings were found higher as compared to non-primed seeds (Dhanya, 2014). There was increase in total protein content and free amino acid content during the diverse growth stages after seed bio-priming with PGPR (Aishwath *et al.*, 2012; Warwate, 2017, Ahmed *et al.*, 2014). PGPR used in bio-priming enriched the production of particular phenolic substances in plants at diverse growth phases (Singh *et al.*, 2003). Moreover to plant growth preferment, seed bio-priming also encourages the production of defense-related enzymes (peroxidase, superoxide dismutase, catalase, chitinase, ammonia lyases, *etc.*) which deals with the plant fitness benefit against biotic and abiotic stress. The respiration, energy metabolism and early reserve mobilization events in crops were controlled by bio-priming (Chen *et al.*, 2013; Paparella *et al.*, 2015). Seed biopriming with biocontrol agents was also found useful in the development of induced systemic resistance in plants. Biocontrol agents, particularly rhizobacteria, have been shown to be effective in defeating nematode disease infection by encouraging a resistance mechanism called "induced systemic resistance" (ISR) in varied crops (Van Loon *et al.*, 1998). Induced resistance is defined as encouragement of plants with enhanced defensive ability of plants against different nematode plant pathogens.

8.8. CONCLUSION

Bio-priming seed treatments can provide a high level of protection against plant parasitic nematodes. This protection was generally equal or superior to the control provided with nematicide seed treatment. So, it could be suggested that bio-priming (combined treatments between seed priming and seed coating with biocontrol agents) may be safely used commercially as substitute for traditional nematicide seed treatments for controlling seed and soil-borne plant pathogens. Besides, bio-priming also improve seed germination, seedling establishment and vegetative growth. This has explored up new dimension of biological control for preventive as well as remedial for seed-borne infection by bioagents. There is a need to undertake considerable research and development to commercialize the successful use of bio-seed priming. Thus, bio-priming can be exploited by seed companies and organic farmers in the sustainable agriculture, which would be more economical and environment friendly.

Chapter - 9

Avermectins: Promising Solution for Phytonematode Management

9.1. INTRODUCTION

Globally, farmers are still depending on chemical nematicides to control PPNs, because of their effectiveness, but the environmental aspects of these synthetic pesticides was drastically devastating (Khalil and Darwesh, 2018). Avermectins are one of new alternatives which proved its activity towards different genera of plant parasitic nematodes (Jansson and Rabatin, 1998; Cabrera *et al.*, 2009; Ibrahim *et al.*, 2013; Bi *et al.*, 2015; El-Tanany *et al.*, 2018; Khalil and Alqadasi, 2019). Avermectins are sub-class of natural products that consisting of a large macrocyclic lactone ring which produced from metabolites of Gram-positive bacterium, *Streptomyces avermectinius* (Khalil and Abd El-Naby, 2018).

The avermectins are a new class of macro-cyclic lactones derived from mycelia of the soil actinomycete, *Streptomyces avermectinius* (soil inhabiting which is ubiquitous in nature). These compounds were reported to be possessing insecticidal, acaricidal and nematicidal properties (Putter *et al.*, 1981). They are commonly distributed in most of the cultivated soils and are in widespread use, especially as agents affecting plant parasitic nematodes, mites and insect pests. The water solubility of avermectin B1 is approximately 6-8 ppb and its leaching potential through many types of soil is extremely low. These physical properties also confer many advantages upon the use of avermectins as pesticides. Their rapid degradation in soil and poor leaching potential suggest that field applications would not result in persistent residues or contamination of ground water.

Avermectins offer an outstanding alternative to any of the available synthetic pesticides. Their novel mode of action, high potency and specific physico-chemical properties makes the avermectins excellent candidates for further insecticidal, acaricidal and nematicidal studies.

Abamectin and emamectin are members of avermectin family which categorized as very effective nematicides which proved capability of reducing PPNs significantly in various crops.

9.2. DISTINGUISHING CHARACTERISTICS OF *STREPTOMYCES AVERMECTINIUS*

» Brownish-grey spore color.

» Smooth spore surface.

» Spiral sporophore structure (Fig 9.1).

» Spores in chain (Fig 9.1).

» Production of melanoid pigments.

» Cultural and carbon utilization patterns.

» Preferred temperatures of 28°C and 37°C for growth.

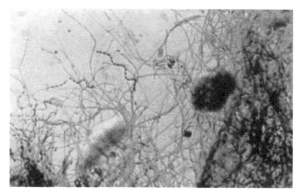

Fig. 9.1. Spiral sporophores of *Streptomyces avermectinius* **and spores in chain.**

9.3. ABAMECTIN

Abamectin is a blend of B_{1a} and B_{1b} avermectins. The chemical structure of abamectin is depicted in Fig. 9.2. Avermectin B1 (abamectin), the major component of the fermentation, also showed potent activity against arthropods in preliminary laboratory evaluations and was subsequently selected for development to control plant nematodes, phytophagous mites and insect pests on a variety of agricultural and horticultural crops worldwide. Major applications for which abamectin is currently registered include uses on ornamental plants, citrus, cotton, pears and vegetable crops at rates in the range of 5 to 27 grams abamectin per hectare as a foliar spray. Abamectin has shown low toxicity to non-target beneficial arthropods which has accelerated its acceptance into integrated pest management programs. Extensive studies have been conducted to support the safety of agricultural uses of abamectin to man and the environment. Abamectin is highly unstable to light and has been shown to photo degrade rapidly on plant and soil surfaces and in water following agricultural applications. Abamectin was also found to be degraded readily by soil microorganisms. Abamectin residues in or on crops are very low, typically less than 0.025 ppm, resulting in minimal exposure to man from harvesting or consumption of treated crops. In addition, abamectin does not persist or accumulate in the environment. Its instability as well as its low water solubility and tight binding to soil, limit abamectin's bioavailability to non-target organisms and, furthermore, prevent it from leaching into groundwater or entering the aquatic environment.

Abamectin, the active ingredient in Avicta seed treatment nematicide, is composed of two molecules that are produced by the soil micro-organism, *Streptomyces avermictinius*. Abamectin interferes with the signal transmission between nerve cells inside the nematode at a novel target site, the GABA receptor protein. Avicta is a contact nematicide, in that the active ingredient kills the nematode immediately upon contact and does not allow the nematode to feed or reproduce.

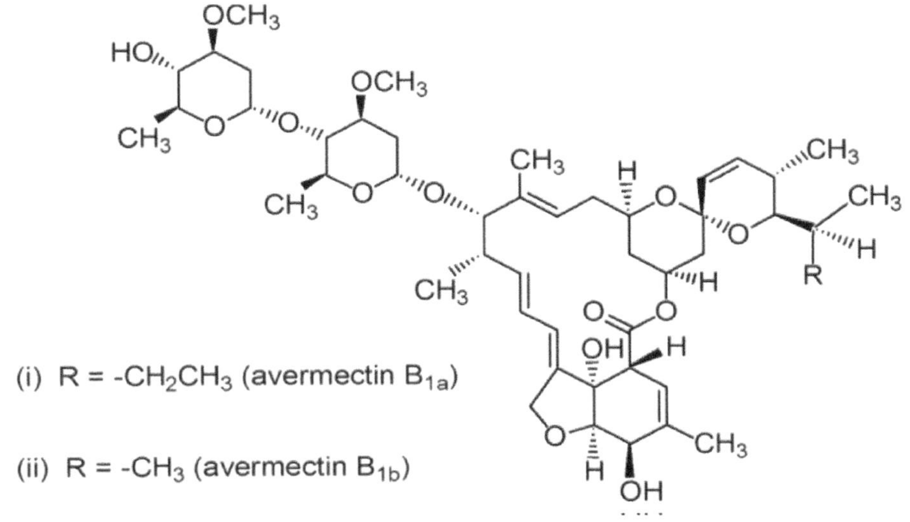

(i) R = -CH₂CH₃ (avermectin B₁ₐ)

(ii) R = -CH₃ (avermectin B₁ᵦ)

Fig 9.2. The chemical structure of Abamectin (Avermectin B$_{1a}$ and B$_{1b}$).

9.3.1. Environmental aspects

The toxicity of abamectin is low towards non-target organisms. This aspect nominated abamectin to join into the integrated pest management (IPM) programs, as well as proved its safety to human beings and environmental components (Khalil, 2013).

Avermectins have relatively shorter residual activities. The stability of abamectin is moderate in environment. The half-life of abamectin under field conditions was about 31 ± 6 days, while the half-life was ranged between 20 and 47 days in soils with 5 - 9 pH. The photo-degradation occurs in thin films (6 hours) and water (12 hours), while it was 21 hours in soil (Boina *et al.*, 2009). Despite its rapid decomposition in various systems, abamectin still provides a relatively long residual activity against target pests in field conditions due to its translaminar activity (Wright *et al.*, 1985).

The systemic activity of abamectin is limited and the water solubility is very low because of its binding with soil particles tightly, resulting in poor movement of the product through the soil profile (Boina *et al.*, 2009; Bull *et al.*,1984; Chukwudebe *et al.*, 1996).

The residues of abamectin are very low in treated plants because of it is highly degraded readily by soil microorganisms. Moreover, the most avermectin degradation products have been reported to pose 1–3 times less toxicity than the parent compound. The temperature coefficient of abamectin is positive which mean that the toxicity increased with the increment of temperature till 37 °C (Boina *et al.*, 2009).

9.3.2. Bio-efficacy on nematodes

The avermectin B1 (Abamectin) proved its ability to manage different pests as an acaricide (Khalil, 2013; Dybas and Green, 1984), insecticides (Lumaret *et al.*, 2012), nematicide (Khalil and Alqadasi, 2019;

Radwan *et al.*, 2019)] and Molluscicide (Abdallah *et al.*, 2015; 2018; Abdelgalil, 2016). Meanwhile, abamectin was applied in many different methods such as soil treatment, seed treatment, injection into plant stem and seedling root dip for the management of plant parasitic nematodes (Table 9.1).

Table 9.1. Effect of abamectin on various plant parasitic nematodes in different crops

Compounds/ Application methods	Crop/Target nematode	Effect on nematode reduction parameters (%)	Reference
Vertemic (1.8% EC) Soil appln.	Rapeseed, *Meloidogyne incognita*	Popn.-94.2, Galls-61.5, Egg masses-92.5	Korayem, *et al.*, 2008
Abamectin (1.8% EC) Soil appln.	Eggplant, *M. incognita*	Galls-61.77, Egg masses-78.82	Shahid *et al.*, 2009
Abamectin Seed treat.	Maize, *Pratylenchus zeae*	Popn.-80	Cabrera *et al.*, 2009
	Sugar beet, *Heterodera schachtii*	Popn.-60	Cabrera *et al.*, 2009
	Cotton, *M. incognita*	Galls-80	Cabrera *et al.*, 2009
Vertemic (1.8% EC) Soil appln.	Tomato, *M. incognita*	Popn.-74.0, Galls-60.7	Khalil, 2012
Vertemic (1.8% EC) Soil appln.	Cabbage, *M. incognita*	Galls-40-88, Egg masses-58-98	Ibrahim *et al.*, 2013
	Cabbage, *H. schachtii*	Popn.-68-92	Ibrahim *et al.*, 2013
Vertemic (1.8% EC) Soil appln.	Tomato, *M. incognita*	Popn.-61.3, Galls-68.0, Egg masses-41.7	Youssef & Lashein, 2015
Tervigo (2% SC) Soil appln.	Tomato, *M. incognita*	Popn.-75.34, Galls-66.69, Egg masses-66.31, Eggs-16.34	Saad *et al.*, 2017
Tervigo (2% SC) Soil appln.	Orange trees, *Tylenchulus semipenetrans*	Popn.-78-87	El-Tanany *et al.*, 2018
	Tomato, *M. incognita*	Popn.-86-91, Galls-67-71, Egg masses-70-78, Eggs-40-63	Radwan *et al.*, 2019
	Tomato, *M. incognita*	Popn.-78.88-81.07, Galls-71.64-72.48	Khalil and Alqadasi, 2019

Avermectin is widely used in the field because of its excellent nematicidal activities; however, it is adsorbed by organic matter near the soil surface, resulting in poor migration in the soil, limiting its field performance. Many organic compound pesticides suffer from the same challenge. Six avermectin B1a (AVB1a) derivatives with phenyl carbamate groups were obtained and found to have increased water solubility and significantly decreased oil–water partition coefficients (K_{ow}) and soil adsorption coefficients (K_f). Two soil mobility experiments verified that all derivatives could move to a farther location compared with AVB1a. These derivatives could reduce nematode viability, though the nematicidal activities were lower by nearly 1.5–5-fold compared with AVB1a. In the field, the six derivatives exhibited excellent efficacy, 20–30% higher than that of the control compound. This research emphasizes the significance of uniform distribution in the soil by hydrophilic modification with phenyl carbamate that increased the delivery efficiency and bioavailability of the agent, thereby improving its efficacy (Fig. 9.3) (Jing *et al.*, 2020).

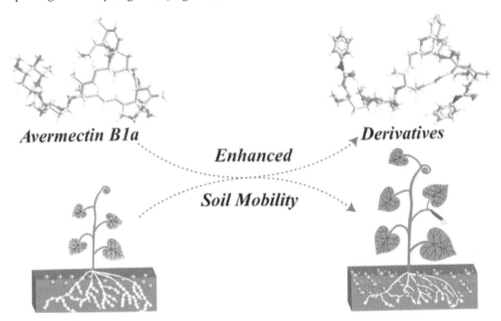

Fig. 9.3. Phenyl isocyanate-modified avermectin B1a improves the efficacy against plant-parasitic nematode diseases by facilitating its soil mobility.

9.4. EMAMECTIN BENZOATE

Emamectin benzoate is a novel avermectin derivative which belongs to the macrocyclic lactones family. Emamectin benzoate was developed as a pesticide by Merck and Company and was classified as a second generation of avermectins which is a derivative of abamectin. Emamectin benzoate is a biological pesticide that contains a mixture of the benzoic acid salt of two structurally complex heterocyclic compounds which were emamectin B1a (> 90%) and emamectin B1b (< 10%). The only differs between emamectin B1a and emamectin B1b is the presence of an additional methylene unit on the side chain at C-25 (Mushtaq *et al.*, 1998).

However, emamectin benzoate is consisting of a 16-membered ring macrolide and a di-saccharide via substitution of an epi-methylamino (-NHCH$_3$) group for a hydroxyl (-OH) group at the 4"-position on the di-saccharide. The limited potentiality of abamectin towards the most of Lepidoptera pushes the scientists to focus and try many prompts to get an alternative compound(s). According to these prompts they discovered of 4"-epi-methylamino -4"-deoxyavermectin B1a and b (emamectin) in 1984 (Fig. 9.4). Emamectin was derived from abamectin via a five step synthesis (Cvetovich *et al.*, 1994).

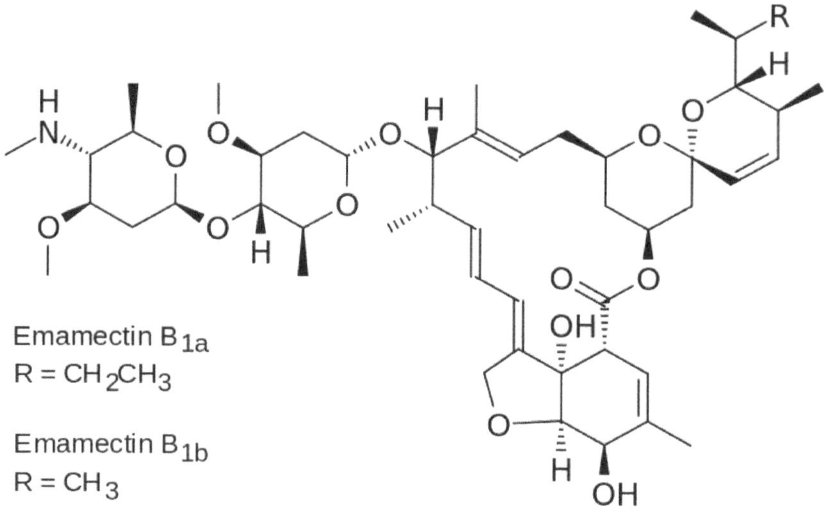

Fig. 9.4. The chemical structure of emamectin benzoate

Many investigations clarified that emamectin benzoate was applied as soil drench to control underground nematodes (Radwan *et al.*, 2019; Aatif *et al.*, 2016; Khalil, 2018) and as injection into the plant trunk/stem (Jansson and Dybas, 1998).

9.4.1. Environmental aspects

Because of that emamectin was derived from soil microbes it offers a promising solution to inhibit environmental pollution and crop losses due to the invasion of pests and diseases. The diversity in structure, activity, biodegradability and the eco-friendly properties make these proposed microbial derivatives, agricultural active agents of the future generation (Birtle *et al.*, 1982). Also, emamectin benzoate is safe towards non-target organisms, as well as it has low residue and environmental pollution (Zhou *et al.*, 2016; Han *et al.*, 2011; Shang *et al.*, 2013; Guo *et al.*, 2015). The both components of emamectin have similar structure; thus, their physicochemical properties, fate, and toxicity profiles are almost to be similar.

Emamectin benzoate showed thermal stability and greater water solubility, which then resulted in a broader spectrum of insecticidal activity than abamectin (Jansson and Dybas, 1998). The half-life of emamectin benzoate in water may reach to 7 days, but would be reduced to as short as one day if the water contained a natural photosensitizer such as humic acid. Meanwhile, the stability of emamectin benzoate's characteristics is depending on pH levels. For example, the solubility of emamectin in water reach to 320 mg/L at pH 5, 93 mg/L at pH 7, and 0.1 mg/L at pH 9. Similarly, its log K_{ow} is 5.0 at pH 7 and 5.9 at pH 9.

Also, the photo-degradation of emamectin benzoate in soils may reach to 22 days under sunlight. While in darkness the degradation process of emamectin is stable in soils (Mushtaq *et al.*, 1998; 1996). Previous studies revealed that photo-degradation may produce many byproducts under UV light (Mushtaq *et al.*, 1998; Feely *et al.*, 1992; Wrzesinski *et al.*, 1996; Crouch and Feely, 1995). The vapor pressure and Henry's constant of emamectin benzoate are low, therefore these suggest that volatility from soil and water will be low.

9.4.2. Bio-efficacy on nematodes

Emamectin benzoate is a novel semi-synthetic derivative antibiotic which has a wide spectrum activity against pests (Zhou *et al.*, 2016; Han *et al.*, 2011). However, certain reports indicated that emamectin benzoate was effective against different genera of plant parasitic nematodes according to the received information (Abbas *et al.*, 2008; Bi *et al.*, 2015; Khalil and Abd El-Naby, 2018), but emamectin was less effective than abamectin. Until now, there is no registered compound of emamectin benzoate to manage PPNs. The obtained data in the following Table 9.2 showed the impact of emamectin benzoate on different plant parasitic nematodes genera in various crops.

Table 9.2. The effect of Emamectin benzoate on various plant parasitic nematodes.

Compounds/ Application methods	Crop/Target nematode	Effect on nematode reduction parameters	Reference
Emamectin benzoate (5% SG), Stem injection	Banana, *Meloidogyne javanica*	Galls-46.67% to 86.67%	Jansson and Rabatin, 1998
	Banana, *Radopholus similis*	Popn.- 64.29% to 76.62% Root dip-81.73%	Jansson and Rabatin, 1998
Emamectin benzoate, Soil application	Tomato, *Meloidogyne incognita*	Egg masses-49.50% to 58.10%, Eggs-22.90% to 26.80%	Rehman, *et al.*, 2009
		Galls-73.4% to 87.7%	Cheng, *et al.*, 2015
	Eggplant, *Meloidogyne* spp.	Galls-43.75%, Egg masses-56.41%	Shahid *et al.*, 2009
		Galls-46.8% (females),	Atif *et al.*, 2016
	Pinewood trees, *Bursaphelenchus xylophilus*	Popn.- 98.03%, Eggs-85.61%	Bi *et al.*, 2015
Emamectin benzoate (Proclaim® 5% WG)	Tomato, *Meloidogyne incognita*	Popn.- 91.81%, Galls-71.65%, Egg masses-76.28%, Eggs-61.71%	Khalil, 2018
Soil application		Popn.- 60% to 63%, Galls-75% to 77%, Egg masses-60% to 62%, Eggs-90% to 93%	Radwan *et al.*, 2019

9.5. MODE OF ACTION

9.5.1. GABA antagonists

The avermectins affect invertebrates by potentiating the ability of neurotransmitters, such as glutamate and/or GABA, to stimulate and influx of chloride ions into nerve cells. The chloride ion flux produced by the opening of the channel into neurons results in loss of cell function and disruption of nerve impulses. Consequently, invertebrates are paralyzed irreversibly and stop feeding. The avermectins do not exhibit rapid knock down effect. The safety of avermectin to mammals is due to (i) the lack of glutamate-gated chloride channels in mammals, (ii) the low affinity of avermectins for other mammalian ligand gated chloride channels, and (iii) their inability to readily cross the blood-brain barrier.

The avermectins have a direct action on neurotransmission. They are antagonists of gamma-amino butyric acid (GABA) which is a neuromuscular transmitter. Avermectins irreversibly block post-synaptic potentials at a neuromuscular junction by reducing muscle-membrane resistance. While the reversible change involves an increase in chloride conductance in cells with GABA-receptors, an irreversible AVM-induces increases in chloride conductance in cells that lack GABA-receptors.

The mode of action of avermectins has been studied on plant parasitic nematodes in terms of their gross effects on the movement and infective behavior of the parasites. Juveniles of *Meloidogyne incognita* exposed to a 120 nM aqueous solution of avermectin B2a-23-ketone i) initially lost movement within 10 min. while being responsive to touch, ii) partially recovered within 30 min. of exposure, and iii) irreversibly lost movement after 120 min. (Wright *et al.*, 1984a). A similar triphasic response was also seen in *M incognita* juveniles exposed to avermectin B1. The initial loss of movement in *M incognita* may be reflective of avermectin's activity as GABA-antagonists as inhibitory synapses. Wright *et al.* (1984b) found that the GABA-antagonists picrotoxin and bicuculline counteract the effects of avermectins on the locomotion of *M incognita* juveniles.

Avermectins potentiate glutamate and GABA gated chloride-channel opening. A number of total syntheses of simpler analogues of the natural products were undertaken with the hope to get access to constructs containing the pharmacophore which would be biologically active, but more economical to prepare. None of these attempts led to a compound with high activity. On the other hand, the search for derivatives of the natural products with improved biological properties turned out to be very successful.

Recently, avermectin-binding proteins were detected. Three specific avermectin-binding polypeptides (53, 47 and 8 K Da) have been identified in membranes of *Caenorhabditis elegans* and one (approximately 47 K Da) in *Drosophila melanogaster*.

9.6. COMMERCIAL PRODUCTS

The commercial products of abamectin are presented in Table 9.3.

Table 9.3. Commercial products of abamectin

Commercial products	Pests controlled	Application/comments
Avid	Spider mites, leaf miners	Many beneficials can be released one week after use.
Avicta	Root-knot and reniform nematodes on cotton and vegetables	Seed treatment
Agri-Mek	Apples and pears as an acaricide/insecticide against European red mite and pear psylla, and spotted tentiform leaf miner.	Foliar spray

The suppliers and branded products of avermectins are presented in Table 9.4.

Table 9.4. Suppliers and branded products of avermectins

Branded products	Suppliers
Akomectin	AAKO BV
Mectinide	AgriGuard Ltd.
Avermk	Agro-Care Chemical Industry Group Limited
Greyhound	ArborSystems
AbaMecK, Transact	Astra Industrial Complex Co Ltd (Astrachem)
Zoro	Cheminova A/S
Temprano	Chemtura AgroSolutions
Abacin	Crystal Phosphates Ltd.
Denka-Flylure	Denka International BV
Fertimectin	Fertiagro Pvt. Ltd.
Gilmectin	Gilmore Marketing & Development Inc
Satin	Hubei Sanonda Co. Ltd.
Laotta	Lainco SA
Abba	Makhteshim Agan Group
Abba, Abba Ultra	MANA - Makhteshim Agan North America Inc
Pilarmectin	Pilar AgriScience (Canada) Corp
Romectin	Rotam CropSciences Inc.
Eagrow	Shandong Kesai Eagrow Co. Ltd.
Agri-Mek, Avicta, Avid, Clinch, Dynamec, Epi-Mek, Varsity, Vertimec, Zephyr	Syngenta
Alba	Wangs Crop-Science Co. Ltd.
Zabamec	Zagro Singapore Pvt. Ltd.

The formulators' and trade names of avermectins are presented in Table 9.5.

Table 9.5. Formulators' and trade names of avermectins

Trade name/s	Formulators
Artig	Agroquimicos Versa SA de CV
Cam-Mek	Cam For Agrochemicals
Bermectin	Chema Industries
Minx	Cleary Chemical LLC
Bentar, Crysabamet, Crysmectin	Dupocsa Protectores Quimicos para el Campo SA
Torpedo	Hektas Ticaret TAS
Eminence, Instar AD, Noflye	Ingenieria Industrial SA de CV
Inimectin	Insecticidas Internacionales, CA

Saddle	Ladda Co. Ltd.
Reaper	Loveland Products Inc.
Medamec	Medmac Agrochemicals
Nagmectin	Multiplex Fertilizers Pvt. Ltd.
Bermectine	Probelte, SA
Arvilmec	SAFA TARIM AS
ManChongGai	Shanghai Nong Le Biological Products Co. Ltd.
Wopro-abamectin	BV Industrie & Handelsonderneming Simonis
Frog	Sinochem Ningbo Chemicals Co. Ltd.
Ieungaechoong	SM-BT Co. Ltd.
Acaritina	Stockton Agrimor AG
Odin, Tinamex	TRAGUSA (Tratamientos Guadalquivir SL)
Arrow, Aviat, Biok	United Phosphorus Ltd.
Abamex, Maysamectin, Vapcomic	VAPCO, Veterinary & Agricultural Products Mfg. Co. Ltd.
Vibamec	Vietnam Pesticide Joint Stock Co (VIPESCO)
Abamine, Akirmactin, Armada, Bitech, Verkotin	Willowood Ltd.

The premix products of avermectins are presented in Table 9.6.

Table 9.6. Premix products of avermectins

Premix products	Trade name/s	Suppliers
Abamectin + *beta*-Cypermethrin	Awei Gaolv	Shanghai Agro-Chemical Industry Co. Ltd.
Abamectin + Bifenazate	Sirocco	OHP Inc.
Abamectin + Bifenthrin	Athena	FMC Corp
Abamectin + Chlorpyrifos	Paragon	BMC Group Co. Ltd.
	Slamfast	
	Vibafos	Vietnam Pesticide Joint Stock Co. (VIPESCO)
	Vinc	Wangs Crop-Science Co. Ltd.
Abamectin + Cyromazine	Locking	Wangs Crop-Science Co. Ltd.
Abamectin + Imidacloprid	Extra Power	Wangs Crop-Science Co. Ltd.
Abamectin + Propargite	Choice	Wangs Crop-Science Co. Ltd.
Abamectin + Thiamethoxam	Acceleron INT-210	Monsanto Co.
	Agri-Flex	Syngenta
	Avicta Duo Corn	Syngenta

	Avicta Duo Cotton	Syngenta
Abamectin + Azoxystrobin + Fludioxonil + Metalaxyl-M + Thiamethoxam	Avicta Complete Corn	Syngenta
Abamectin + Azoxystrobin + Thiamethoxam	Avicta Complete Cotton	Syngenta
Abamectin + Fludioxonil + Thiamethoxam	Avicta Complete Beans	Syngenta

9.7. NEMATODE MANAGEMENT

The available literature on avermectins indicate that *Meloidogyne incognita*, *M. javanica*, *M. arenaria* and *Tylenchulus semipenetrans* were the commonly used test organisms to assess the nematicidal activity of these metabolites on tomato and citrus (Garabedian and Van Gundy, 1983).

While avermectins B1 and B2a have shown high toxicity against *M. incognita* in greenhouse conditions when incorporated into soil, avermectin B2a was found to be more potent than B1 (Putter *et al.*, 1981).

In a microplot study, avermectin B1 at 0.17 kg a i per ha was equal to ethoprop, aldicarb, fenamiphos, oxamyl and carbofuran at 6.7 kg a i per ha (Nordmeyer and Dickson, 1981).

Abamectin is being used as a seed treatment to control plant-parasitic nematodes (*M incognita and Rotylenchulus reniformis*) on cotton and some vegetable crops. Using an assay of nematode mobility, LD_{50} values of 1.56 µg/ml and 32.9 µg/ml were calculated based on 2 hr. exposure for *M incognita* and *R. reniformis*, respectively. There was no recovery of either nematode after exposure for 1 hr. Mortality of *M incognita* continued to increase following a 1 hr. exposure, whereas *R. reniformis* mortality remained unchanged at 24 hr. after the nematodes were removed from the abamectin solution. Sublethal concentrations of 1.56 to 0.39 µg/ml for *M. incognita* and 32.9 to 8.2 µg/ml for *R. reniformis* reduced infectivity of each nematode on tomato roots. The toxicity of abamectin to these nematodes was comparable to that of aldicarb (Faske and Starr, 2006).

9.7.1. Banana nematodes, *Meleidogyne javanica, Radopholus similis*

Injections (1 ml) of ≥ 100 µg a.i. /plant of abamectin into banana (*Musa acuminata* cv Cavendish) pseudo stems were effective in controlling *M javanica* and *R similis*, and were comparable to control achieved with a conventional chemical nematicide, fenamiphos, in a protectant assay. Abamectin injections of 250 and 500 µg a.i. /plant were effective at reducing nematode infections 28 to 56 days after inoculation. Abamectin was more effective than ivermectin in controlling nematodes after their establishment in banana roots. Injections of between 100 and 1000 µg a.i /plant were effective in controlling nematodes for at least 56 days after treatment. These studies demonstrated that abamectin has potential for controlling nematode parasites on banana when injected into the pseudo stem (Jansson and Rabatin, 1997).

9.7.2. Citrus nematode, *Tylenchulus semipenetrans*

In citrus, a monthly rate of 1.1 kg a.i. per ha of avermectins for 7 months gave maximum increase in yield and reduction of *T. semipenetrans* population (Garabedian and Van Gundy, 1983).

9.7.3. Potato cyst nematode, *Globodera pallida*

In glasshouse experiment, abamectin at 18.0 and 36.0 µg/mL significantly reduced number of eggs, juveniles, cyst/g soil and reproduction rate of *G. pallida* in comparison to both untreated control and fosthiazate nematicide treatment. Soil applications of abamectin provided significant *G. pallida* control with LD_{50} and $LD_{99.9}$ of 14.4 and 131.3 µg/mL, respectively. These results indicate the efficacy of abamectin to control *G. pallida* on potato crops and its potential use in organic agriculture or in an integrated pest management program (Sasanelli *et al.*, 2019).

9.7.4. Tomato root-knot nematode, *Meloidogyne incognita*

The per cent root infection by *M. incognita* juveniles in tomato seedlings raised in seed pans treated with aqueous solution of avermectins (100 ml of 0.001% solution w/v) were lowest as compared to carbofuran (2 g a.i. /seed pan) or neem cake (250 g/seed pan) (Parvatha Reddy and Nagesh, 2002).

The aqueous solution of avermectins (250 ml of 0.001%/m² nursery bed) significantly reduced root galls of tomato seedlings raised in root-knot infested nursery beds compared to carbofuran (10 g a.i. /m² nursery bed) or neem cake (1 kg/m² nursery bed). The treated tomato nursery beds yielded robust and healthy seedlings with no root-knot nematode infection (Fig 9.5) (Parvatha Reddy and Nagesh, 2002).

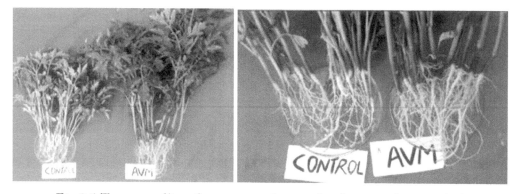

Fig. 9.5. Tomato seedlings from avermectin treated and untreated nursery beds

Talc and charcoal formulations of *S. avermitilis* at 100 g/m² nursery bed significantly controlled *M. incognita* infection in tomato and were superior to carbofuran. The above treatments also gave vigorous and root-knot free tomato seedlings (Parvatha Reddy and Nagesh, 2002).

In a root-knot infested micro-plot study, avermectins at 0.015 kg/ha effectively controlled *M. incognita* on tomato and increased yield by 11 and 8% compared to the untreated and carbofuran treated plots, respectively.

All sublethal concentrations greater than 0.39 µg abamectin/ml inhibited (P < 0.05) infection of tomato roots by *M. incognita*. No reduction of root galls occurred with abamectin concentration less than 0.15 µg/ml (Faske and Starr, 2006).

Avermectins B1 and B2a applied to soil through drip irrigation systems at 0.093 to 0.34 kg a.i. /ha applied as a single dose or 0.24 kg a.i. /ha applied as 3 doses each at 0.08 kg a.i. /ha on tomatoes against *M. incognita* were as effective as oxamyl and aldicarb at 3.36 kg a.i. /ha. There was no significant difference in efficacy between B1 and B2a (Garabedian and Van Gundy, 1983).

9.7.5. Tomato reniform nematode, *Rotylenchulus reniformis*

Sub-lethal concentrations greater than 8.2 µg abamectin/ml lowered ($P < 0.05$) the number of *R. reniformis* females observed per root (Faske and Starr, 2006).

9.7.6. Brinjal and chilli root-knot nematode, *Meloidogyne incognita*

Avermectins effectively reduced root-knot nematode infection in chilli and brinjal under field conditions (Parvatha Reddy and Nagesh, 2002).

9.7.7. Cucumber root-knot nematode, *Meloidogyne incognita*

Under soil-free conditions, avermectin B2a-23-ketone reduced the invasion of cucumber roots by *M. incognita* larvae and their further development to a saccate stage at concentrations (0–0003 µg 1 cm^{-3}) much lower than were needed to immobilize the juveniles. Wright *et al.* (1984a) proposed that avermectins might affect the behavioral sequence preceding invasion, a mode of action also suggested for organophosphorus and carbamate nematicicdes.

9.7.8. Garlic stem and bulb nematode, *Ditylenchus dipsaci*

Abamectin at 10-20 ppm as the 20-minute hot dip (49 °C) or as a 20-minute cool dip (18 °C) following a 20-minute hot-water dip was highly effective in controlling *D. dipsaci* and was non-injurious to garlic seed cloves. This treatment was not as effective as a hot water-formalin dip and was non-eradicative, but showed high efficacy on heavily infected seed cloves relative to non-treated controls. Abamectin was most effective as a cool dip. These abamectin cool-dip (following hot-water dip) treatments can be considered as effective alternatives to replace formalin as a dip additive for control of clove-borne *D. dipsaci* (Roberts and Matthews, 1995).

9.7.9. Carnation and gerbera root-knot nematode, *Meloidogyne incognita*

Avermectins (0.001%) applied as post-plant treatment at 250 ml/m^2 at two intervals (6 and 12 months after planting) effectively controlled root-knot nematodes (*M. incognita*) in carnation and gerbera in commercial polyhouses (Parvatha Reddy and Nagesh, 2002).

9.7.10. Tobacco root-knot nematode, *Meloidogyne incognita*

In a small-plot field trial, soil incorporation of the avermectins B1, B2a and B2a-23-ketone were equally effective in inhibiting root-gall development and reproduction of *M. incognita* on tobacco. At application rates ranging from 0.168 to 1.52 kg a.i. /ha, these avermectins were as effective as ethoprop and fenamiphos at 6.73 kg a.i. /ha (Sasser *et al.*, 1982).

In another microplot study, avermectins B1a at 0.17 kg a.i. /ha was equal to ethoprop, aldicarb, fenamiphos, oxamyl and carbofuran at 6.7 kg a i/ha in significantly increasing yields of tobacco infested with *M. incognita* or *M. arenaria* (Nordmeyer and Dickson, 1985).

9.7.11. Cotton root-knot nematode, *Meloidogyne incognita*

In greenhouse tests, 35 days after planting (DAP), plants from seed treated with abamectin were taller than plants from non-treated seed, and root galling severity and nematode reproduction were lower where treated seed were used. The number of second stage juveniles that had entered the roots of plants from seed treated with 100 g abamectin/kg seed was lower during the first 14 DAP than with non-treated seed. In microplots tests, seed treatment with abamectin and soil application of aldicarb at 840 g/ha reduced the numbers of juveniles penetrating seedling roots during the first 14

DAP compared to the non-treated seedlings. In field plots, population densities of *M incognita* were lower 14 DAP in plots that received seed treated with abamectin at 100 g/kg seed than where aldicarb (5.6 kg/ha) was applied at planting (Montfort *et al.,* 2006).

The number of galls caused by *M incognita* race 3 in cotton was also reduced more than 80% with abamectin at 0.1 mg a.i. seed⁻¹ (Cabrera *et al.,* 2009).

Based on the position of initial root-gall formation along the developing tap root from 21 to 35 DAP, infection by *M. incognita* was reduced by abamectin seed treatment. Penetration of developing tap roots by nematode species was suppressed at tap root length of 5 cm by abamectin-treated seed, but root penetration increased rapidly with tap root development. Based on an assay of nematode mobility to measure abamectin toxicity, the mortality of *M. incognita* associated with 2-day-old emerging cotton radical was lower than mortality associated with the seed coat, indicating that more abamectin was on the seed coat than on the radical. Thus, the limited protection of early stage root development suggested that only a small portion of abamectin applied to the seed was transferred to the developing root system (Fig 9.6) (Faske and Starr, 2007).

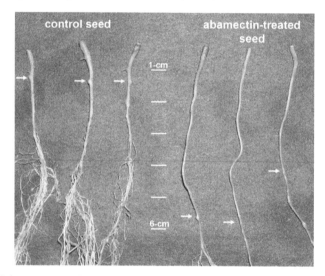

Fig. 9.6. Effect of abamectin seed treatment on the position of initial gall formation by *Meloidogyne incognita* **on cotton tap roots. To better visualize galling, secondary roots were removed 2 cm past the initial gall for both control and abamectin-treated seed.**

9.7.12. Sugar beet cyst nematode, *Heterodera schachtii*

Penetration of *Heterodera schachtii* in sugar beets was reduced over 60% when sugar beet seeds were treated with abamectin at a concentration of 0.3 mg a.i. seed⁻¹ (Cabrera *et al.,* 2009).

9.7.13. Cereal cyst nematode (CCN) on wheat, *Heterodera avenae*

Greenhouse experiment and field trials showed that soil applications of abamectin provided significant CCN control (46.9 to 65.4%) and higher straw dry weights and wheat grain yields. There was an 8.5 to 19.3% yield increase from the various abamectin treatments compared with the control. The results of this study demonstrated that abamectin exhibited a high nematicidal activity to *Heterodera avenae* and adequate performance to enhance wheat crop yields (Zhang *et al.,* 2017).

9.7.14. Maize lesion nematode, *Pratylenchus zeae*

Penetration of *Pratylenchus zeae* was reduced more than 80% in maize with abamectin at a dose of 1.0 mg a.i. seed^{-1}.

9.8. INTEGRATED NEMATODE MANAGEMENT

Under pot conditions, the application of *Syncephalastrum racemosum* fungus alone or combined with the nematicide, avermectin significantly increased the cucumber plant vigor index by 31.4% and 10.9%, respectively compared to the *M. incognita*-inoculated control. All treatments reduced the number of root galls and juvenile nematodes compared to the untreated control (Huang *et al.*, 2014).

Under greenhouse conditions, all treatments reduced the nematode severity and enhanced fruit yield compared to the untreated control. Fewer nematodes infecting plant roots were observed after treatment with avermectin alone, *S. racemosum* alone or their combination compared to the *M. incognita*-inoculated control. Among all the treatments, application of avermectin or *S. racemosum* combined with avermectin was more effective than the *S. racemosum* treatment (Huang *et al.*, 2014). Application of *S. racemosum* combined with avermectin not only reduced the nematode number and plant disease severity but also enhanced plant vigor and yield. The combination of *S. racemosum* with avermectin could be an effective biological component in integrated management of RKN on cucumber (Huang *et al.*, 2014).

The RGI control efficacy of the combination at 337.5 g a.i. /ha of fluopyram (41.7% SC) and 450 g a.i. /ha of abamectin (1.8% EC) gave effective control of *M. incognita* on tomato and 1: 5 combination at 450 g a.i. /ha increased tomato yields by 24.07 and 23.22% compared to the control in field trials during two successive years. This is because fluopyram promotes abamectin diffusion during the process of transport and uptake from the soil and roots, and abamectin has high toxicity to RKN. The combination of fluopyram and abamectin provides good nematode measure, and it can increase tomato yields. It provides an effective solution for the integrated management of southern RKN (Fig. 9.7) (Li *et al.*, 2020).

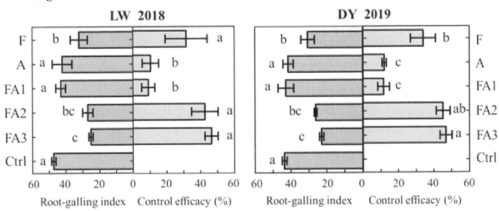

Fig. 9.7. Effect of Fluopyram, Abamectin and different combinations on the root-galling index and control efficiency in field trials at the end of tomato production.
Different lower case letters indicate significant differences (p < 0.05).
(LW - Laiwu in Jinan City, DY - Daiyue in Tai'an City, F - Fluopyram A - Abamectin F: A -Fluopyram: Abamectin (Mass ratio of a.i.) 1: 1, 1: 3, 1: 5.

9.9. CONCLUSION

Avermectins offer an outstanding alternative to any of the available synthetic pesticides as they showed excellent insecticidal, nematicidal and acaricidal action. Their novel mode of action, high potency and specific physico-chemical properties makes the avermectins excellent candidates for further insecticidal, acaricidal and nematicidal studies. Mobility of avermectins may be increased by formulating with suitable surfactants which compete for adsorption sites (Morton, 1986) or by forming micelles around the avermectin molecules which will reduce adsorption. Some other approaches such as using slow release and encapsulated formulations (Morton, 1986) can also be applied to advance research with the avermectins.

Viable and effective formulations need to be developed for wide scale application. The talc, charcoal and coir pith formulations examined were effective for short-term/immediate use for pest control. Keeping the cost of commercial production, they can be recommended for high value and polyhouse grown crops. Considering its moderate persistence in the environment, low toxicity to non-target beneficial organisms (Lumaret *et al.*, 2012), the degradability by soil microorganisms, its poor leaching potential not resulting in ground water contamination, abamectin could be a potential bio-nematicide to use in integrated pest management programs and organic farming.

Chapter - 10

Natural Genetic and Induced Nematode Resistance

10.1. INTRODUCTION

All cultivated plant species are subjected to biotic stresses that can endanger crop yields and cause relevant economic losses. Genetically based resistance may be a simple and efficient solution to protect plants against pathogens and pests. There is a need to introduce innovative strategies and reduce cropping system reliance on pesticides. The current generally accepted perspective is that exploitation and management of plant genetic resistance must be the key elements of a durable Integrated Pest Management (IPM). The basic understanding of crop-pest systems is crucial for a durable exploitation of plant genetic resistance, which, however, should not be considered as the single solution but a major component of a global IPM approach aiming at designing disease suppressive growing systems and reducing chemical inputs.

Use of resistant cultivars and rootstocks has been prioritized as a major goal for nematode management (Barker *et al.*, 1984). Resistance occurs naturally in wild relatives and can be transferred to crop cultivars through conventional breeding methods. In this chapter, issues involving natural genetic resistance introduced in crop cultivars by classical breeding and its present practical use will be addressed.

Induced resistance is mainly based on treatment of plants with natural or synthetic chemicals able to trigger systemic acquired resistance (SAR). SAR is widely diffused in plants, is systemically activated by necrogenic pathogen attacks, induces a broad-spectrum disease resistance, and is mediated by salicylic acid (SA) (Kessmann *et al.*, 1994). SAR elicitors do not exhibit any direct antimicrobial activity and seem to be environmentally benign, unlike traditional pesticides. However, less investigation has been carried out to date on induced resistance to plant parasitic nematodes. Most of the available reports refer to induced resistance against RKNs in tomato (Cooper *et al.*, 2005; Javed *et al.*, 2007; Molinari and Baser, 2010; Oka *et al.*, 1999; Vavrina *et al.*, 2004).

As cyst and root-knot are the most damaging groups of nematodes, most of the efforts to study parasitism and find sources of resistance and other control strategies have been focused on them.

10.2. PPN LIFE CYCLES

PPNs are divided into three major groups according to feeding behavior: ectoparasitic, semi-endoparasitic, and endoparasitic (Decraemer and Hunt, 2013; Palomares-Rius *et al.*, 2017; Smant *et al.*, 2018).

10.2.1. Ectoparasitic nematodes

Ectoparasites take up nutrients from plant cells without invading the plant root. Some ectoparasites such as needle nematodes (*Longidorus* spp.) and dagger nematodes (*Xiphinema* spp.) induce the formation of nurse cells which extends the period of feeding. Ectoparasitic nematodes spend their entire life cycle outside of the host, with the only physical contact being the insertion of a long and rigid feeding stylet (Fig. 10.1A) (e.g. *Belonolaimus* spp. and *Xiphinema* spp.).

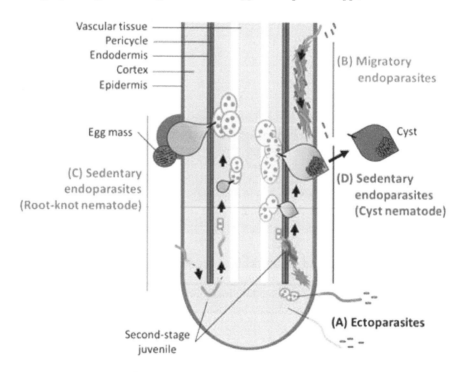

Fig. 10.1. Infection strategies of PPNs.

10.2.2. Endoparasitic nematodes

Endoparasitic nematodes completely enter the root and feed on internal tissues. This feeding type is further divided into either migratory or sedentary lifestyles. Migratory endoparasites move through inside of the root tissues causing destruction en route and feed on plant tissues. Migratory endoparasites include the root-lesion nematodes *Pratylenchus* spp., and the burrowing nematodes *Radopholus* spp.) which migrate through root tissues to feed on plant cells, causing damage to tissues as they migrate (Fig. 10.1B).

Sedentary endoparasites include the root-knot nematodes (RKNs), *Meloidogyne* spp. and the cyst nematodes (CNs), including *Globodera* spp. and *Heterodera* spp. (Fig. 10.1C). Second-stage RKN juveniles enter the root near the root-tip then migrate intercellularly to the vascular cylinder where they reprogram root tissues into giant cells. After establishment of giant cells, RKN juveniles become sedentary and take up nutrients and water through a feeding stylet. Adult RKN females form an egg mass on or below the root surface. Second-stage juveniles of the CNs move inside of the root intracellularly, causing destruction of plant tissues as they move, and establish syncytia in the vascular tissues as feeding cells (Fig. 10.1D). CN juveniles also become sedentary and start feeding from syncytia. Adult CN females retain eggs inside of the body, which forms a cyst after death. Sedentary endoparasites move into the vascular cylinder and induce redifferentiation of host cells into multinucleate and hypertrophic feeding cells. RKNs and CNs are the most devastating nematodes in the world (Jones *et al.*, 2013).

10.2.3. Semi-endoparasitic nematodes

Semi-endoparasitic nematodes penetrate roots to feed, with its posterior part remaining in the soil. The examples of semi-endoparasitic nematodes include the citrus nematode (*Tylenchulus semipenetrans*) and the reniform nematode (*Rotylenchulus reniformis*) (Fig. 10.2).

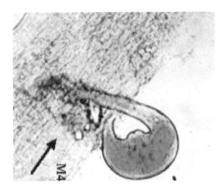

Fig. 10.2. Semiendoparasitic nematode, *Rotylenchulus reniformis.*

10.3. NATURAL GENETIC RESISTANCE

Nematode resistance can be dominant, recessive or additive in expression, and can be conferred by single major genes or by combinations of two or more genes or quantitative trait loci (QTLs). Most of the characterized genes (R-genes) confer resistance to the three most economically important nematode groups, the root-knot nematodes (*Meloidogyne* spp.) attacking most of the cultivated plants, the potato cyst nematodes (*Globodera* spp.), and another cyst nematode family attacking a wide range of annual crops (*Heterodera* spp.). Thus, most of the information available on genetic resistance to nematodes and its use as a control strategy refers to sedentary endoparasitic nematodes. An extensive list of genes conferring resistance to RKNs in annual and perennial crops, describing their inheritance/expression, has been reported (Williamson and Roberts, 2009). Relatively fewer resistance genes against cyst nematodes have been isolated and characterized (Williamson and Kumar, 2006). On the contrary, of the six genes cloned, only one, the Mi-1.2 tomato gene, confers resistance against the 3 most diffused RKN species, *M. incognita*, *M. javanica*, and *M. arenaria*. Mi-1.2 also confers resistance to specific isolates of the potato aphid, *Macrosiphum euphorbiae* (Rossi *et al.*, 1998) and to two biotypes of the white fly, *Bemisia tabaci* (Nombela *et al.*, 2003), being the only known R-gene that confers resistance against such different groups of pests. The other cloned genes are Hs1pro^{-1} conferring resistance in sugar beet to *H. schachtii*, Gpa2 in potato to *G. pallida*, Gro1-4 in potato to *G. rostochiensis*, Hero A in *Solanum pimpinellifolium* (a wild relative of cultivated tomato) against a broad range of *G. pallida* and *G. rostochiensis* pathotypes, and rhg1/rhg 4 in soybean to the pathotype 0 of *H. glycines* (Fuller *et al.*, 2008).

Most of our information on signaling pathways and defense responses of resistant plants upon a nematode attack is based on RKN-tomato interactions. Tomato resistance is expressed by a hypersensitive reaction (HR), leading to a rapid and localized cell death, whose earliest visible indications can be seen about 12 hr. after inoculation of roots with J2, while they attempt to establish a feeding site (Paulson and Webster, 1972). Mi-mediated resistance seems to be regulated by a salicylic acid (SA)-dependent defense pathway (Branch *et al.*, 2004; Molinari, 2007; Molinari and Loffredo, 2006), as it has generally been found in most R-gene-mediated defenses (Glazebrook, 2005). A specific oxidative burst with enhanced generation of reactive oxygen species (ROS) has been proved to occur early in incompatible RKN-tomato interactions (Melillo *et al.*, 2006). The increase of ROS in root cells may be caused by a specific and very early inhibition of H_2O_2 - degrading enzymes, such as catalase (Molinari, 2001; Molinari and Loffredo, 2006). Overproduction of SA has been reported in resistant tomato attacked by *M. incognita* (Vasyukova *et al.*, 2003), and a possible indirect effect of enhanced SA levels may result in an impaired mitochondrial phosphorylation efficiency, thus leading to the necrosis of the cells involved in HR (Molinari, 2007; Molinari *et al.*, 1990).

10.3.1. Plant genetic resistance against nematodes

Resistance has been proved to be an effective management tool that improves crop yields, lowers nematode population densities, and favors the developments of effective rotation systems (Starr *et al.*, 2002). Although the most common definition of nematode resistance is based on a measurable restriction of the pest reproduction, the primary aim of resistance used as a control strategy is to protect yield potential. Therefore, an applicable use of resistance must provide minimal crop damage and be associated with tolerance of nematode attack that may be distinct from the recognized ability of resistant plants to restrict pest reproduction. However, resistant plants are generally tolerant of nematode attack, even though resistance is typically a post-infection process. Suppression of reproduction is a valuable tool to lower population densities in soil, thus protecting subsequent susceptible crops, although yield must be considered the priority and population density management an additional benefit (Starr and Roberts, 2004). For instance, decrease of nematode density is an important effect of the use of Mi-1.2 gene in tomato against RKNs (Roberts and May, 1986).

The primary benefit of the use of resistance is that it is economically convenient for managing nematodes in both high- and low-value agricultural systems, as direct cost to the growers is minimal. Effectiveness and profitability of the use of resistant tomato has been proved in a plastic house naturally infested by *M. javanica* (Sorribas *et al.*, 2005). Growth of resistant tomato increased profits by 30,000- and 88,000-euros ha^{-1} in non-fumigated and methyl bromide fumigated soils, respectively, compared with growth of susceptible tomato. Furthermore, resistant crops are environmentally compatible, do not require specialized applications, additional cost input or deficit, and are amenable to integration with other management systems that may promote resistance durability or provide additional protection when resistance is not sufficient (Roberts, 1993).

The potential for success in releasing resistant cultivars seems very promising in light of the numerous available sources of resistance to nematodes in a broad range of plant families, and the rapid advances in techniques such as in-embryo rescue, somatic hybridization, and direct gene transfer that should promote a more efficient genetic transfer across conventionally difficult biological barriers (e.g., sexual incompatibility, polyploidy, unacceptable gene linkages). Most of the resistance factors in the major annual and perennial crops, currently available for farmers, are summarized in Table 10.1. It should be noted that, in potato, resistance is available to only some pathotypes of *G. rostochiensis* and *G. pallida*, although the resistance in the cultivars used in Europe against *G. pallida* is only partial (Whitehead and Turner, 1998). Recently, also grafting susceptible high-yielding tomatoes onto resistant rootstocks has been tested as a control measure against RKNs, although response ranged from highly resistant to fully susceptible (López-Pérez *et al.*, 2006).

Table 10.1. Major annual and perennial crops carrying resistance to root-knot, cyst, and other families of nematodes currently used for nematode pest management (Roberts, 1992)

Crops	Root-knot nematode resistance	Cyst nematode resistance	Resistance to other nematodes
Beans	*Meloidogyne incognita,* *M. javanica*	---	*Pratylenchus scribneri*
Carrot	*M. incognita,* *M. javanica*	---	---
Soybean	*M. incognita,* *M. javanica*	*Heterodera glycines*	*Rotylenchulus reniformis*
Tobacco	*M. incognita,* *M. arenaria*	*Globodera* spp.	---
Tomato	*M. incognita, M. javanica,* *M. arenaria*	---	---
Alfalfa	*M. incognita*	---	*Ditylenchus dipsaci*
Cotton	*M. incognita*	---	---
Cowpea	*M. incognita*	---	---
Potato	*Meloidogyne* spp.	*Globodera* spp.	---
Small grains (wheat, barley, oat)	---	*H. avenae*	*D. dipsaci*
Prunus Nemaguard rootstock (almond, nectarine, peach, plum)	*M. incognita, M. javanica,* *M. arenaria*	---	---
Citrus	---	---	*Tylenchulus semipenetrans*
Grape	*Meloidogyne* spp.	---	*Xiphinema index*

Resistance acts against "target" nematode species, or only one species, and, sometimes, even a subspecific race or pathotype. Resistance to restricted targets present in field populations may result in a competitive advantage to the species that are not controlled by the resistance. For example, while cultivars of cotton with resistance to *M. incognita* have been developed and used in USA, all are highly susceptible to *R. reniformis* (Robinson *et al.*, 1999), and fruit tree rootstocks that are resistant to RKNs are attacked by different genera, such as *Pratylenchus, Xiphinema, Helicotylenchus, Criconemella* (Nyczepir and Becker, 1998). Also, the selective pressure exerted by intensive use of potato cultivars containing the H1-gene for controlling *G. rostochiensis* in UK has caused a spread of the sibling species *G. pallida*, for which available resistance is much less effective (Cook and Evans, 1987).

10.3.2. Durability of resistance

Another serious concern is the selection of virulent populations, induced by an intensive use of R-genes, that break resistance and may make the use of resistant cultivars ineffective at specific locations. Therefore, durability of R-genes must be taken into account when resistance is to be used as a control strategy against nematodes. Thus far, durability of resistance to sedentary plant parasitic nematodes can be considered generally high. The potato gene H1 has been used against *G. rostochiensis* over 30 years in UK without the development of virulent populations (Fuller *et al.*, 2008), the tomato gene Mi-1 has been used for more than 20 years in California against RKNs with only very few isolated cases of resistance breakdown (Kaloshian *et al.*, 1996), and the resistance to *Meloidogyne* spp. in the *Prunus* rootstock Nemaguard has not been endangered by virulent populations during a 50-year use in commercial orchards (Williamson and Roberts, 2009).

Currently, at least among RKNs exposed to resistant crop cultivars or rootstocks, occurrence of virulent populations appears widespread and widely distributed geographically. Selection of virulent populations has been reported on the resistant tomato cultivar Sanibell, widely used in Florida (Noling, 2000), and, in North Carolina, on resistant soybean attacked by *H. glycines* (Starr and Roberts, 2004). The extensive use of Rk-carrying cowpea cultivars has resulted in several fields in California with virulent *M. incognita* populations (Petrillo *et al.*, 2006), and virulence in *M. chitwoodi* to the resistance gene Rmc1(b/b) in potato puts in danger the use of such resistance in the US Pacific Northwest (Mojtahedi *et al.*, 2007). Virulent populations of *M. incognita* and *M. javanica* were found on resistant tomato cropped in Morocco and Greece (Eddaoudi *et al.*, 1997; Tzortzakakis and Gowen, 1996), as well as of *Meloidogyne* spp. from Spain and Uruguay on resistant bell pepper (Robertson *et al.*, 2006). More recently, virulence has been detected in *M. incognita* and *M. javanica* populations collected from resistant tomato in Tunisia, and in *M. incognita* populations collected from resistant pepper grown in Hungarian greenhouses. Moreover, the Mi-1.2 gene in tomato is not effective against the species *M. hapla* (Liu and Williamson, 2006) and *M. enterolobii* (a senior synonym of *M. mayaguensis*) (Brito *et al.*, 2007). This latter tropical species (*M. enterolobii*), due to its potential to become a quarantine pest, was placed on the European and Mediterranean Plant Protection Organization (EPPO) alert list, and is of great importance as it displayed virulence also against resistant N-carrying bell pepper and is considered particularly aggressive (Brito *et al.*, 2007; Kiewnick *et al.*, 2009).

Studies on approaches to enhance resistance durability to RKNs are in progress (Djian-Caporalino *et al.*, 2010), based also on investigations on gene specificity and pathogenic potential of populations selected for virulence (Molinari, 2010a). Virulent populations of RKNs selected on Mi-1.2-carrying tomato did not reproduce either on different tomato genes (Mi2–Mi9) or resistant bell pepper; comparably, populations virulent on pepper containing the Me3-gene were not able to develop on bell pepper containing the Me1-gene as well as on resistant tomato (CastagnoneSereno *et al.*, 1996; Molinari, 2010a; Williamson, 1998). Moreover, selection for virulence resulted into adverse fitness costs, which were detected by decreased reproduction and damage potentials of virulent populations, either on susceptible or resistant tomato, with respect to field starting populations. Additional data showed that the lines selected to overcome the Mi-1.2 resistance gene could have a compromised ability to compete with other lines on susceptible cultivars, as they would reproduce less efficiently (Castagnone-Sereno *et al.*, 2007). Accordingly, it has been reported that field populations of *M. incognita* may comprise a mixture of virulent and avirulent lineages, in which the virulent forms had reduced reproductive potentials (Petrillo and Roberts, 2005). In such cases, reproduction on susceptible host plants in a rotation may result in a decline in the overall level of virulence in the field (Williamson and Roberts, 2009). Alternatively, the subsequent use of single R-genes in rotation or the mixture of lines bearing different R-genes may lower the emergence of virulent populations and enhance the sustainability of cropping resistant varieties. An additional approach to extend resistance durability is pyramiding multiple resistance genes into the same cultivar. In cowpea, resistance to *Meloidogyne* spp. based on genes Rk plus rk3 may be more durable than resistance conferred by gene Rk alone (Elhers *et al.*, 2000); moreover, another gene, Rk2 , is being bred into cultivars near release to manage Rk-virulent populations (Williamson and Roberts, 2009). On the other hand, in tomato, other genes (Mi2–Mi9), conferring resistance to *Meloidogyne* spp., have been proved difficult to introgress from wild relatives and are still unavailable for tomato growers. Therefore, an increased availability of different introgressed resistance genes in commercial varieties will broaden the chances of growers to address their problems of nematode-infested soils by using resistance as the main control strategy, or preventing some initial drawbacks of its use. For this purpose, also transgenic transfer of already cloned or novel resistance genes may speed up the process of introgression of additional efficient resistance factors into high-yield crop cultivars and prolong resistance durability. Combinatorial transgenic resistance would enhance the durability of any resistance deployed against nematodes (Atkinson *et al.*, 2009). In the future, the use of resistant varieties in nematode-infested soils should be preceded by an accurate analysis of the virulence potential of the local populations on the available resistant cultivars, to account not only for the theoretical durability of the R-genes involved but also for the more practical sustainability of the cropping system.

10.3.3. Assessment of crop resistance

The assessment of crop resistance to the most important sedentary endo-parasitic (*Meloidogyne, Heterodera, Globodera* spp.) and migratory ecto-parasitic (*Xiphinema* spp.) nematodes is mainly based on nematode reproduction (Starr *et al.*, 2002). Within the nematode groups to which resistance is a current strategy of control, only the infestation of root-knot nematodes (*Meloidogyne* spp.) induces specific and clearly visible symptoms on roots, such as the typical root knots or galls, which result from expansion of the cortical tissue surrounding the giant cells used by nematodes to feed at the infection site (Williamson and Hussey, 1996). Evaluation of genotypes for root-knot nematode resistance has relied thus far on determination of the degree of galling, egg mass number, or total eggs collected from the root system (Hussey and Janssen, 2002; Molinari, 2009a). An accurate assessment of the resistance of a cultivar or breeding line can only result from the measurement of all the aforementioned indicators, as, in most cases, the extent of correlation between galling and nematode reproduction cannot be predicted. For galling and egg mass evaluations, 0-5 indexes have been developed associated with the percentage of galled roots (gall index, GI) and numbers of egg masses per root system (egg mass index, EI), respectively (Ammati *et al.*, 1985; Hoedisoganda and Sasser, 1982). The number of egg masses per root system indicates the number of individuals, among those present at sowing or planting (initial population density, Pi), which were able to enter the root, develop into gravid females and reproduce, but does not give information on the reproduction rate of the nematode population. Reproduction rates are usually evaluated by a reproduction index (RI) which refers to the total number of eggs and J2 produced on the test roots and is expressed as the percentage of the total number of eggs on the roots of a fully susceptible cultivar of the same species, taken as a reference (Roberts and Thomason, 1986). In field and greenhouse tests, the multiplication rates may also be calculated as Pf / Pi ratio, where Pf is the nematode population in the soil at the end of the crop cycle or of the test time (Ornat *et al.*, 2001). Decreases in the multiplication rate are correlated to increases in initial population density because of the concomitant effects of higher competition for food and root damage; a similar inverse correlation exists between initial population density and crop yield (Greco and Di Vito, 2009). Nematode-resistant cultivars usually yield more than susceptible cultivars when planted in fields with population densities exceeding the damage threshold, thus indicating to be more tolerant across a range of initial nematode densities (Starr and Roberts, 2004). Both nematode reproduction and gall index were used by Canto-Sa´enz (1985) for a qualitative classification of plant response to RKN as susceptible (high reproduction and galling), tolerant (high reproduction, low galling), resistant (low reproduction and galling), and hyper-susceptible (low reproduction, high galling).

10.4. INDUCED RESISTANCE TO NEMATODES BY CHEMICALS

Induced resistance can be broadly divided into systemic acquired resistance (SAR) and induced systemic resistance (ISR). SAR develops locally or systemically in response to pathogen infection or treatment with certain chemicals and is mediated by a SA-sensitive pathway (Durrant and Dong, 2004). Conversely, ISR develops as a result of colonization of plant roots by plant-growth-promoting rhizobacteria (PGPR) and is mediated by a jasmonate- or ethylene-sensitive pathway (van Loon *et al.*, 1998). Resistance to pathogens can be chemically induced by applying to plants SA and compounds which can mimic the action of SA, such as acibenzolar-S-methyl (ASM) and 2,6-dicholoroisonicotinic acid (INA); indeed, ASM is the first synthetic chemical developed as a SAR activator and is marketed in Europe as BION and in USA as ACTIGARD (Walters *et al.*, 2005). These chemicals induce SAR in plants and do not show any antimicrobial activity *in vitro*. INA and ASM have generally been proved to control crop diseases caused by fungi, bacteria, and viruses in field experiments (Vallad and Goodman, 2004). Conversely, relatively few and contrasting data have been reported thus far on the effectiveness of SAR elicitors in restricting nematode infestation (Chinnasri and Sipes, 2005; Molinari, 2008; Nandi *et al.*, 2003; Oka *et al.* 1999; Sanz-Alfe´rez *et al.*, 2008). After a test involving several chemicals that induce resistance to many pathogens, only DL-b-amino-n-butyric acid (BABA) was found to be effective in inducing resistance to *M. javanica* in tomato, either by foliar spray or soil drench (Oka *et al.*, 1999). Also, jasmonic acid and neem (*Azadirachta indica*) formulations have been reported to promote a restraint of RKN infestation on tomato plants (Cooper *et al.*, 2005; Javed *et al.*, 2007). Recently, an extensive study reported the effect of SA, methyl-salicylic acid (MetSA), INA, and ASM on RKN infestation to tomato (Molinari and Baser, 2010). In this study, SAR elicitors have been tested as inducers of resistance to RKNs, taking into account a number of variables, such as the effect of different chemical concentrations, of different application methods, and of soil composition. Moreover, it was proved that the inhibitory action of a SAR inducer may be explicated on different stages of nematode infestation process, i.e. penetration, establishment of a successful feeding site, development into gravid females by the invading juveniles, root galling, reproduction, invasion by successive nematode generations, *etc*. The determination of as many as possible different infestation factors is crucial for a full comprehension of how an inducer should be used, i.e. at which dosage, the type, and numbers of treatments, the growth stages of plants to be treated, *etc*. The conclusions were that SA and ASM, correctly applied at the most effective dosages, can be used for nematode management in conventional and organic tomato protected cultivation, better if included in integrated management programs. The use of SA as such to induce resistance to nematodes is particularly interesting as it is a natural compound and non-phytotoxic at the proper dosages (Molinari, 2008); SA has also been proved to be nematicidal, a strong attractant for *M. incognita*, and an irreversible inhibitor of hatch (Wuyts *et al.*, 2006). However, further investigation is still to be done to verify whether SA and its analogs can be applied to a larger variety of crops and in the more complex field conditions.

One of the most widely researched chemical inducers for controlling phytonematodes is acibenzolar-S-methyl (ASM). Sprayed onto vine leaves seven days prior to inoculation with *M. javanica* and *M. arenaria*, ASM reduced the number of galls and eggs by 40 to 80%, compared to untreated plants (Owen *et al.*, 1998). Similar results have been obtained by prior application to the *M. incognita*-tomato pathosystem (Silva *et al.*, 2004). ASM is thought to interfere in the formation of giant cells via a protein essential to this process and to affect nematode reproduction (Silva *et al.*, 2002). A significant increase in the activity of β-1, 3- glucanase in the roots has also been observed five days after application of ASM (Owen *et al.*, 1998).

Other chemical resistance inducers have been researched, such as salicylic acid, potassium phosphite and jasmonic acid. Molinari and Baser (2010) evaluated salicylic, methyl-salicylic acid and ASM in the control of *M. incognita* in tomato roots and observed that salicylic acid reduced 50% of the egg masses, and the reproduction by 57%. Methyl-salicylic acid was ineffective in reducing the variables assessed; this result was attributed to the low concentration used in the study since it was phytotoxic when applied at high concentrations. Kempster (1998) confirmed induction of resistance to *Heterodera trifolii* in bioassays on clover (*Trifolium repens*), applying salicylic acid and benzothiadiazole. Resistance was evidenced by a reduction in nematode fecundity, non-viability of cysts and fewer eggs in the cysts.

The use of methyl jasmonate and potassium silicate reduced the number of *M. incognita* per gram root and increased peroxidase enzyme activity in sugarcane. Furthermore, the parasitism by *M. incognita* increased peroxidase activity at 14 and 21 days after inoculation (Guimarães *et al.*, 2010). Jasmonic acid plays a fundamental role in signaling the expression of plant defenses. Some defense genes are controlled by the jasmonate pathway, including those coding for proteinase inhibition. Proteinases are crucial for impeding nematode feeding and development (Gheysen and Fenoll, 2002). However, in the *Meloidogyne*-tomato pathosystem, jasmonic acid and methyl jasmonate did not induce resistance when sprayed onto the leaves or applied to the soil (Oka *et al.*, 1999).

The capability of phosphites to trigger plant defense mechanisms, including the production of phytoalexins, was reported by Dercks and Creasy (1989). In a study conducted by Dias-Arieira *et al.* (2012), potassium phosphite was effective in decreasing the population of *Pratylenchus brachyurus* in maize. Similar results have been obtained for other nematode species (Oka *et al.*, 2007). Although phosphites affect microorganisms directly (Guest and Grant, 1991), in a study conducted by Dias-Arieira *et al.* (2012), phosphite was applied to the aerial part, that is, spatially separated from the nematode, proving its capability of triggering plant defense mechanisms, which include the phytoalexins production (Dercks and Creasy, 1989). This hypothesis is backed up by the result that in a study carried out by Salgado *et al.* (2007), potassium phosphite increased hatching of *M. exigua*, but did not kill juveniles, that is, did not directly affect the parasite. Furthermore, Oka *et al.* (2007) observed that potassium phosphite applied to the aerial part was effective in controlling *Heterodera avenae* and *Meloidogyne marylandi* in wheat and oats. This result can be ascribed to potassium phosphite's capability of translocating through the xylem and phloem (Quimette and Coffey, 1990).

Silicon has also been used to induce resistance in various pathosystems, and although the mechanism by which silicon activated resistance in plants has not yet been elucidated; its deposition on the cell walls has resulted in the hypothesis of a possible physical barrier (Terry and Joyce, 2004). The silicon absorbed by plants is rapidly translocated to the aerial part and, during transpiration, the dissolved silicon becomes supersaturated and polymerizes, forming solids that are incorporated into the cell walls enhancing rigidity (Epstein and Bloom, 2004). Although most of the silicon is polymerized or solidified, its role in resistance to disease is mainly due to the silicon fraction in solution within the plant, which would suggest that it helps produce defensive compounds by activating the synthesis of substances such as phenols, lignin, suberin and callose in the cell wall (Rodrigues and Datnoff, 2005).

Dutra *et al.* (2004) observed that applying calcium silicate caused a decrease in the number of galls and eggs of various species of *Meloidogyne* in common bean, tomato and coffee. The authors attributed the induced and enhanced resistance to the silicon was thought to stimulate the production of enzymes and substances related to defense mechanisms. Researching the biochemical resistance response of coffee to *M. exigua* mediated by silicon, Silva *et al.* (2010) submitted evidence that the reproductive capability of nematodes in coffee roots supplied with silicon was impaired. The response was associated with the production of lignin and increased activity of peroxidase, polyphenol oxidase and phenylalanine ammonia-lyase, especially in the susceptible cultivar studied. According to Guimarães *et al.* (2008), potassium silicate was effective in inducing resistance to *M. incognita* in sugarcane, since it reduced the number of nematode eggs in the RB867515 and RB92579 varieties. However, it did not affect aerial part biomass in the RB867515 and RB863129 varieties, nor the population density of *Pratylenchus zeae* in the soil and roots, 100 days after transplanting.

In practice, applying the inducer in advance restricts nematode control, since the nematodes are already present in the soil when the seeds germinate. On the other hand, if the inducer impairs the multiplication of the nematode in subsequent cycles, the damage that they cause can be minimized. But to be ascertaining of this hypothesis, researchers should work over a complete crop cycle, or at least with a longer observation period. Studies are still necessary on resistance inducers in seed treatments to find out whether host protection can be achieved in early stage of germination. Investigations are also required on the treatment of plants germinated on trays or nursery and subsequently transplanted in the field. Resistance induction in plants has proved to be a method of control with potential to attenuate the severity of nematodes, but it is important to increase our knowledge of the mechanisms by which resistance inducers operate.

10.5. RESISTANCE MECHANISMS

Multiple plant immune responses against plant parasitic nematodes include:

» Plants secrete anti-nematode enzymes such as papain-like cysteine proteases (PLCPs) and chitinases into the apoplast to attack PPNs (Fig. 10.3A).

» Resistant plants produce a wide range of secondary metabolites in response to PPN infection (Fig. 10.3B). Some metabolites inhibit egg hatching, suppress the motility of migrating PPNs, arrest growth and development, or kill nematodes. Plants may also reduce chemoattraction by secreting less amounts of attractants or more repellents.

» Plants reinforce their cell walls by accumulating lignin, suberin, and callose, which strengthen the physical barrier to PPNs (Fig. 10.3C).

» PPN infection induces the production of ROS, which may be directly toxic to PPNs (Fig. 10.3D). Hydrogen peroxide plays a role in cell wall cross-linking. ROS may also work as a transducing signal to activate immune responses and to control HR-cell death.

» Nitric Oxide (NO) production is induced upon PPN infection and may play a role in JA-mediated defense responses, possibly through the production of protease inhibitor 2 (Fig. 10.3E).

» HR-cell death is crucial for limiting PPN movement and completing the life cycle (Fig. 10.3F). HR-cell death occurs during penetration and migration of PPNs in cortical and epidermal tissues, contributing to inhibition of migration (Fig. 10.3F-1). HR-cell death is induced in cells infected by RKNs or CNs, which inhibit the formation of feeding cells (Fig. 10.3F-2). HR-cell death is also induced in cells surrounding feeding cells, often resulting in degeneration of feeding cells (Fig. 10.3F-3). Even if some feeding cells survive, the nutrient transport from surrounding tissues to the feeding cells is limited, causing a reduction in the number of eggs, and production of relatively more males. Some resistant plants induce the deterioration of feeding cells without any HR-cell death of surrounding cells.

10.5.1. Secretion of anti-nematode enzymes into the apoplast

The fact that the PPN effector Gr-VAP1 inhibits RCR3pim, a PLCP, implies that its enzymatic activity is important in immunity against PPNs (Fig. 10.3A). Indeed, the absence of RCR3pim homologs in *Arabidopsis* results in enhanced susceptibility to CN (Lozano-Torres *et al.*, 2014). In addition to Gr-VAP1, Mc1194, an effector of RKN *Meloidogyne chitwoodi* targets another PLCP, RD21A in *Arabidopsis* (Davies *et al.*, 2015b). Lack of RD21A leads to hyper-susceptibility to *M. chitwoodi*, showing that this PLCP also plays a positive role in immunity against RKN. However, it is not yet known how these PLCPs inhibit PPN infection.

Chitinases are also potentially important apoplastic enzymes in immunity against PPNs (Fig. 10.3A). Upon fungal infection, plants often secrete chitinases, which degrade chitin in the fungal cell walls (Kumar *et al.*, 2018). In nematodes, chitin is the main component of the egg shell (Perry and Trett, 1986) and makes up part of the pharyngeal lumen walls of *Caenorhabditis elegans* (Zhang *et al.*, 2005), suggesting that chitinases may have antinematodal activity and thus contribute to immunity against PPNs. Consistent with this idea, chitinase activity and transcript levels are upregulated after PPN infection in resistant plants (Bagnaresi *et al.*, 2013). However, there is currently no genetic evidence connecting plant chitinases to resistance against PPNs.

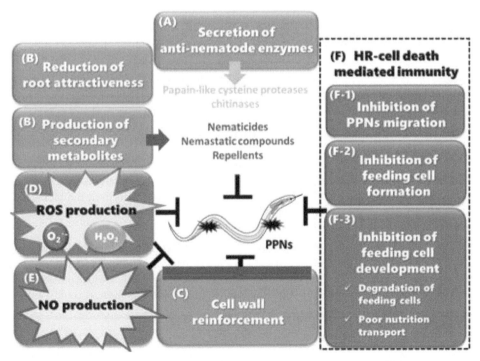

Fig. 10.3. Multiple plant immune responses against PPNs.

10.5.2. Production of anti-nematode compounds

Plants produce secondary metabolites in response to PPN invasion (Fig. 10.3B). For instance, chlorogenic acid, a phenolic compound, is produced in various plants including solanaceous plants (Pegard *et al.*, 2005), carrots (Knypl *et al.*, 1975), and rice (Plowright *et al.*, 1996), suggesting a common defense response against PPN infection. Although the production of chlorogenic acid is well-correlated with PPN resistance levels, chlorogenic acid itself is only weakly nematicidal for *M. incognita* (D'Addabbo *et al.*, 2013) with moderate activity against *Nacobbus aberrans*, a false root-knot nematode (López-Martínez *et al.*, 2011). One possible explanation for this lack of correlation between response and effectiveness is that metabolized products of chlorogenic acid have higher nematicidal activity in the target organism, but those compounds may be unstable or highly toxic in plants. Chlorogenic acid can be hydrolyzed to quinic acid and caffeic acid, with the latter being further oxidized to orthoquinone, which is toxic to PPNs (Mahajan *et al.*, 1985). However, the roles of caffeic acid and orthoquinone in resistance against PPNs need to be further established.

Another phenolic compound, phenylphenalenone anigorufone accumulates at the infection sites of the burrowing nematode, *Radopholus similis* in a resistant banana cultivar (*Musa* sp.) (Dhakshinamoorthy *et al.*, 2014). Anigorufone has high nematicidal activity because of the formation of large lipid–anigorufone complexes in the bodies of *R. similis*.

Some flavonoids (mostly belong to the classes of flavonols e.g., kaempferol, quercetin, myricetin, isoflavonoids, and pterocarpans) play important roles in PPN resistance by functioning as nematicides, nemastatic compounds (which do not kill but inhibit their movement), repellents, or inhibitors of egg hatching (Chin *et al.*, 2018). Kaempferol inhibits egg hatching of *R. similis* (Wuyts *et al.*, 2006b). Kaempferol, quercetin, and myricetin are repellents and nemastatic to *M. incognita* juveniles (Wuyts *et al.*, 2006b), and medicarpin also inhibits the motility of *Pratylenchus penetrans* in a concentration-dependent manner (Baldridge *et al.*, 1998). Similarly, patuletin, patulitrin, quercetin, and rutin are nematicidal for infective juveniles of *Heterodera zeae*, a CN (Faizi *et al.*, 2011).

Apart from phenolic compounds, other nematicidal chemicals are produced by several nematode-antagonistic plants, such as marigold and asparagus, which have been used for reducing nematode populations in soil. Marigold roots secrete α-terthienyl (Faizi *et al.*, 2011), an oxidative stress-inducing chemical that effectively penetrates the nematode hypodermis and exerts nematicidal activity (Hamaguchi *et al.*, 2019). Similarly, asparagus produces asparagusic acid, which inhibits hatching of two important CNs, *Heterodera glycines* and *G. rostochiensis* (Takasugi *et al.*, 1975).

In Brassicaceae family plants, the broad spectrum antimicrobial isothiocyanates and indole glucosinolates are considered as anti-PPN compounds. Isothiocyanates effectively inhibit hatching of CNs and RKNs (Yu *et al.*, 2005) and also have toxicity to RKNs and the semi-endoparasitic nematode *Tylenchulus semipenetrans* (Zasada and Ferris, 2003).

Ethylene, which is normally produced after wounding as well as during pathogen invasion, reduces PPN attraction to the root (Marhavý *et al.*, 2019). An ethylene-overproducing *Arabidopsis* mutant is less attractive for PPNs. PPN infection induces ethylene production, which possibly prevents secondary PPN invasion by reducing attractiveness. The reduced attractiveness could be due to a reduction in attractant secretion or an increase in repellents.

10.5.3. Reinforcement of cell wall as a physical barrier

Since all PPNs must penetrate the cell wall for feeding, reinforcement of cell wall structure has been implicated as an effective defense as a physical barrier (Fig. 10.3C). For instance, PPN infection often induces accumulation of lignin in resistant plants (Dhakshinamoorthy *et al.*, 2014). Moreover, *Arabidopsis* mutants with increased levels of syringyl lignin have reduced *M. incognita* reproduction rates (Wuyts *et al.*, 2006a). These results suggest that lignin accumulation in roots is an effective antagonist to PPN infection.

The effectiveness of lignin accumulation for suppressing nematode infection is also supported by plant immune inducers such as β-aminobutyric acid (BABA), thiamin, and sclareol. BABA is a non-protein amino acid. Treatment with BABA inhibits RKN invasion, delays giant cell formation, and retards RKN development. Interestingly, BABA induces lignin accumulation in roots, and callose accumulation in galls (Ji *et al.*, 2015). Thiamin (vitamin B1) treatment also induces lignin accumulation in roots; enhances the expression of phenylalanine ammonia-lyase, a key enzyme of the phenylpropanoid biosynthesis pathway; reduces PPN penetration; and delays PPN development (Huang *et al.*, 2016). An inhibitor of phenylalanine ammonia-lyase suppresses thiamin-mediated immunity, indicating that activation of the phenylpropanoid pathway with subsequent lignin accumulation is important for thiamin-mediated immunity against nematodes.

Callose deposition and suberin accumulation may also reinforce cell walls and contribute to immunity against PPNs. The RKN *Meloidogyne naasi* induces callose deposition at an early infection stage, and suberin accumulation at a later stage in the resistant grass plant *Aegilops variabilis* (Balhadère and Evans, 1995b). Overexpression of the transcription factor RAP2.6 in *Arabidopsis* leads to enhanced callose deposition at syncytia and results in higher resistance to CN (Ali *et al.*, 2013).

Lignin and suberin in suberin lamellae and casparian strips at the endodermis are also important basal physical barriers to RKNs. RKNs are not able to directly cross the endodermis because of the reinforcement of cell walls by suberin lamellae and casparian strips (Abad *et al.*, 2009).

10.5.4. Reactive oxygen species (ROS)

The rapid production of ROS, such as superoxide anion and hydrogen peroxide, is a conserved signaling response across kingdoms, and in plants, it is induced at an early stage of PPN infection (Fig. 10.3D). ROS have direct antimicrobial properties but also serve as signaling molecules to activate additional and complementary immune outputs such as strengthening cell walls by cross-linking polymers, amplifying and propagating intra and intercellular defense signals, and regulating HR-cell death (Kadota *et al.*, 2015). Resistant tomato plants carrying the Mi-1.2 gene respond to RKN infection with a strong and prolonged induction of ROS. Similarly, strong ROS production is induced in *Arabidopsis* roots during incompatible interactions with the soybean CN *H. glycines* (Waetzig *et al.*, 1999).

10.5.5. Nitric oxide (NO) and protease inhibitor-based immunity

NO is an essential signaling molecule that has multiple functions in plants (Scheler *et al.*, 2013) (Fig. 10.3E). After infection with *M. incognita*, resistant tomato plants carrying Mi-1.2 produce more NO than susceptible cultivars (Melillo *et al.*, 2011). Application of an exogenous NO donor, sodium nitroprusside (SNP), to susceptible tomato plants significantly enhances immunity against RKNs (Zhou *et al.*, 2015). Treatment with SNP reduces the number of egg masses and restores the growth inhibition associated with PPNs, suggesting that NO plays a positive role in immunity. Because both JA- and SNP-induced RKN defense responses are compromised by silencing protease inhibitor 2 (PI2), the NO- and JA-pathways likely converge to induce immunity against PPNs (Zhou *et al.*, 2015).

Interestingly, heterologous expression of various protease inhibitors, including trypsin inhibitors and cysteine protease inhibitors, confer resistance against PPNs, showing the effectiveness of protease inhibitor-based immunity against PPNs (Urwin *et al.*, 2003).

10.5.6. HR-cell death-based inhibition of nematode development

HR-cell death, a type of programed cell death that is induced after the invasion of avirulent pathogens to prevent the spread of biotrophic pathogens (Huysmans *et al.*, 2017), also plays a crucial role in PPN immunity (Fig. 10.3F). HR-cell death has been observed at three different phases of PPN infection in resistant plants: (i) in the cortex and epidermis during PPN penetration and migration (Davies *et al.*, 2015a), (ii) in vascular tissues during the initiation of feeding cell formation (Melillo *et al.*, 2006), and (iii) in cells adjacent to developing feeding cells (Ye *et al.*, 2017).

The resistance of soybean to SCN is based on a measure of nematode reproduction termed the female index (FI). The FI is calculated by taking an average cyst count across replicates of the test line and dividing by the average cyst count across replicates of the susceptible check and multiplying by 100. If the female index is less than 10% on a soybean line, the cultivar is regarded as resistant. In resistant soybean cultivars, infective juveniles are unable to establish a syncytium. Instead, the syncytium becomes necrotic and degenerates shortly after it is initiated in a manner similar to the hypersensitive response (HR) that leads to localized cell death in response to pathogens (Ghezzi *et al.*, 1996). Since the syncytium provides SCN nutrition for growth and reproduction, the collapse of the syncytium will lead to the demise of nematode due to starvation. What differs among resistant soybean cultivars is the timing of the HR (Acedo *et al.*, 1984). In Peking, syncytium degradation occurs around 2 days post-infection (Ghezzi *et al.*, 1996). In contrast, in PI 209332, the HR response occurs at 8 to 10 days after SCN infection (Acedo *et al.*, 1984). In both scenarios, the SCN penetrates the root but is unable to complete its life cycle (Fig. 10.4). Whether the differences in the timing of HR have any biological impact on plants is still unknown.

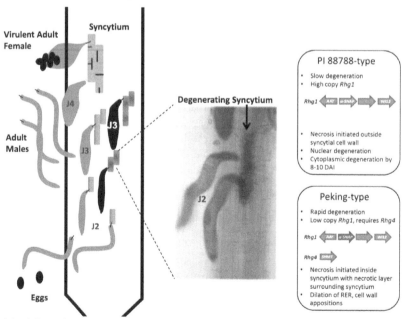

Fig. 10.4. The life cycle of the soybean cyst nematode (SCN) in a Peking-type resistant soybean plant. The two black boxes highlight the features of two types of resistance to SCN Hg Type 0 (Kim *et al.*, 1987; 2010b; 2012). J2 – Second stage juvenile, J3 – Third stage juvenile, J4 – Fourth stage juvenile.

10.6. RESEARCH NEEDS

Key needs in the area of genetic resistance and plant breeding include cooperation with, and support for, classical plant breeding programs, and research financial support in the following areas:

>> Natural and novel gene source identification and development.

>> Classical and novel gene transfer techniques.

>> Genetic studies on HPR, tolerance traits, nematode parasitism, and virulence factors.

>> Bioengineering studies to facilitate molecular approaches to gene study.

>> Transfer and expression of foreign genes in plants.

>> Certain other potential areas such as microtoxin use *in vitro* and immunization systems.

Development of a state-wide computerized database is also needed to allow access to information on plant susceptibility to nematodes and on management techniques. Longer-term research needs include the development of resistant plants using germplasm from wild species, development of transgenic plants, understanding nematode genetics and its impact on stability of HPR, and understanding the mechanisms of HPR.

A considerable amount of on-going research in molecular biology is directed toward understanding the nature of control and inheritance of HPR, but it is still unclear how the resistance traits are expressed relative to the nematode and how they are integrated into the physiology of the whole plant.

>> What is the effect of the expressed gene on nematode behavior, feeding, and reproductive habits?

>> How does expression of the gene affect the growth, fruitfulness, or longevity of the plant?

>> When is the gene expressed?

>> What environmental conditions influence or constrain the expression of the gene?

For example, the expression of resistance conferred by the *Mi* gene of tomato is not thermostable. The resistance breaks down at soil temperatures above 28 $^{\circ}$C, which could be a serious constraint in warm environments or at certain times of the year.

In addition to understanding the ramifications of HPR in whole plants, it is also necessary to understand the subtleties of its expression under field conditions. Critical to this understanding is genetic variability of nematode populations, which is not readily measured and is not well documented.

Some interesting lessons have been learned already about the impact of employing single gene resistance against nematodes in whole fields of plants. For example, single genes have been incorporated into soybeans to confer resistance to the soybean cyst nematode, *H. glycines*. Repeated culture of the resistant varieties selects for variants in the nematode population so that resistance-breaking biotypes become predominant. Consequently, the resistant cultivar is no longer effective in that field, and new sources of resistance are needed.

Similar examples exist for the cyst-nematode parasites of potatoes and cereals in Europe. In another example, a gene for resistance to the root-knot nematode, *M. incognita*, has been incorporated into tobacco varieties grown in the south-eastern United States. As a result, *M. arenaria* (another root-knot nematode more virulent than *M. incognita*) has now been selected for in that region.

The solution to the problem of selection for aggressive biotypes or species of a nematode pest by using resistant cultivars may lie in what Vanderplank (1984) described as "stabilizing selection." The underlying principle assumes that genes for aggressiveness to a resistant cultivar may not confer any advantage to the biotype of the nematode in the absence of the selection pressure imposed by the resistance. In other words, the original biotype or species of the pest was probably predominant because it was well adapted to that environmental situation and ecological niche, and was probably a better competitor for resources than the variants. Consequently, removal of the selection pressure, by growing a susceptible cultivar, will remove any advantage provided for the new biotype or species and allow selection for the better-adapted, original genotype.

The superior competitive abilities of the original genotype should result in a decline of the new genotype. Obviously, this could lead to reversion to the original problem, so a delicate balance of rotating resistant and susceptible cultivars must be achieved to keep both genotypes below damaging levels.

The implementation of stabilizing selection to preserve the longevity and utility of HPR, based on single-gene sources, requires both the study and measure of a number of epidemiological and population genetics parameters:

» What is the frequency of the new genotype in the population?
» How rapidly does the frequency of the new genotype increase and the frequency of the old genotype decrease when the resistant cultivar is grown?
» Should the same question be addressed in reverse for the susceptible cultivar?
» What is the appropriate rotation of resistant cultivars, susceptible cultivars, and non-host crops to minimize crop damage, maximize yields and profits, and to preserve the usefulness of the resistant cultivar for that field?

Clearly, in addressing these questions, the planning horizon for nematode management is extended from a single crop season to multiple growing seasons. Also, the perception of management of genotype frequencies of nematode populations in a field over time is introduced, as opposed to merely controlling the numbers of nematodes to a low level prior to planting.

A relatively untapped area for plant-parasitic nematodes is that of induced resistance. Can the plant's defensive forces, its "immune system," be activated by some stimulus so that it is better protected against nematode parasitism? Considerable research in this area, in both plant pathology and entomology, suggests that infection of a plant by a minor pathogen, or feeding by a relatively undamaging insect species, may trigger defensive responses in the plant that render it less susceptible to subsequent infection or feeding.

Several questions are suggested:

» Could such principles be applied to nematode parasitism?
» Is it possible that inoculation of a plant with a fungus, nematode, or homogenate of their products might elicit defensive mechanisms?
» How might such defensive mechanisms be induced to translocate downward into the root system?

These and other questions will provide fruitful avenues for additional research in this area.

10.7. CONCLUSION

Resistance, when available, can generally be considered as the best option for nematode management basically because it is cost effective and environmentally benign. However, unlike the use of nematicides that is not dependent on biological specificity of action among nematodes, natural host resistance can be applied only after an investigation of the specificity of the resistance traits and their targets among the wide variety of nematode species and subspecies. Despite the great potential of this management tactic, resistance appears to have been underutilized thus far, probably because many of its problems are still to be properly addressed. Most of these impediments will be overcome or minimized with additional research, breeding efforts, effective grower education programs, and by a closer collaboration between nematologists and plant breeders. One of the major problems to solve is the poor availability of resistance traits in many important crops and/or resistant traits to a wider variety of nematode groups, especially to migratory endo- and ectoparasites. Screening of the available germplasm resources for nematode resistance is still limited because bioassay protocols are very time-consuming as they should be capable of reliably evaluating the possible hundreds of genotypes of a germplasm collection. The same problems are present in a breeding program for nematode resistance in which the genotypes to be screened may be thousands. Moreover, assessment of the level of resistance to nematodes is not a trivial task, as described earlier. Technical difficulties in selecting nematode-resistant plant accessions or progenies in a breeding program could be minimized by adopting more rapid and objective methods of genetic identification, such as marker-assisted selection (MAS), where molecular markers come mainly from DNA polymorphism. Molecular marker techniques are beginning to be integrated into nematode resistance breeding programs, although apparently limited to RKNs and soybean cyst nematodes thus far (Hussey and Janssen, 2002; Young and Mudge, 2002).

Spread of virulent populations and resistance durability is an issue which is of late raising increasing concern. At least with RKN populations virulent on resistant tomato and pepper, it is becoming clear that selection, probably at its early stage, implies adverse fitness costs and that virulence is strictly specific to the gene on which selection occurs (Djian-Caporalino *et al.*, 2010; Molinari, 2010a). The cases in which reduced fitness has not been observed may be explained by the capacity of well-established nematode populations to compensate and restore fitness in time. This is the reason why assessment of resistance breaks and actions to support resistance durability should be promptly or preventively applied. In this case, spatial heterogeneity of resistant/susceptible varieties can significantly decrease pest population density and the rate of spread of new virulent populations. Considering the specificity of virulence, random spatial patterning of differential monogenic resistances can reduce pest density and plant damage at the same level as multigenic resistance. Finally, it may be possible in the future to pyramid combinations of novel transgenic resistance and natural resistance genes to develop broad-based and durable resistance.

Chapter - 11

Vegetable Grafting for Improved Nematode Resistance

11.1. INTRODUCTION

Recently, growers and researchers have begun examining vegetable grafting as an integrated pest management tool for successful vegetable production. Research has focused on grafted seedling production, use, and economics (Kubota *et al.*, 2008; Rivard *et al.*, 2010b); grafting as an alternative to methyl bromide in field production (Freeman *et al.*, 2009); and the use of resistant rootstocks for controlling RKN and other soil-borne diseases such as bacterial wilt (*Ralstonia solanacearum*), fusarium wilt, and southern blight (*Sclerotium rolfsii*). With the phase-out of methyl bromide for soil fumigation and the continued rise in demand for organic produce, the need for alternative disease control methods that do not rely on synthetic biocides has increased.

Grafting can be defined as the natural or deliberate fusion of plant parts so that vascular continuity is established between them and the resulting genetically composite organism functions as a single plant. Growing grafted vegetables was first launched in Japan and Korea in the late 1920s by grafting watermelons to gourd rootstocks. Even though grafting has been a mere common practice in fruit trees since ages, vegetable grafting is of recent popularization on a commercial scale (Sakata *et al.*, 2007). Continuous cropping is unavoidable in greenhouses, which again leads to reduction in yield and quality of the produce due to the menace of soil-borne pathogens and nematodes. Since soil sterilization can never be complete, grafting has become an essential technique for the production of repeated crops of vegetables grown in both greenhouse and open field conditions.

At first the cultivation of grafted plants in vegetables started to challenge the serious crop loss caused by infection of soil-borne diseases aggravated by successive and intensive cropping. After the first trial, the cultivated area of grafted vegetables, as well as the kinds of vegetables being grafted, has been consistently increased. At present, most of the watermelons (*Citrullus lanatus*), musk melons (*Cucumis melo*), Oriental melons (*Cucumis melo* var. *makuwa*), greenhouse cucumbers (*Cucumis sativus*), and several solanaceous crops in Korea and Japan are grafted before being transplanted in the field or in greenhouse (Ito, 1992; Kurata, 1992). The purpose of grafting also has been greatly expanded, from reducing infection by soil-borne diseases caused by pathogens such as *Fusarium oxysporum* (Itagi, 1992; Yamakawa, 1983) to increasing low-temperature (Tachibana, 1989), salt and wet-soil tolerance (Tachibana 1989; Park 1987), enhancing water and nutrient uptake (Heo, 1991; Jang *et al.*, 1992), increasing plant vigor, and extending the duration of economic harvest time (Itagi, 1992; Ito, 1992), among other purposes (Ali *et al.*, 1991; Dole and Wilkins, 1991; Matsuzoe *et al.*, 1991). Growing grafted vegetables, compared to growing grafted trees, was seldom practiced in the United States or in other western countries where land use is not intensive, i.e., proper crop rotation is being practiced. However, it is highly popular in some Asian and European countries where land use is very intensive and the farming area is small (Hartmann and Kester, 1975). *Since grafting gives increased disease tolerance and vigor to crops, it will be useful in the low-input sustainable horticulture of the future.*

11.2. OBJECTIVES

The main objective of vegetable grafting is to avoid soil-borne diseases such as Fusarium wilt in Cucurbitaceae (Cucumber, melons etc.), bacterial wilt in Solanaceae (Tomato, pepper etc.), and root-knot nematodes (*Meloidogyne* spp.) on several vegetable crops. However, research has shown that this technique can be effective against a variety of fungal, bacterial, viral, and nematode diseases (King *et al.*, 2006).

Grafting's early purpose was to avoid or reduce the soil-borne disease caused by *F. oxysporum*, but the reasons for grafting, as well as the kinds of vegetables grafted, have increased dramatically (Fig. 11.1). Watermelons, other melons (*Cucumis* spp.), cucumbers, tomatoes (*Solanum lycopersicum*) and eggplants (*Solanum melongena*) are commonly grafted to various rootstocks. Numerous rootstocks also have been developed. Watermelons are commonly grafted to gourds (*Lagenaria siceraria*) or to inter-specific hybrids (*C. maxima x C. moschata*). Cucumbers are frequently grafted to Figleaf gourd (*C. ficifolia*) or inter-specific hybrids (*C. maxima x C. moschata*).

Details of the objectives of grafting for each vegetable crop are shown in Table 11.1.

Table 11.1. Objectives of grafting fruit-bearing vegetables

Species	Objective/s
Watermelon	Tolerance to Fusarium wilt (*F. oxysporum*), root-knot nematodes (*Meloidogyne* spp.), low temperature, wilting due to physiological disorders, and drought tolerance.
Cucumber	Tolerance to Fusarium wilt, root-knot nematodes (*Meloidogyne* spp.), low temperature, and tolerance to *Phytophthora melonis*.
Muskmelon	Tolerance to Fusarium wilt (*F. oxysporum*), root-knot nematodes (*Meloidogyne* spp.), low temperature, wilting due to physiological disorders, and tolerance to *Phytophthora* disease.
Tomato	Tolerance to bacterial wilt (*Ralstonia solanacearum*), *Fusarium oxysporum*, *Verticillium dahlia*, *Pyrenochaeta lycopersici*, and root-knot nematodes (*Meloidogyne* spp.).
Egg plant	Tolerance to bacterial wilt (*Ralstonia solanacearum*), *Verticillium albo-atrum*, *Fusarium oxysporum*, root-knot nematodes (*Meloidogyne* spp.), low temperatures, and greater vigor.

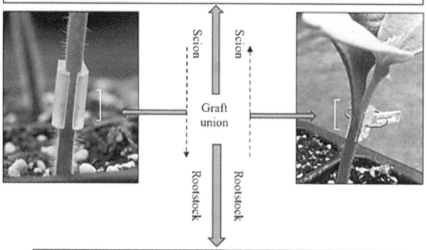

Desirable aboveground (shoot) traits intrinsic to scion cultivar (e.g., fruit yield, quality attributes, and resistance/tolerance to foliar diseases)

Characteristics as a result of rootstock-scion interactions (e.g., increased growth vigor, hormonal status modification, fruit quality modification, improved tolerance to low temperature and soil salinity, and changes in virus susceptibility)

Scion

Scion

Graft union

Rootstock

Rootstock

Desirable belowground (root) traits intrinsic to rootstock cultivar (e.g., resistance/tolerance to soilborne diseases, vigorous root systems, and tolerance to low temperature and soil salinity)

Characteristics as a result of rootstock-scion interactions (e.g., improved nutrient and water use efficiency, root architecture and morphology)

Fig. 11.1. A general concept of vegetable grafting (Photo by Xin Zhao).

11.3. AREA UNDER SELECTED GRAFTED VEGETABLES SEEDLINGS

Table 11.2 shows the statistics of cultivated area under greenhouse conditions using grafted seedlings, and the estimated number of grafted vegetable seedlings in Korea and Japan. About 766.3 million grafted seedlings are planted annually in Korea and 721.3 million in Japan in greenhouses. About 95% of the watermelons in both countries are grafted. The majority of greenhouse cucumbers (75%) are grafted. Most of the Oriental melons are grafted to squash (*Cucurbita* spp.) rootstock (Ito, 1992; Jang *et al.*, 1992). It is likely that these quantities will be increased soon with the introduction of new uses of grafting for other vegetables and the development of rootstock having desirable characteristics.

Table 11.2. Cultivated area under greenhouse of some vegetables being grafted and the number of grafted seedlings needed annually in Japan and Korea (2005) (Lee *et al.*, 2010).

Vegetables	Cultivated area (ha)	No. of grafted seedlings/ha (x 1000)	% use of grafted seedlings	No. of grafted seedlings required* (millions)
Republic of Korea				
Watermelon	23,179	6-9	95	198.2
Melon[1]	13,000	7-10	90	117.0
Cucumber	5,853	20-30	75	131.7
Tomato	6,749	20-30	25	50.6
Eggplant	933	10-20	20	3.7
Pepper[2]	67,023	20-40	10	268.2
Total				766.3
Japan				
Watermelon	13,400	6-9	92	111.0
Melon[1]	10,400	7-10	30	31.2
Cucumber	13,400	20-30	75	301.5
Tomato	13,000	20-30	40	156.0
Eggplant	10,400	10-20	55	114.4
Pepper[2]	3,620	20-40	5	7.2
Total				721.3

[1] Including net melons, cantaloupes, oriental melons.

[2] Including hot peppers for dry and fresh uses.

Vegetable grafting is also actively practiced in other countries like China, Spain, USA, Taiwan, Italy, and France. Grafted vegetable seedlings were distributed in North America to the extent of 40–45 million in 2005 (Kubota *et al.*, 2008). Spain is by far the leading European country in using grafted vegetable transplants (129.8 million in 2009) (Hoyos Echeverria, 2010) followed by Italy (47.1 million) (Morra and Bilotto, 2009) and France (about 28 million). It was estimated that about 20% of China's watermelons and cucumbers are grafted.

11.4. GRAFTING METHODS

Grafting involves the union of suitable varieties of greenhouse crops used as scions over the resistant rootstocks usually from the same family. Trials have been conducted on grafting of tomato seedlings over the available rootstocks from different families. Although, the success has been achieved in grafting of greenhouse tomato over interfamilial plants, however, tomato and cucurbits only gave good results. Technology is underway and is proving to be promising for greenhouse cultivation in India.

Various grafting methods have been developed and growers must choose their favorite methods based on experiences and preferences. Grafting cucurbitaceous crops is commonly done when scion and rootstock seedlings are young, i.e., before the outgrowth of the first true leaf between the cotyledons.

Grafting methods vary considerably with the type of crops being grafted, and the sowing time for scion and stock seeds vary with grafting method and crop. For example, "hole insertion grafting" would be convenient for watermelons because of their small seedling size compared to the size of stock seedlings, such as gourd and squash. In cucumbers, however, "tongue approach grafting" has been used widely, mainly because of their large seedling size, including hypocotyl length and diameter, and grafting ease. Some of the most widely practiced grafting methods are as follows:

11.4.1. Hole insertion grafting (HIG), terminal or top insertion grafting

This method is mostly used for cucurbits, mainly for watermelon. Scion seedling size needs to be smaller than the rootstock seedlings. In this method, squash or bottle gourd are used as rootstock. Watermelon seeds are sown 7–8 days after the sowing of gourd rootstock seeds or 3–4 days after sowing squash rootstock seeds. Grafting is made 7–8 days after the sowing of watermelon seeds. The true leaf including the growing point should be carefully and thoroughly removed with a scoping motion. A hole is made with a bamboo or plastic gimlet or drill at a slant angle to the longitudinal direction in the removed bud region.

The hypocotyl portion of the watermelon scion is prepared by slant cutting to a tapered end for easy insertion into rootstock hole (Fig. 11.2). After that the grafted plant is placed into a healing chamber.

Fig. 11.2. Hole insertion grafting.

11.4.2. Tongue approach grafting (TAG)

It is usually done in watermelons, cucumbers, and melons. Scion and rootstock seedlings need to be of similar height and stem diameters. In this method, seeds of cucumber are sown 10-13 days before grafting and pumpkin seeds 7-10 days before grafting, to ensure uniformity in the diameter of the hypocotyls of the scion and rootstock. The growing point of the rootstocks should be carefully removed so that the shoot cannot grow. The grafting cut for rootstock should be made in a downward direction and the scion cut in an upward direction at an angle, usually 30°–40° to the perpendicular axis in such a way that they tongue into each other (Fig. 11.3). Grafting clips are placed to secure graft position at the graft union site. The hypocotyl of the scion is left to heal for 3-4 days and crushed between fingers and later the hypocotyl is cut off with a razor blade 3 or 4 days after being crushed.

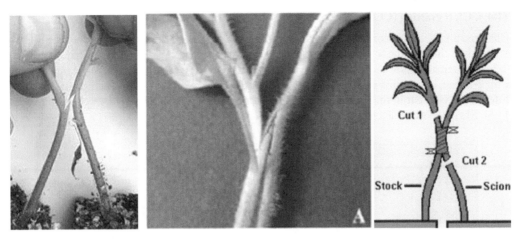

Fig. 11.3. Tongue approach grafting

11.4.3. Splice grafting (SG), tube grafting (TG), and one cotyledon splice grafting (OC-SG)

This grafting method is mainly used for the vegetable seedlings which are grown in plug trays i.e. cucumber and watermelon. This method is two or three times faster than the conventional method. The rootstock should be grafted when cotyledons and the first true leaf start to develop (about 7 to 10 days after sowing). One cotyledon and the growing tip are removed. The seedling is cut at a slant from the base of one cotyledon to 0.8–1.0 cm below the other cotyledon, removing one cotyledon and the growing tip. The length of cut on the scion hypocotyl should match that of the rootstock and should be at a 35° to 45° angle. The scion is attached to the rootstock and fixed tightly by a grafting tube or clip (Fig. 11.4). OCG is called tube grating when the joined plants are held together with a length of tube instead of a grafting clip.

Fig. 11.4. Splice grafting

11.4.4. Cleft grafting (CG)

Cleft grafting is usually confined to solanaceous crops. The seeds of the rootstock are sown 5-7 days earlier than those of the scion. The stem of the scion (at four leaf stage) are cut at right angle with 2-3 leaves remaining on the stem. The rootstock (at the four to five leaf stages) is cut at right angles, with 2-3 leaves remaining on the stem. The stem of the scion is cut in a wedge, and the tapered end fitted into a cleft cut in the end of the rootstock. The graft is then held firm with a plastic clip (Fig. 11.5). Move the tray filled with grafted plants to proceed for healing up.

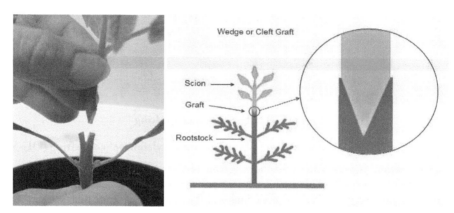

Fig. 11.5. Cleft grafting

11.4.5. Pin grafting (PG)

Pin grafting is basically the same as the splice grafting. Instead of placing grafting clips to hold the grafted position, specially designed pins are used to hold the grafted position in place. The cotyledons of the rootstock and scion are cut horizontally and a ceramic pin about 15 mm long and 0.5 mm in diagonal width of the hexagonal cross-section is inserted into the cut surface this helps align and secure the joined sections (Fig. 11.6). This is very easy method, reducing labor cost, but ceramic pins are expensive, and a special environmentally controlled chamber is needed to acclimatize the grafted plants.

Fig. 11.6. Horizontal-pin grafting

The different rootstocks suitable and various methods used for grafting of cucurbitaceous and solanaceous vegetable crops are listed in Table 11.3.

Table 11.3. Grafting methods and rootstocks used in cucurbitaceous and solanaceous vegetable crops

Scion plant	Suitable rootstocks	Grafting methods
Eggplant	*Solanum torvum, S. sissymbrifolium Solanum khasianum*	Tongue and cleft method, Cleft method, Both tongue and cleft methods
Tomato	*L. pimpinellifolium, S. nigrum*	Only Cleft method, Tongue and cleft methods
Cucumber	*C. moschata, Cucurbita maxima*	Hole insertion and tongue method, Tongue method
Water melon	*Benincasa hispida, C. moschata, C. melo, C. moschata × C. maxima, Lagenaria siceraria*	Hole insertion and cleft method, Hole insertion and cleft method, Cleft method, Hole insertion method, Splice Grafting
Bitter gourd	*C. moschata, Lagenaria siceraria*	Hole insertion and tongue method, Hole insertion

11.5. POST GRAFT CARE

In the past, grafting was routinely carried out by the growers themselves, but it is now rapidly shifting to cooperative operations owing to the efficiency, ease of post-graft care, and recent expansion of the commercial seedling industry (such as sales of plug-grown seedlings). Even though the single-edged razor blade is still the most widely used grafting knife among farmers today, many other devices, such as specifically designed knives, clips, tubes, or glue, have now been developed for easier grafting and post-graft care (Kurata, 1994). Semi-automatic grafting machines and/or fully automatic grafting machines using robots (Itagi, 1992; Ito, 1992) are being developed. By using the most sophisticated machine, grafting efficiency could be significantly increased from the present 150 seedlings/hour per expert. For this purpose, it is important to define the grafting timeline (Fig. 11.7).

The grafted seedlings are conditioned (hardened) in the dark and cool shade nets prepared specially for the purpose before planting in seedling trays and finally their transplantation in the greenhouses (Fig. 11.8). Grafting robots are being increasingly used for the commercial production of healthy nursery.

The labor required for intensive post-graft care, mostly 7 to 10 days of careful management, could be markedly reduced by using specifically designed conditioning chambers (Ito, 1992).

Grafting an indeterminate tomato scion onto a vigorous rootstock makes it possible to extend the harvest period when environmental conditions are adequate. Grafted tomato plants are increasingly used also in soilless cultivation systems, where, under controlled conditions and with crop cycles extended up to one year, grafted plants can reach their full potential (Fig. 11.9).

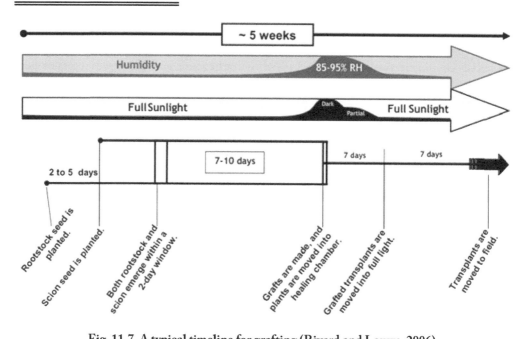

Fig. 11.7. A typical timeline for grafting (Rivard and Louws, 2006).

Fig. 11.8. Nursery production of grafted vegetables in greenhouse

Fig. 11.9. Double-stem grafted tomato plants grown in a greenhouse soilless system in a year round crop cycle (Photo by Francesco Di Gioia).

11.6. NEMATODE MANAGEMENT

Under continuous cropping, soil-borne nematode problems are likely to increase. Since soil sterilization can never be complete, grafting has become an essential technique for the production of repeated crops of fruit-bearing vegetables.

Grafting as a technique is gaining wide attention throughout the world, especially for greenhouse cultivation of vegetable crops, mainly the Solanaceous (tomato, eggplant and sweet pepper) and Cucurbitaceous (cucumber, watermelon and musk melon) ones, from the view point of resistance against the soil-borne nematode pathogens in addition to obtaining better yield and quality.

Many rootstocks having distinctive characteristics are available (Yamakawa, 1983), and growers select the rootstocks, they think are the most suitable for their growing season, cultivation methods (field or greenhouses), soil type, and the type of crops and cultivars (Lee, 1989).

The vigorous roots of the rootstock exhibit excellent tolerance to serious soil-borne diseases, such as those caused by *Fusarium, Verticillium, Ralstonia*, and root-knot nematodes, even though the degree of tolerance varies considerably with the rootstock. The mechanism of disease resistance, however, has not been intensively investigated. The disease tolerance in grafted seedlings may be entirely due to the tolerance of stock plant roots to such diseases.

However, in actual plantings, adventitious rooting from the scion is very common (Lee, 1989). Plants having the root systems of the scion and rootstock are expected to be easily infected by soil-borne diseases.

Disease resistance researchers around the world have demonstrated that grafting can be effective against a variety of soil-borne fungal, bacterial, viral, and nematode diseases. In Morocco and Greece, grafting is used to control root-knot nematodes (*Meloidogyne* species) in both tomatoes and cucurbits. Researchers have proposed using grafted plants instead of methyl bromide to manage soil-borne nematode diseases in these regions of the world.

11.6.1. Cucurbitaceae

In cucurbits, resistance to *Meloidogyne incognita* was identified in *Cucumis metuliferus, C. ficifolius*, and bur cucumber (*Sicyos angulatus*) (Fassuliotis, 1970; Gu *et al.*, 2006). Using *C. metuliferus* as a rootstock to graft RKN-susceptible melons, led to lower levels of root galling and nematode numbers at harvest (Siguenza *et al.*, 2005). Moreover, *C. metuliferus* showed high graft compatibility with several melon cultivars (Trionfetti Nisini *et al.*, 2002). Cucumbers grafted on the bur cucumber rootstock exhibited increased RKN resistance (Zhang *et al.*, 2006). Promising progress has also been made in developing *M. incognita*-resistant germplasm lines of wild watermelon (*Citrullus lanatus* var. *citroides*) for use as rootstocks (Thies *et al.*, 2010). However, at present, cucurbit rootstocks with resistance to RKN are not commercially available (Thies *et al.*, 2010).

Several new rootstocks are being developed. For example, bur-cucumber (*Sicyos angulatus*) collected near Andong, Korea, showed good compatibility with cucumbers and watermelons for early summer growth and good resistance to root-knot nematodes (Lee, 1992; Lee *et al.*, 1992).

Diseases controlled by grafting in different Cucurbiaceous vegetable crops due to fungi, bacteria, viruses and nematodes are listed in Tables 11.4 and 11.5.

Table 11.4. Various rootstocks used for grafting some vegetable crops

Vegetables	Popular rootstock species
Watermelon	Interspecific hybrids, *Cucurbita maxima* × *C. moschata* Bur-cucumber, *Sicyos angulatus*
Cucumber	Wild cucumber, *Cucumis pustulatus* Bur-cucumber, *Sicyos angulatus*
Musk melon	African horned cucumber, *Cucumis metuliferus* Wild cucumber, *Cucumis pustulatus*

Table 11.5. Major types of rootstocks used for melon and watermelon grafting and their key characteristics

Type of rootstock	Rootstock cultivar example	Major beneficial characteristic
Cucurbita moschata (squash)	'Marvel' (American Takii)	Resistant to Fusarium wilt; possibly tolerant to root-knot nematodes; tolerant to low temperature
Cucurbita maxima × *C. moschata* (interspecific hybrid squash)	'Carnivore', 'Super Shintoism' (Syngenta); 'Cobalt' (Risk Swan); 'RST-04-109- MW' (DP Seeds); 'Flexifort' (Enza Zaden)	Resistant to Fusarium wilt; possibly resistant to melon necrotic spot virus; tolerant to low and high temperatures; tolerant to root-knot nematodes; vigorous growth
Citrullus lanatus var. *citroides* (wild watermelon)	'Ojakkyo' (Syngenta)	Resistant to Fusarium wilt and root-knot nematodes
Cucumis metuliferus (African horned cucumber)	No commercial rootstocks available	Resistant to Fusarium wilt and root-knot nematodes

Adapted from Davis *et al.* (2008) and Louws *et al.* (2010).

11.6.1.1. Watermelon: Commercial watermelon 'Fiesta' (diploid seeded) (*Citrullus lanatus* var. *lanatus*) plants grafted on rootstock from the wild watermelon germplasm line RKVL 318 (*C. lanatus* var. *citroides*) had significantly less (P< 0.05) root galling (11%) due to *M. incognita* than non-grafted 'Fiesta' watermelon (36%) or plants with the squash hybrid or bottle gourd rootstocks (Fig. 11.10). Wild watermelon germplasm lines derived from *C. lanatus* var. *citroides* were identified that may be useful as resistant rootstocks for managing root-knot nematodes in water-melon.

11.6.1.2. Muskmelon: Melon scion cvs. 'Honey Yellow' and 'Arava' grafted onto *C. metuliferus* rootstock exhibited significantly lower galling and reduced RKN population densities in the field (Table 11.6) (Guan *et al.*, 2014). *Cucumis metuliferus* has resistance to *M. incognita* race 1 and 3. *C. metuliferus* has also been reported resistant to *M. arenaria* and *M. javanica* (Walters *et al.*, 1993). Fassuliotis (1970) found no hypersensitive reaction to infection by *M. incognita*, but the development of J2 to adult was delayed in *C. metuliferus*.

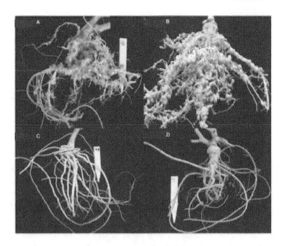

Fig. 11.10. Effect of rootstocks on root-knot nematode, *Meloidogyne incognita* resistance in watermelon. A - 'Emphasis' bottle gourd (*Lagenaria siceraria*) rootstock exhibiting severe root galling caused by root-knot nematodes. B - 'Strong Tosa' squash hybrid (*Cucurbita moschata × C. maxima*) rootstock exhibiting severe root galling. C - RKVL 301 wild watermelon (*C. lanatus* var. *citroides*) rootstock resistant to root-knot nematodes. D - RKVL 302 wild watermelon (*C. lanatus* var. *citroides*) rootstock resistant to root-knot nematodes.

Table 11.6. Effect of rootstocks on root gall index (GI) and numbers of *Meloidogyne javanica* **second-stage juveniles (J2) in melon under field conditions.**

Treatment	Gall index (0-10 scale)	*Meloidogyne* density (J_2/ 100 cm³ of soil)
Non-grafted 'Honey Yellow'	7.14 a	378.2 a
Self-grafted 'Honey Yellow'	6.70 a	515.6 a
'Honey Yellow' grafted on *C. metuliferus*	0.08 b	1.2 b
Non-grafted 'Arava'	5.20 a	200.2 a
Self-grafted 'Arava'	4.45 a	140.2 ab
'Arava' grafted on *C. metuliferus*	0.15 b	4.8 b

11.6.1.3. Watermelon, muskmelon and cucumber: Wild cucumber rootstock, *Cucumis pustulatus* gave effective control of root-knot nematodes on watermelon, muskmelon and cucumber and increased fruit yield (Liu *et al.* 2014) (Fig. 11.11 and Table 11.7).

Table 11.7. Management of root-knot nematodes in Cucurbits using wild cucumber rootstock, *Cucumis pustulatus*

Treatment	Gall index	Yield in kg/plant
Cucumber grafted on *C. pustulatus*	1	2.032
Cucumber non-grafted	5	1.234
Muskmelon grafted on *C. pustulatus*	1	2.276

Muskmelon non-grafted	5	1.612
Watermelon grafted on *C. pustulatus*	1	3.901
Watermelon non-grafted	5	1.650

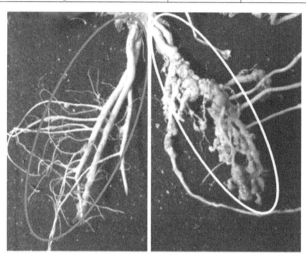

Fig. 11.11. Effect of rootstocks on root-knot nematode, *Meloidogyne incognita* **resistance in watermelon (***Left* **– Cucumber grafted on resistant rootstock,** *Right* **– Non-grafted).**

11.6.2. Solanaceae

Different popular rootstocks used in various solanaceous vegetable crops for the management of diseases and nematodes are listed in Table 11.8.

Table 11.8. Popular rootstock species for grafting of solanaceous vegetables

Vegetables	Popular rootstock species
Tomato	Multifort (*S. lycopersicum* x *S. habrochaites*), Survivor (*S. lycopersicum*), Nematex, Maxifort, Beaufort (*Mi gene*), *Solanum pimpinellifolium*, and *S. hirsutum*
Eggplant	*Solanum integrifolium, Solanum tarvum, S. sisymbriifolium,* Accessions EG 195, EG 203
Sweet pepper	AR-96023 (*Capsicum annuum*)

11.6.2.1. Tomato: The *Mi* gene, which provides effective control against RKN in tomato, has been introgressed into cultivated tomatoes and rootstock cultivars (Louws *et al.*, 2010). Grafting of susceptible tomato cultivars on RKN-resistant rootstocks was effective in controlling RKN in fields naturally infested with RKN (Rivard *et al.*, 2010). However, as a result of temperature sensitivity of the *Mi* gene, such resistance may not be uniformly stable (Cortada *et al.*, 2009).

Solanum torvum wild brinjal germplasm identified as resistant against root-knot nematode was used as rootstock to graft with scions of promising tomato varieties Kashi Aman, Kashi Vishesh and Hisar Lalit. Grafted plants were compatible between rootstock and scion and also showed significant resistance against root-knot nematode by reducing soil population, reproduction and gall index (Gowda *et al.*, 2017).

Diseases controlled by grafting on different rootstocks in tomato due to fungi, bacteria, virus and nematode are listed in Table 11.9.

Table 11.9. Commonly used rootstocks for tomato and their disease resistance/susceptibility and vigor.

Rootstock	TMV	Fusarium Wilt		Verticillium Wilt Race 1	Bacterial Wilt	Root-knot Nematodes	Vigor*
		Race 1	Race 2				
Anchor[1]	R	R	R	R	R	R	5
Arnold	R	R	R	N/A	N/A	R	N/A
Beaufort[2]	R	R	R	R	S	R	3
Body[3]	R	S	R	R	S	R	5
Estamino	R	R	R	R	S	R	N/A
Maxifort[2]	R	R	R	R	S	R	5
Multifort	S	R	R	S	S	R	N/A
Survivor[1]	R	R	R	R	M	R	5
Aegis[1]	R	R	R	R	M	R	4
Robusta[3]	R	S	R	R	M	R	3
RT-04-105T	S	R	R	R	R	R	N/A
RT-04-106T	R	R	R	S	R	S	N/A

R – Resistant, M – Moderate, S – Susceptible, N/A – Not characterized.

* Vigor is measured on a scale of 1 to 5, where 1 represents poor and 5 represents excellent

Seed supplier [1] Takii Seeds, [2] deRuiter Seeds, [3] Bruinsma Seeds

The hybrid rootstocks i.e., interspecific tomato hybrid rootstock 'Multifort' (*S. lycopersicum* x *S. habrochaites*) and tomato hybrid rootstock 'Survivor' (*S. lycopersicum*) significantly reduced root galling compared with the non-grafted and self-grafted scions by ~ 80.8%. The root galling reduction by 'Survivor' (97.1%) was significantly greater than that by 'Multifort' (57.6%) compared with non-grafted scions (Table 11.10). In general, tomato plants grafted onto 'Multifort' tended to be more vigorous than all other treatments. 'Brandywine' grafted to 'Multifort' resulted in significantly higher (P<0.05) yields than the non- and self-grafted 'Brandy-wine' treatments (Fig. 11.12). Grafting with appropriate rootstocks may play an effective role in RKN management during the transition to organic production when high populations of nematodes are present (Barrett *et al.*, 2012).

Table 11.10. Effect of grafting treatments on root-knot nematode galling ratings on tomato cultivars.

Treatment	Root galling (1-10 scale) in cultivar	
	Brandywine	Flamme
Non-grafted	9.30 a	8.06
Self-grafted	7.30 b	6.12
Grafted on 'Multifort' rootstock	3.88 c	3.48
Grafted on 'Survivor' rootstock	0.54 d	0.00

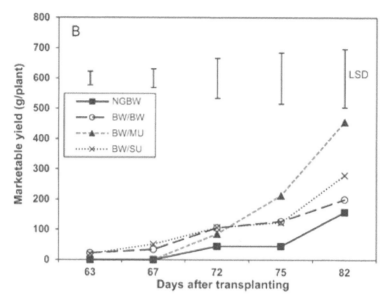

Fig. 11.12. Cumulative marketable yield for non-grafted and grafted tomato cultivar Brandywine.

Most eggplant lines utilized will graft successfully with tomato lines. The Asian Vegetable Research and Development Centre (AVRDC) recommends eggplant accessions EG195 and EG203 that are resistant to flooding, bacterial wilt, root-knot nematode (*Meloidogyne incognita*), *Fusarium* wilt (*Fusarium oxysporum* f. sp. *lycopersici*), and southern blight (*Sclerotium rolfsii*) (Black *et al.*, 2003).

Owusu *et al.* (2016) reported that the tomato scion cvs. Tropimech and Big Power (locally grown nematode-susceptible cultivar) grafted on resistant rootstocks such as Big Beef, Celebrity, and Jetsetter (resistant to *Verticillium* wilt, *Fusarium* wilt, nematodes, and tobacco mosaic virus) exhibited the least nematode populations in the greenhouse. In field experiments, nematode population levels were lower in Big Power that had been grafted on Celebrity, Jetsetter, and Big Beef rootstocks, compared to self-grafted or non-grafted Big Power (Fig. 11.12). Fruit yields were also higher in the grafted plants utilizing resistant rootstocks than non-grafted plants.

Tomato grafting onto a resistant rootstock of wild brinjal (*S. sisymbriifolium*) under farmers' field conditions at Hemza of Kaski district against root-knot nematodes showed that the root system of the grafted plants was free from gall formations. However, non-grafted plants had an average of 7.5 gall index (GI) (Figs. 11.13 and 11.14). Fruit yields significantly ($P > 0.05$) increased by 34.64% in the grafted plants compared with the non-grafted plants (Table 11.11) (Baidya *et al.*, 2017).

Table 11.11. Tomato on 'Wild Brinjal' rootstock

Treatment	Yield (Kg/plant)	RKI
Non-grafted	8.980	8
Grafted on *Solanum sisymbriifolium*	12.064	0
Yield increase over non- grafted	3.104	---
% yield increase	34.64	---

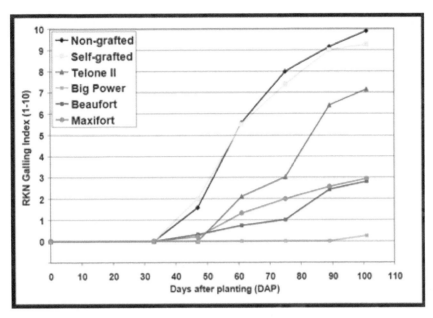

Fig. 11.13. Effect of rootstocks on root galling on tomato

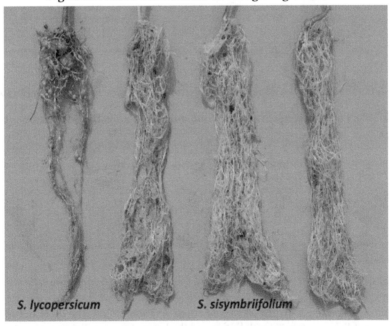

Fig. 11.14. The root of tomato (*Solanum lycopersicum*) on the left shows severe galling compared to three healthy root systems of *S. sisymbriifolium* after infection with *Meloidogyne enterolobii*.

Among eight wild *Solanum* rootstocks and two tomato hybrids were screened against root-knot nematode infection; *S. sisymbriifolium*, *Physalis peruviana*, and *S. torvum* had the least galls per 10 g root (6, 5, 5) and females per g root (2, 2, 2), respectively, and showed the highest level of expression of phenolics and defense-related enzymes viz., peroxidases, polyphenol oxidases, phenylalanine ammonia lyase, and acid phosphatase from leaf samples, compared to the susceptible tomato scion (US-618) (Dhivya *et al.*, 2014).

In grafted tomatoes, the rootstocks 'Aloha' and 'Multifort' both reduced root galling (Table 11.12 and Fig. 11.15).

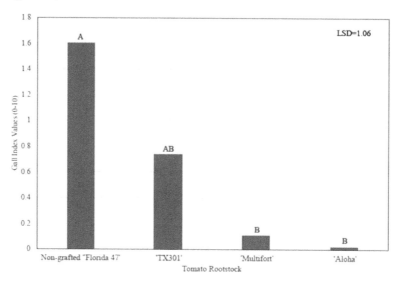

Fig. 11.15. The rootstocks 'Aloha' and 'Multifort' both reduced root galling in tomato

Table 11.12. Effect of 'Aloha' and 'Multifort' rootstocks on root galling of tomato

Rootstock	*M. incognita* (J_2/g root)
Non-grafted 'Florida 47'	2.16
'TX301'	1.52
'Multifort'	0.10
'Aloha'	0.10

In a previous study, two garden eggplant rootstocks *S. torvum* and *S. aethiopicum* were found to be poor hosts of *M. javanica* and *M. incognita* (Ioannou, 2001).

11.6.2.2. Brinjal: The rootstocks *Solanum torvum* and *S. sisymbriifolium* showed resistant reaction against root-knot nematode, *Meloidogyne incognita*. Three cultivated eggplant varieties viz., Sufala, Singnath, and Kazla were grafted on *S. torvum* and *S. sisymbriifolium*. The highest grafting success was 95% in case of *Solanum torvum* with Sufala and the lowest (85%) in *Solanum sisymbriifolium* with Singnath. The success of grafting was not affected significantly due to the effect of scion and or of rootstocks. The grafted plants showed resistant reaction against *M. incognita* while the scion plants showed susceptibility in the root-knot nematode sick beds. The grafted plants also showed resistant reaction against *M. incognita* under the field conditions. The grafted plants also out-yielded compared to the scion plants. The grafting combination *Solanum torvum* with Sufala gave the highest yield compared to other grafting combinations and non-grafted plants, while the grafting combination *S. sisymbriifolium* + Sufala gave least number of galls (Table 11.13) (Rahman *et al.*, 2002).

Table 11.13. Effect of different rootstocks on root-knot nematode resistance in three eggplant cvs.

Treatments	Fruit yield (tons/ha)	No. of galls/g root
S. torvum + Singnath	35.3 bcd	10.40
S. sisymbriifolium + Singnath	33.3 d	6.42
S. torvum + Sufala	39.2 a	7.80
S. sisymbriifolium + Sufala	38.0 ab	5.56
S. torvum + Kazla	36.5 abc	6.20
S. sisymbriifolium + Kazla	36.4 cd	7.92
Singnath	13.4 f	36.20
Sufala	15.6 f	40.62
Kazla	21.4 e	39.42
CD (P = 0.05)	7.36	---

In a column, means having a common letters do not differ significantly at 5% level.

Grafting of eggplant onto root-knot nematode resistant rootstock, usually *Solanum torvum*, is used in high value production systems in some European and Asian countries. This method is an effective way to prevent nematode damage and build-up.

Grafting of eggplant cv. Aristotle onto root-knot nematode resistant rootstocks like Charleston Hot, Carolina Wonder, Charleston Belle, Mississippi Nemaheart and Carolina Cayenne significantly reduced population of root-knot juveniles and root galling (Table 11.14).

Table 11.14. Effects of grafting Brinjal cv. Aristotle on root-knot nematode resistant rootstocks.

Treatment	*M. incognita*/g root	RKI
Aristotle non-grafted	16.73 b	3.28 b
Aristotle self-grafted	43.29 a	5.18 a
Aristotle grafted on Charleston Hot	8.75c	0.18 c
Aristotle grafted on Carolina Wonder	1.97 c	0.02 c
Aristotle grafted on Charleston Belle	0.22 c	0.24 c
Aristotle grafted on Mississippi Nemaheart	1.72 c	0.10 c
Aristotle grafted on Carolina Cayenne	1.11 c	0.12 c

11.6.2.3. Sweet pepper: Sweet pepper (*Capsicum annuum*) cultivars possessing the *N* gene, which controls resistance to RKNs (*M. incognita*, *M. arenaria*, and *M. javanica*), have been effective as rootstocks to control RKNs (Table 11.15) (Oka *et al.*, 2004; Thies and Fery, 1998, 2000).

Table 11.15. Rootstock for root-knot resistance in sweet pepper

Scion/rootstock	GI (1-10 scale)	Fruit yield (kg/6m²) 20 plants
Celia non-grafted	7.8a	20.2b
Celia grafted on AR-96023 (*C. annuum*)	0.7b	43.8a

11.7. ADVANTAGES AND DISADVANTAGES

Major advantages and disadvantages of using grafted vegetable transplants are presented in Table 11.16.

Table 11.16. Major advantages and disadvantages of using grafted vegetable transplants

Advantages	Disadvantages
» Yield increase	» Additional seeds for rootstocks
» Shoot growth promotion	» Experienced labor needed
» Disease tolerance	» Wise selection of scion/rootstock combinations
» Nematode tolerance/resistance	» Different combinations for cropping season
» Low temperature tolerance	» Different combinations for cropping methods
» High temperature tolerance	» Increased infection of seed-borne diseases
» Enhanced nutrient uptake	» Excessive vegetative growth
» Enhanced water uptake	» Fruit harvesting may be delayed
» High salt tolerance	» Inferior fruit quality (taste, color and sugar contents)
» Wet soil tolerance	
» Heavy metal and organic pollutant tolerance	» Increased incidence of physiological disorders
» Quality changes	» Symptoms of incompatibility at later stages
» Extended harvest period	» Different cultural practices should be applied
» Multiple and/or successive cropping allowed	» Higher prices of grafted seedlings
» Convenient production of organic wastes	
» Ornamental values for exhibition and education	

Various problems are commonly associated with grafting and cultivating grafted seedlings. Major problems are the labor and techniques required for the grafting operation and post-graft handling of grafted seedlings for rapid healing for 7 to 10 days. An expert can graft 1200 seedlings per day (150 seedlings per hour), but the numbers very with the grafting method. Similarly, the post-graft handling method depends mostly on the grafting methods. The problems could be minimized or easily overcome by careful cultural management and wise selection of scion and rootstock cultivars.

11.8. ROBOTIC GRAFTING

The introduction of mechanization and automation technology will help to address large-scale production issues. A fully automated grafting robot for cucurbits, which was recently developed based on the 1993 version of Iseki's GR800 semi-automated robot. The robot has scion and rootstock feeders, which pick, orient, and feed the scion and the rootstock shoots to the grafting processor

(Fig. 11.16), performing 750 grafts per hour with a 90% success rate. The robotic grafting process is illustrated in Fig. 11.17.

Even though there are many problems associated with cultivating grafted vegetable seedlings, the demand for successful grafted seedlings is growing rapidly. Breeding multipurpose rootstocks and developing efficient grafting machines and techniques will undoubtedly encourage increased use of grafted seedlings in many countries. Large-scale commercial production of vegetable seedlings is expanding rapidly in many developed countries and this will lead to an increased commercial supply and use of grafted vegetable seedlings throughout the world.

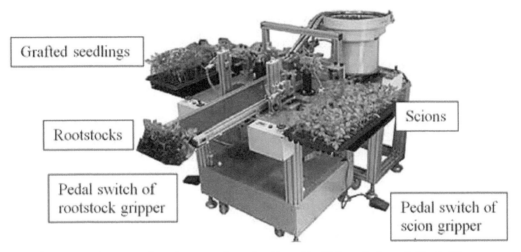

Fig. 11.16. Automated grafting machine developed by Helper Robotech Co. in Korea

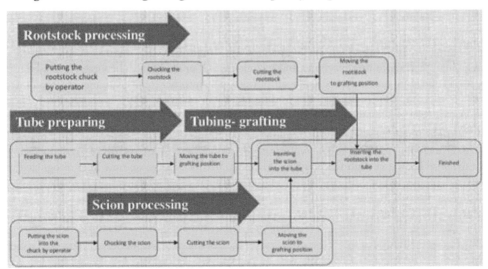

Fig. 11.17. The illustration of robotic grafting process

A fully automated grafting robot can produce 500 to 1,200 grafts/hour depending on the machine used and crop involved (Table 11.17).

Table 11.17. The capacity of production of grafted plants by different automated grafting robots.

Grafting robot	Company/ Country	Crop(s)	Efficiency
CCG	Iam Brain, Japan	Cucumber	500/hr.
G892	Iam Brain, Japan	Cucumber	1200/hr.
G710	Nasmix Co. , Japan	Cucurbits	600-800/hr.
G720	Nasmix Co. , Japan	Solanaceous	600-800/hr.
AG1000	Yanma, Japan	Solanaceous	1000/hr.

11.9. CONCLUSION

Grafting as a technique is gaining wide attention throughout the world, especially for greenhouse cultivation of vegetable crops, mainly the Solanaceous and Cucurbitaceous ones, from the view point of resistance against the soil-borne pathogens including root-knot nematodes in addition to obtaining better yield and quality. Cucurbits are commonly grafted to gourds or to inter-specific hybrids. Grafting methods, grafting for management of soil-borne pathogens, problems commonly associated with grafted plants, and grafting-conferred defense mechanisms are discussed.

Grafting is a valuable nematode disease management tactic for heirloom tomato growers. This practice originated as a way to ensure fruit quality while keeping disease resistance high for melon production systems with soil-borne disease pressure. This same principle lends itself well to heirloom tomato production. Furthermore, research worldwide has demonstrated increased yields from grafted vegetable plants in comparison to non-grafted plants. As South-Eastern growers realize its potential benefits and relative ease, grafting will become a more popular disease management technique in the United States.

Even though there are many problems associated with cultivating grafted vegetable seedlings, the need for successfully grafted seedlings is growing rapidly. Breeding multipurpose rootstock and developing efficient grafting machines and techniques will undoubtedly encourage increased use of grafted seedlings not only in Korea and Japan, but also in many other countries. Large-scale commercial production of vegetable seedlings is expanding rapidly in many developed countries, and this will lead to an increased commercial supply and use of grafted vegetable seedlings throughout the world.

Introduction of excellent rootstocks possessing multiple disease resistance and efficient grafting machines including grafting robots will greatly encourage the extended use of grafted vegetables over the world. Even though the benefits of using grafted seedlings are now fully recognized over the world, production of uniform, healthy grafted seedlings at reasonable prices is the key point for wider use, especially in those countries with limited experience.

Chapter - 12

Strategies for Transgenic Nematode Resistance

12.1. INTRODUCTION

Resistance in plants is an attractive approach for controlling nematode populations. The resistance may be either occurring naturally and transferred to crop cultivars from wild relatives or breeding lines through conventional breeding methods, or engineered through molecular techniques. Biotechnology offers several benefits for nematode control in an integrated management strategy, such as reducing risks to the environment and to human health, accessibility for food producers in the developing world, and the possibility of achieving durable, broad-spectrum nematode resistance (Thomas and Cottage, 2006).

Research efforts must be expanded in the isolation and cloning of desirable traits of HPR and tolerance to facilitate understanding of mechanisms of resistance and host-parasite recognition. Cloned genes will enable direct transfer of resistance into crop plants that are unrelated to the gene donor plant. Success in transfer will depend upon adequate, non-disruptive expression of the desired HPR trait in the transgenic plants. The use of novel transfer techniques, including tissue culture, cell and protoplast fusions, embryo cultures, and embryo cloning techniques may help to overcome incompatible plant barriers.

In conjunction with transfer of natural resistance genes, the use of novel sources of resistance (derived through induced mutations and somatic and somaclonal variations) can be explored, as well as incorporation of toxin-producing genes or inhibitors into plants.

In the past three decades, many researchers have published regarding the transgenic resistance in model plants, as well as the crop species using natural resistance (*R*) genes, proteinase inhibitors and RNA interference of nematode effector genes (Gheysen *et al.*, 2010; Atkinson *et al.*, 2012).

Genetic engineering has evolved as a promising field in the management of plant pathogenic nematodes by the means of gene cloning and gene modification of host plants. Various transgenics plants have been developed so far against plant pathogenic nematodes. The key objective of the genetic manipulation would be to control all possible physiological and biological activities of nematode due to the counter effect of host plants by possessing resistance gene/s on the basis of gene for gene concept. There are several proteinase inhibitors genes which have been identified and transferred into host plants to create resistance against pathogenic nematodes. Nematicidal proteins are also considered as "anti-nematode proteins" can directly inhibit the multiplication of pathogenic nematodes. Protein from *Bacillus thuringiensis*, lectins and some antibodies are regarded as nematicidal proteins. Similarly, other house-keeping genes have been manipulated through RNA interference technique. This chapter focuses on the most recent literature on PPNs with the emphasis on the use of different transgenic approaches to manage PPNs in different plant species of economic importance.

12.2. TRANSGENIC NEMATODE RESISTANCE

Biotechnology has a role to play in incorporation of resistance against nematodes and biological control of plant nematodes. A number of genes that mediate nematode resistance have now been or soon will be cloned from a variety of plant species. Nematode resistance genes are present in several crop species and are an important component of many breeding programs including those for tomato, potato, soybeans, and cereals (Trudgill, 1991). Several resistance genes have been mapped for chromosomal locations or linkage groups and some of them have been cloned. The first nematode resistance gene to be cloned was *Hs1pro-1*, a gene from a wild relative of sugar beet conferring resistance to *Heterodera schachtii* (Cai *et al.*, 1997). The cDNA, under the control of the CaMV 35S promoter, was able to confer nematode resistance to sugar beets transformed with *Agrobacterium rhizogenes* in an *in vitro* assay (Cai *et al.*, 1997). The *Mi* gene from tomato conferred resistance against a root-knot nematode and an aphid in transgenic potato (Rossi *et al.*, 1998). The gene *Mi* is a true *R* gene, characterized by the presence of NBS and LRR domains. *Gpa2* gene that confers resistance against some isolates of the potato cyst nematode, *Globodera pallida* was identified. This gene shares extensive homology with the Rx1 gene that confers resistance to potato virus X suggesting a similarity in function (Van Der Vossen *et al.*, 2000).

Biotechnology offers sustainable solutions to the problem of plant parasitic nematode control. There are several possible approaches for developing transgenic plants with improved nematode resistance; these include anti-invasion and migration strategies, feeding-cell attenuation, and anti-nematode feeding and development strategies. The essential elements of an effective control strategy are (i) genes that encode an anti-nematode effector protein, peptide or interfering RNA, and (ii) promoters that direct a specific pattern of expression for that effector. There are essentially three approaches for engineering resistance: transgenic expression of natural resistance genes in heterologous species; targeting and disruption of the nematode; and feeding site attenuation (Thomas and Cottage, 2006).

Based on growing molecular knowledge of plant parasitic nematodes and their infection cycle, gene technology offers a much wider toolbox to engineer nematode resistance into plants. This technology is not only useful for key nematode problems in developed countries (in crops such as potato and soybean) but certainly also in developing countries (in crops such as banana and rice). Recent developments indicate that countries in Asia and Africa are less hesitant to test and cultivate these crops than some European countries.

The strategy to improve plant defenses to plant parasitic nematodes has centered on the use of proteinase inhibitors (PIs). Cysteine proteinases were selected as the target of the defense as they represent the predominant proteinase activity in plant parasitic nematodes and are not involved in mammalian digestion. Furthermore, inhibitors of cysteine proteinases (cystatins) are not toxic and are already present in the human diet (e.g. maize, rice grain, and sunflower seeds). Plant cystatins are not normally produced in the roots where the nematodes attack, but plants can be modified by the use of root specific promoters driving the expression of cystatin in areas of the plants where the nematode infests.

The approach provides resistance to a wide range of nematodes including *Globodera pallida,
Heterodera schachtii, Meloidogyne incognita and Rotylenchulus reniformis* and can be used in many
important crops, such as potato. Furthermore the novel resistance is additive and capable of ensuring
that a naturally partially resistant potato cultivar provides a high level of resistance to potato cyst
nematodes.

Recent advancements in biotechnological approaches have made it possible to incorporate
and express indigenous and heterologous proteins from one organism to another which has brought
about new era in crop improvement. Genetic engineering has led to the exceptional enhancement
of nematode resistance in different crop plants. Different methods of engineering resistance genes
in crop plants to suppress the nematode infection and populations in the soil below the threshold
level are discussed.

In the past, several transgenic strategies have been used for enhancement of nematode resistance
in plants. The resistance genes from natural resources have been cloned from numerous plant species
and could be transferred to other plant species, for instance, *Mi* gene from tomato for resistance
against *M. incognita*, *Hs1^{pro-1}* from sugar beet (*Beta vulgaris*) against *H. schachtii*, *Gpa-2* from potato
against *Globodera pallida* and *Hero A* from tomato against *G. rostochiensis* (Fuller *et al.*, 2008). The
overexpression of different protease inhibitors (PIs) such as cowpea trypsin inhibitor (CpTI), PIN2,
cystatins, and serine proteases have been used for producing nematode resistant plants (Lilley *et
al.*, 1999). Another main strategy was the targeted suppression of important nematode effectors in
plants using RNA interference (RNAi) approach. Unlike these strategies, some recent researches
have suggested that nematode resistance could be enhanced in plants by modifying the expression of
particular genes in syncytia (Ali *et al.*, 2013a, b). An overview of various transgenic strategies aimed
at nematode resistance is provided in Table 12.1.

Table 12.1. Various transgenic strategies used for nematode resistance in plants.

Molecular strategy	Name of gene	Source	Transgenic plant	Effective against	Resistance response	Reference
Natural resistance genes	*Mi-1.2*	Wild tomato (*S. peruvianum*)	Tomato (*Solanum lycopersicum*)	*M. incognita*	Triggering of HR before significant establishment of giant cells on roots	Milligan *et al.*, 1998
	Gpa-2	Potato	Potato (*Solanum tuberosum*)	*G. pallida*	Development of stagnated & translucent female nematodes on plant roots	van der Vossen *et al.*, 2000
	Hero A	Tomato	Potato	*G. pallida* *G. rostochiensis*	HR response after the initiation of syncytia which become abnormal & necrotic due to degeneration of surrounding cells	Sobczak *et al.*, 2005
	Gro1-4	Potato	Potato	*G. rostochiensis* pathotype Ro1	Unknown	Paal *et al.*, 2004
	Rhg1	Soybean (*Glycine max*)	Soybean	*H. glycines*	HR response leading to abnormal syncytia & necrosis due to degeneration of cells surrounding syncytia	Kandoth *et al.*, 2011

	Rhg4	Soybean	Soybean	H. glycines	HR response following the initiation of syncytia, which become abnormal & necrotic due to degeneration of surrounding cells this HR breaks down with the passage of time	Matthews et al., 2013
	Cre loci	Aegilops spp.	Wheat (Triticum aestivum)	H. avenae	Unknown	Safari et al., 2005
Proteinase/ protease inhibitors	CpTISpTI-1	Cowpea (Vigna unguiculata) Sweet potato (Ipomoea batatas)	Potato Sugar beet	G. pallida M. incognita H. schachtii	Effect on the sexual fate of newly established G. pallida & reduce the fecundity of females without inhibition of growth & development of female H. schachtii	Cai et al., 2003
	PIN2	Potato	Wheat	H. avenae	Unknown	Vishnudasan et al., 2005
	Oc-IΔD86	Rice (Oryza sativa)	Potato	G. pallida M. incognita	Reduced reproductive success of PPNs	Lilley et al., 2004
			Rice	M. incognita	-do-	Vain et al., 1998
			Cavendish dessert bananas (Musa acuminata)	Radopholus similis	-do-	Atkinson et al., 2004
			Eggplant (Solanum melongena)	M. incognita	-do-	Papolu et al., 2016
	CeCPI	Taro (Colocasia esculenta)	Tomato	M. incognita	Interferes with nematode ability of sex determination & gall formation	Chan et al., 2010, 2015
	CCII	Maize (Zea mays)	Plantain (Musa spp.)	R. similis, Helicotylenchus multicinctus, Meloidogyne spp.	Anti-feedant, reduces the reproductive success of nematodes	Tripathi et al., 2015
Lectins	lectin GNA	Snowdrop (Galanthus nivalis)	oilseed rape (Brassica napus), potato	G. pallida, Pratylenchus bolivianus, M. incognita	Decrease in number of females & galls developed on plant roots	Ripoll et al., 2003
Bt toxins	Cry6A, Cry5B	B. thuringiensis	Tomato	M. incognita	Significant reduction in nematode reproduction	Li et al., 2008
Anti- invasion peptides	ACHE-I-7.1	Synthetic	Potato	G. pallida	Inhibits nematode acetylcholinesterase (ACHE) leading to disorientation of invading J2s	Lilley et al., 2011

	LEV-I-7.1	Synthetic	Potato	*G. pallida*	Results in chemodisruption of J2s and avoids invasion	Liu *et al.*, 2005
	nAChRbp	Synthetic	Plantain (*Musa* spp.)	*R. similis* *H. multicinctus* *Meloidogyne* spp.	Disrupts chemosensory function of invading J2s	Tripathi *et al.*, 2015
Dual resistance	*Oc-IΔD86 +* nAChRbp	Rice + Synthetic	Potato	*G. pallida*	Reduced reproductive success of PPNs coupled with disruption of chemosensory function of invading J2s	Green *et al.*, 2012
	CeCPI + *PjCHI-1*	Taro and *Paecilomyces javanicus* fungus	Tomato	*M. incognita*	Reduced chitin content & retardation in embryogenesis in the nematode eggs	Chan *et al.*, 2015
	CCII + *nAChRbp*	Maize + Synthetic	Plantain (*Musa* spp.)	*R. similis* *H. multicinctus* *Meloidogyne* spp.	Anti-feedant & an anti-root invasion plants with reduced reproductive success of PPNs coupled with disruption of chemosensory function of invading J2s	Tripathi *et al.*, 2017

12.2.1. Plant natural resistance genes

Natural resistance genes could exist in both polygenic manner and single dominant nature. The resistance conferred by host plant single dominant resistance genes, the *R* genes from plants, interacts specifically with corresponding avirulence (*Avr*) genes in the nematode, resulting in a so-called 'gene-for-gene' interaction. This type of interaction initiates a cascade of defense responses in the plants. A short summary of natural nematode resistance genes is reviewed by Fuller *et al.* (2008) to provide the basis for this kind of resistance in plants.

Several natural host resistance genes have been cloned from some plant species and could be transferred to other plant species. For instance, *Mi-1.2* from tomato against *M. incognita* (Milligan *et al.*, 1998), *Hs1^{pro-1}* from *Beta procumbens* against beet cyst nematode *H. schachtii* (Cai *et al.*, 1997), *Gpa-2* from potato against potato cyst nematode (PCN, *Globodera pallida*) (van der Vossen *et al.*, 2000), *Hero A* from tomato against *G. pallida* and *G. rostochiensis* (Sobczak *et al.*, 2005), and *Cre* loci from *Aegilops* spp. against cereal cyst nematodes in wheat (Safari *et al.*, 2005) are some examples that could be used in future to develop cyst nematode resistance in crop plants.

Transgenic expression of resistance proteins also induces the expression of PR (pathogenesis related) proteins to establish nematode resistance in plants. The potato roots expressing *Hero A* gene showed high levels of several salicylic acid (SA)-dependent PR genes in the incompatible interaction with PCN at 3 dpi (Uehara *et al.*, 2010). They confirmed that SA inducible PR-1(P4) was a hallmark for the cultivar resistance conferred by *Hero A* against PCN and that nematode parasitism resulted in the inhibition of the SA signaling pathway in the susceptible cultivars. Similar effects were found in resistant line of hexaploid wheat carrying *Cre2* gene, which showed upregulation of ascorbate peroxidase coding gene in response to cereal cyst nematode (*H. avenae*) when compared with the expression in the susceptible lines (Simonetti *et al.*, 2010).

Another example is *Gro1-4*, the constitutive expression of which has increased resistance in potato plants against *G. rostochiensis* pathotype Ro1 (Paal *et al.*, 2004). *Rhg1* is another natural resistance gene identified in soybean against soybean cyst nematode (SCN), *H. glycines* (Kandoth *et al.*, 2011). A recent study has shown that map-based cloning of a gene at the *Rhg4* locus, which is a major quantitative trait locus (QTL), contributing resistance against SCN (Liu *et al.*, 2012). In that study, the *Rhg4* mutant was analyzed through transgenic complementation, which revealed that this gene confers resistance in soybean by encoding a serine hydroxymethyltransferase. This was further confirmed through overexpression of serine hydroxymethyltransferase in soybean roots that demonstrated 45% decrease in the number of mature cyst nematode (Matthews *et al.*, 2013). In the same study, overexpression of nine other putative resistance genes (including short chain dehydrogenase, ascorbate peroxidase, lipase, β-1, 4-endoglucanase, calmodulin, DREPP membrane protein, and three proteins with unknown function) resulted in more than 50% decrease in the number of adult females in soybean roots. Similarly, *Hero A* gene gives almost complete resistance (>95%) against cyst nematode (*G. rostochiensis*), while it provides around 80% of resistance against *G. pallida* (Ernst *et al.*, 2002).

Transgenic expression of *Hs1pro-1* (resistance gene against *H. schachtii*) from *B. procumbens* into sugar beet led to nematode resistance, but unluckily was linked with the genes that were negatively correlated with beet yield (Panella and Lewellen, 2007). Moreover, the *R* genes are typically effective against one or limited range of nematode species/pathotypes. Another limitation with this strategy is the development of different nematode pathotypes, which have the effectors (*avr* genes) that would not be recognized by the *R* genes (Jung *et al.*, 1998).

12.2.2. Proteinase inhibitor coding genes

Proteinase inhibitors (PIs) are molecules, mostly protein in nature, which inhibit the function of proteinases/proteases released by the pathogens. After the attack of herbivores and wounding, a variety of proteinase inhibitors are produced into the plants. In case of PPNs, these PIs become active against all the four classes of proteinases from nematodes, i.e., serine, cysteine, metalloproteinases, and aspartic. The PIs used for nematode resistance studies are CpTI (Hepher and Atkinson, 1992), sweet potato (*Ipomoea batatas*) serine PI (sporamin or SpTI-1) (Cai *et al.*, 2003), PIN2 (Vishnudasan *et al.*, 2005), rice (*Oryza sativa*) cystatin (Oc-IΔD86) (Urwin *et al.*, 1998), and some others cystatins from maize (*Zea mays*), taro (*Colocasia esculenta*), and sunflower (*Helianthus annuus*) (Fuller *et al.*, 2008; Chan *et al.*, 2015).

The anti-nematode potential of plant PIs was first described in transgenic potato expressing the serine PI, the cowpea (*Vigna unguiculata*) trypsin inhibitor (CpTI) against PCN (*G. pallida*) (Hepher and Atkinson, 1992). Moreover, the combinations of different PIs could be helpful to couple specificity with wide range of resistance. Transgenic expression of two proteinase inhibitors, CpTI and Oc-I1Δ86, as a translational fusion protein in Arabidopsis resulted in an additive effect against *G. pallida* and *H. schachtii* (Hepher and Atkinson, 1992; Urwin *et al.*, 1998). Cystatin Oc-IΔD86 was effective against different nematode species in various plants species (Urwin *et al.*, 2000, 2003; Atkinson *et al.*, 2004; Lilley *et al.*, 2004). Other important proteinase inhibitors are sporamin (SpTI-1) (Cai *et al.*, 2003) and PIN2 (Vishnudasan *et al.*, 2005), which have shown good resistant response in plants.

As compared to other proteinase inhibitors, cystatins from different plant species have met the promise of enhancement of nematode resistance in a variety of crop plants (Chan *et al.*, 2015; Green *et al.*, 2012; Tripathi *et al.*, 2015; Papolu *et al.*, 2016). Heterologous expression of a taro cystatin established considerable degree of resistance in tomato against *M. incognita* infection by interference with nematode ability of sex determination and gall formation (Chan *et al.*, 2010). Recently, a dual strategy has been used to control PCN without affecting the soil quality (Green *et al.*, 2012). In the first approach, a peptide was precisely targeted under control of a root tip-specific promoter that disrupts chemoreception of nematodes and suppresses root invasion without a lethal effect in both containment as well as under field trial (Lilley *et al.*, 2011). In addition to this chemoreception disruptive peptide, OcIΔD86 cystatin from rice was incorporated to control the invading larvae that are able to cross the barrier of chemodisruptive peptide. This approach establishes that a combination of these genes offers distinct bases for the transgenic plant resistance to *G. pallida* without harmful impact on the non-target nematode soil community (Green *et al.*, 2012). This rice cystatin has also shown good control in lily (*Lilium longiflorum* cv. 'Nellie White') against lesion nematode (*P. penetrans*) (Vieira *et al.*, 2015).

Recently, an anti-feedant maize cystatin and an anti-root invasion synthetic peptide were transformed into plantain (*Musa* spp., cv. Gonja Manjaya), individually and in combination (Tripathi *et al.*, 2015). The field trials of the best transgenic event containing the peptide only demonstrated 186% more yield in addition to 99% control against *R. similis*, *H. multicinctus*, and *Meloidogyne* spp. as compared to non-transgenic control. Moreover, transgenic expression of proteinase inhibitor from maize and synthetic chemodisruptive peptide resulted in enhanced yield and nematode resistance in plantain. Similarly, dual overexpression of a taro cysteine proteinase inhibitor (CeCPI) and a fungal chitinase (PjCHI-1) in tomato under the control of a synthetic promoter, pMSPOA (with NOS-like and SP8a elements), had negative effects on reproduction of *M. incognita* (Chan *et al.*, 2015). This study further revealed that dual gene transformation had more inhibition of nematodes than plants transformed with a single gene. This advocated that the use of gene pyramiding could be employed for developing and improving nematode resistance in plants (Chan *et al.*, 2015; Tripathi *et al.*, 2017).

Recently, a modified rice cystatin (Oc-IΔD86) was expressed in the roots of eggplant (*S. melongena*) under the control of the root-specific promoter, TUB-1 (Papolu *et al.*, 2016). Five putative transformants containing this cystatin exhibited detrimental effects on both the development and the reproduction of *M. incognita* in eggplant. In that study, a single copy transgenic event showed 78.3% reduction in reproductive success of *M. incognita*. This concludes that proteinase inhibitors are potential candidates for induction of nematodes resistance in a variety of crop plants to increase crop yield and minimize the damage caused by these parasitic worms.

12.2.3. Nematicidal proteins

These proteins could be characterized as anti-nematode proteins because these are directly involved in inhibiting the nematode development on the plants. Lectins, some antibodies, and *Bt* Cry proteins are some examples of these proteins.

12.2.3.1. Lectins: The toxicity of lectins is characterized by their ability to obstruct intestinal function of organisms exhibiting or ingesting them (Vasconcelos and Oliveira, 2004). The defense mechanism conditioned by the lectins is vital as several lectins bind with glycans (Peumans and van Damme, 1995). Overexpression of a snowdrop (*Galanthus nivalis*) lectin GNA driven by cauliflower mosaic virus promoter (CaMV35S) has been exploited to exhibit anti-nematode activity in several plants, i.e., Arabidopsis, oilseed rape (*Brassica napus*), and potato, in response to RKNs, CNs and lesion nematodes (Burrows *et al.*, 1998; Ripoll *et al.*, 2003).

12.2.3.2. Plantibodies: Plantibodies are the antibodies expressed in plants and also potential candidates for the development of nematode resistance. These are important because the establishment of a compatible plant–nematode interaction engages a series of processes against which plantibodies may be directed. RKNs and CNs depend on secretions of their pharyngeal glands to mimic re-differentiation of plant cells into specialized nematode feeding sites like giant cells or syncytia. Direction of plantibodies opposite to the active proteins from these secretions could be attenuated to suppress the parasitic ability of the nematode. There are only a few reports available in the literature regarding the use of plantibodies for nematode resistance (Fioretti *et al.*, 2002; Sharon *et al.*, 2002).

12.2.3.3. Bt toxins (Cry proteins): In addition to lectins and plantibodies, different variants of *Bt toxins* (Cry proteins) derived from *Bacillus thuringiensis* have shown promise to induce plant resistance against nematodes. *Bt* toxin was first used as an anti-nematode protein by Marroquin *et al.* (2000), when *C. elegans* was exposed to Cry5B and Cry6A which resulted in the reduction in nematode fertility and viability. The PPNs use feeding tube at the stylet orifice while feeding on the plants roots. The feeding tube operates as molecular sieve, allowing the uptake of certain molecules and excluding the others. The ultrastructure of feeding tubes from root-knot and cyst nematode differ which is based on the observation that root-knot nematodes are able to ingest larger proteins compared with cyst nematodes (Li *et al.*, 2007). Transgenic expression of 54 kDa Cry6A and Cry5B proteins in tomato hairy roots affected the reproduction of root-knot nematode *M. incognita* (Li *et al.*, 2007, 2008). Western blotting technique showed that the 54 kDa Cry6A protein was shown to be ingested by *M. incognita*. On the other hand, this large protein could not be ingested by cyst nematodes (i.e., *H. schachtii*) due to small orifice of the feeding tube having the size limit up to approximately 23 kDa (Urwin *et al.*, 1998). This limitation severely restricts the agronomic application of these toxins against PPNs.

12.2.3.4. Chemodisruptive peptides: Plant parasitic nematodes are highly dependent on their chemoreceptive neurons to sense distinct chemical stimuli for invasion into the plants. Nematodes use acetylcholinesterase (*AChE*) and/or nicotinic acetylcholine receptors for proper functioning of the nervous system. Chemodisruptive peptides are another important strategy to minimize the invasion of PPNs into the plant roots. Two peptides have been shown to bind with these receptors to inhibit their proper function (Winter *et al.*, 2002). Both of these peptides disrupted nematode ability of chemoreception by blocking their reaction to chemical signal at the very minute concentrations of up to 1 nm. Transgenic potato plants expressing a secreted peptide inhibited nematode *AChE* leading to disorientation of invading nematode *G. pallida*, which resulted in a 52% decline in the number of female nematodes (Liu *et al.*, 2005). The peptide is considered effective after its uptake from chemoreceptor sensillae through retrograde transport along nematode neurons to cholinergic synapses. Costa *et al.* (2009) have demonstrated that cyst nematode acetylcholinesterase gene (*AChE*) is expressed in chemo-and mechano-sensory neurons of *C. elegans*, which further support this hypothesis. Similarly, Wang *et al.* (2011) reported indirect evidence to support the mechanism by which such peptide disrupts chemosensory function in cyst nematodes. The peptide exhibits disulphide-constrained 7-mer with the amino acid sequence CTTMHPRLC that binds to nicotinic acetylcholine receptors. Incubation in the peptide solution or root-exudate from transgenic plants that secrete the peptide disrupted normal orientation of infective cyst nematodes to host root diffusate.

Moreover, chemosensory disruptive peptide that inhibits *AChE* has recently been expressed under the control of the constitutive CaMV35S promoter and the root tip-specific promoter in Arabidopsis and potato plants, where it confers resistance against *H. schachtii* and *G. pallida* (Lilley *et al.*, 2011). This root tip-specific promoter from Arabidopsis gene (*MDK4-20*; *At5g54370*) directed expression of the nematode repellent peptide only at the sites of cyst nematode invasion and has shown strong level of resistance against PCN. This strategy has now been combined with the transgenic expression of a rice cystatin in potato to maintain high level of resistance against PCN without affecting soil quality (Green *et al.*, 2012). By using the same technique, the International Institute of Tropical Agriculture (IITA), in partnership with the University of Leeds, UK, developed transgenic plantain for nematode resistance using maize digestive protease inhibitor cystatin and synthetic nematode repellent peptides (Roderick *et al.*, 2012; Tripathi *et al.*, 2013). Furthermore, pyramiding of cystatins and chemodisruptive peptide into different crop plants has shown high degree of nematode resistance and enhanced crop yields against root-knot nematodes (Chan *et al.*, 2015; Tripathi *et al.*, 2017).

12.2.4. Utilization of RNA interference to suppress nematode effectors

RNA interference has emerged as a very useful tool for gene-silencing aimed at functional analysis of different genes by suppressing their expression in a wide variety of organisms including PPNs. In this strategy, the nematodes uptake double-stranded RNA (dsRNA) or short interfering RNAs (siRNAs) from the plants expressing these RNAs, which elicit a systemic RNAi response in nematodes. A schematic diagram elaborating the mechanism of *in planta* RNAi is shown in Fig. 12.1. The double-stranded RNA (dsRNA) is processed by the plant dicer enzyme (A) into (B). Once the dsRNA is uptaken by the nematode from the plant cell while feeding (C), the processing from dsRNA to short interfering RNA (siRNA) can be executed by the nematode dicer. Then, the siRNA is recognized by the RNA-induced silencing (RISC) complex of the nematode (D) and it's unwinding into sense and antisense strands take place. A proportion of the RISC complex loaded with the antisense strand interacts with the corresponding mRNA in the nematode (E) as a result the mRNA is cleaved by the RISC (F) and subsequently degraded (G). Moreover, the targeted mRNA can be made double-stranded after binding of the siRNA, and this dsRNA is then processed to produce additional siRNAs, intensifying the initial silencing signal (Gheysen and Vanholme, 2007). The transgenic expression of dsRNA targeting a specific nematode effector gene could be handful to suppress the expression of that effector gene, which is crucial for infection process. There are several recent review articles emphasizing the usefulness and application of RNAi technology to induce nematode resistance in plants by silencing the expression of nematode effectors mainly (Fuller *et al.*, 2008; Maule *et al.*, 2011; Tamilarasan and Rajam, 2012).

In a review article most of the aspects of the RNAi application in nematode resistance are reviewed (Lilley *et al.*, 2012). Lilley *et al.* (2012) have reviewed various features ranging from *in vitro* assays with *C. elegans* to delivering RNAi *in planta* as an important strategy for crop protection against cyst nematodes.

Youssef *et al.* (2013a) have tested this approach by silencing the *H. glycines* gene *HgALD*, encoding fructose-1, 6-diphosphate aldolase important in the conversion of glucose into energy and actin-based motility during parasite invasion into its host. Transgenic soybean roots expressing an RNAi construct targeted to silence *HgALD* revealed 58% reduction of females formed by *H. glycines*. Recently, Tripathi *et al.* (2017) have reviewed the application of RNAi for the enhancement of nematode resistance by suppression of important effector proteins. Nevertheless, RNAi has become an established experimental tool for the enhancement of resistance against PPNs and also offers the prospect of being developed into a novel control strategy when delivered from transgenic plants.

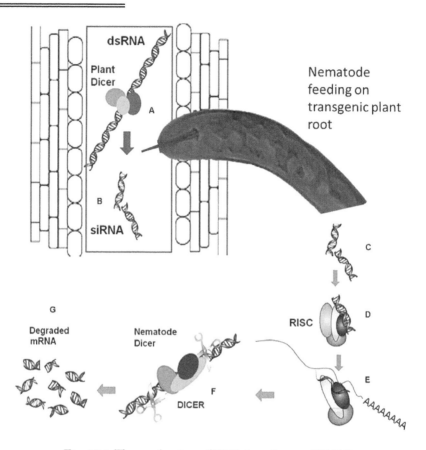

Fig. 12.1. The mechanism of RNA interference (RNAi).

12.2.5. Other strategies

12.2.5.1. Overexpression or suppression of genes: The manipulation of expression of genes from the plants could also be useful for induction of resistance against cyst nematodes. Transgenic plants that overexpress genes correlated with resistance or have silenced genes which are important to syncytium development and nematode success are two obvious areas to explore (Ali, 2012). This could be achieved by using precise delivery of transgene into the feeding sites as the constitutive overexpression or suppression of a particular gene could have detrimental effects on the growth and development of the plants (Ali *et al.*, 2013b). For this purpose, promoters that are specifically expressed in the feeding sites of PPNs, i.e., *Pdf2.1* or *MIOX5*, could be used (Siddique *et al.*, 2011). Although constitutive promoters can deliver the gene of interest or suppress the genes which are important for compatible interaction (Ali and Abbas, 2016), however, it could be lethal for the plants to silence the genes which are also important for other physiological processes in addition to establishment of the nematodes. The promoters of the genes like *Pdf2.1* or *MIOX5* are specially and highly expressed in syncytia and several studies have shown the utility of these promoters to overexpress and silence the genes of interest specially into the feeding sites (Siddique *et al.*, 2011; Ali *et al.*, 2014). In addition to syncytia specific promoters, root and root tip-specific promoters have also been used to drive site specific expression of proteinase inhibitors and nematode chemodistruptive peptides in several plant species (Lilley *et al.*, 2011; Atkinson *et al.*, 2012; Green *et al.*, 2012).

12.2.5.2. Upregulation of genes: The transcriptomes of various plant species infected with different nematode species demonstrated the upregulation of genes important for development of nematode feeding structures in the plant roots. The knockout mutants of two endo-1, 4-β-glucanases, which were highly upregulated in syncytia, revealed less susceptibility in Arabidopsis in response to beet cyst nematode (Wieczorek *et al.*, 2008). Likewise, an ATPase gene from Arabidopsis (*At1g64110*) was reported to be induced in syncytia caused by *H. schachtii* (Ali *et al.*, 2013b). The knocking down of this gene using syncytia specific promoters (prom.AtPDF2.1 and prom.AtMIOX5) supported less number of nematodes.

12.2.5.3. Downregulation of genes: Conversely, the nematodes are smart enough to suppress the defense mechanisms of the plants as most of the genes involved in defense related pathways were downregulated in the feeding sites revealed by plant transcriptomes in response to nematode infection (Kyndt *et al.*, 2012; Ali *et al.*, 2015). One strongly downregulated group comprised peroxidase gene family, as out of top 100 differentially expressed genes with the strongest decrease in expression, 14 were peroxidases (Szakasits *et al.*, 2009). Similarly, ethylene responsive transcription factor from Arabidopsis, AtRAP2.6, was one of the strongly downregulated transcripts in syncytia. This gene was driven through the constitutive promoter CaMV35S to overexpress in Arabidopsis at the global level, which resulted in reduced susceptibility in overexpression lines (Ali *et al.*, 2013a). This also supports the debate of use of CaMV35S promoter to expression gene of interest in the syncytia induced by beet cyst nematodes in Arabidopsis roots (Ali and Abbas, 2016).

12.2.5.4. Camalexin and callose synthesis: Another strategy is the expression of the genes involved in the defense pathways like camalexin and callose synthesis in plants (Mao *et al.*, 2011; Birkenbihl *et al.*, 2012). Recently, expression of *AtPAD4* under the control of FMV-sgt promoter has resulted in enhanced resistance against soybean cyst and root-knot nematodes in soybean (Youssef *et al.*, 2013b). Expression of *AtPAD4* in soybean roots decreased the number of mature *H. glycines* females and *M. incognita* galls up to 68 and 77%, respectively. The pathway of camalexin synthesis in plant–nematode interactions was recently dissected based on infection assays of *AtWRKY33* and *AtPAD3* mutants and overexpression lines (Ali *et al.*, 2014). The syncytia specific overexpression of *WRKY33* resulted in the suppression of susceptibility in Arabidopsis. Similarly, the overexpression of a soybean salicylic acid methyltransferase (*GmSAMT1*) gene is found to confer resistance to SCN (Lin *et al.*, 2013).

12.2.5.5. Genome editing: Genome editing is a group of techniques that enable us to change an organism's DNA, including insertion, replacement or removal of DNA from a genome with a high degree of specificity. Several approaches to genome editing have been developed. One of the most widely used is known as CRISPR-Cas9, which is short for clustered regularly interspaced short palindromic repeats and CRISPR-associated protein 9 (Fig. 12.2). CRISPR-Cas9 was adapted from a naturally occurring genome editing system in bacteria. The bacteria capture snippets of DNA from invading viruses and use them to create DNA segments known as CRISPR arrays. The CRISPR arrays allow the bacteria to "identify" the viruses and target the viral DNA, the bacteria then use Cas9 or a similar enzyme to cut the DNA apart, which disables the virus. Gene editing has a significant potential to improve specific plant traits (and quality of plant raw material), or confer resistance to pests and diseases. Use of CRISPR-Cas9 technology has revolutionized non-GM breeding since its adoption in 2012.

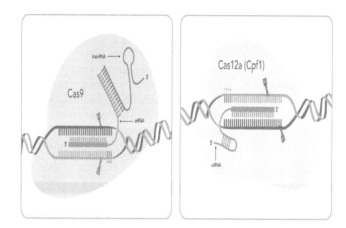

Fig. 12.2. CRISPR genome editing using Cas9

Start-up Company Phytoform Labs Ltd. (phytoformlabs.com) aims to commercialize gene editing technologies for the horticulture sector and speed up the development of new varieties with enhanced pest, disease, and nematode resistance, or improved quality traits. They are developing a new unique breeding platform based on the use of genomics, gene editing tools and proprietary automated screening technology. If proven to work, this platform could significantly reduce the breeding cycle and provide significant cost savings for plant breeders.

12.3. NEMATODE RESISTANT GM CROPS

12.3.1. Cotton root-knot nematode, *Meloidogyne incognita*

Root-knot nematode, *Meloidogyne incognita* is one of the most economically damaging cotton plant parasites with populations widespread throughout the US Cotton Belt. Currently, control options in cotton are limited to planting genetically tolerant varieties and/or early season control by seed treatments. Using marker-assisted breeding techniques, Monsanto is introducing resistance into elite genetics to develop varieties that could potentially increase the lint yield by an average of 8 to 10 percent under root-knot nematode infestations (Fig. 12.3). Genetic markers for the genes responsible for resistance to root-knot nematode in upland cotton are located on chromosomes 11 and 14. This information should help breeders develop new varieties of nematode-resistant cotton.

Fig. 12.3. Root-knot nematode resistance in cotton

To date increased transcript and protein levels of MIC3 in galls of resistant plants remains the only example of a gene whose expression is correlated with the onset of RKN resistance. In this report, this correlation was further validated via over-expression of MIC3 in the RKN-susceptible line Coker 312. A MIC3 overexpression cassette driven by the CaMV35S promoter was constructed using the binary vector pBI121. Transgenic cotton lines harboring this cassette were created using *A. tumefaciens*. Six homozygous T2 lines were identified that showed a range of elevated MIC3 transcript and protein levels in roots and leaves compared to non-transgenic controls. For the RKN assays, data from two independent experiments showed that both high and low levels of MIC3 overexpression affected RKN egg production. It was found that the transgenic line 11-1-1Top, which showed the highest level of MIC3 transcript in uninfected roots, reduced RKN eggs/plant by 70% compared to the non-transgenic susceptible control Coker 312.

12.3.2. Cotton reniform nematode, *Rotylenchulus reniformis*

Reniform nematodes (*Rotylenchulus reniformis*) reduce cotton plant growth and lint yield, and the increasing severity and incidence of infestations is causing concern in the central US Cotton Belt. A new, high throughput screening method is being utilized to accelerate the incorporation of resistance into elite genetics. The product could potentially increase lint yield by an average of 10 to 15 percent under infestation conditions (Fig. 12.4). Resistance to reniform nematode in a wild *Gossypium barbadense* line is governed by more than one gene, and they have identified markers linked to these genes on chromosomes 21 and 18.

Fig. 12.4. Reniform nematode resistance in cotton. *Left* – **Susceptible,** *Right* - **Resistant (transgenic).**

12.3.3. Soybean cyst nematode, *Heterodera glycines*

Soybean cyst nematode (SCN) (*Heterodera glycines*) is responsible for significant reduction in soybean yields. This nematode enters the plant root and starts to feed (Fig. 12.5). Consequently, soybean plant losses from this pathogen can reach 30% (Ichinohe, 1988). Resistant cultivars are the most widely used approach to SCN management and result in increased soybean yields concomitant with decreased nematode populations.

Resistance to SCN is an oligogenic quantitative trait (Young, 1996). Restriction fragment length polymorphism (RFLP) mapping and linkage analyses have shown that at least three quantitative trait loci (QTLs) exist among the sources of resistance studied (Concibido *et al.*, 1997). Among these, a resistance locus on linkage group G was found to act in a race independent manner and to account for more than 50% of total expressed phenotypic resistance to SCN (Concibido *et al.*, 1997). Further mapping and analysis has indicated the presence of a gene coding for resistance to SCN (*rhg1*) in this genomic region.

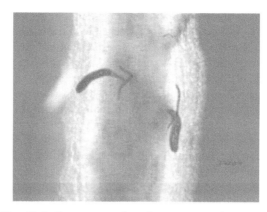

Fig. 12.5. Cyst nematode infection on soybean root

A. rhizogenes, the causative organism of hairy root disease (Nilsson and Olsson, 1997), produces large numbers of transgenic hairy roots in soybean (White *et al.*, 1985) and is cultivar independent (Owens and Cress, 1985). *A. rhizogenes* induces the formation of transgenic hairy roots by transforming host plant cells with the T-DNA of the Ri plasmid (Nilsson and Olsson, 1997). In addition, *A. rhizogenes* also introduces novel genes into hairy roots from the T-DNA of binary vector plasmids (Hamill *et al.*, 1987).

Seedling cotyledons of the SCN-susceptible cultivars, Agassiz and Parker, and SCN-resistant Bell and Faribault were infected with *A. rhizogenes* strain K599 transformed with T-DNA binary vectors containing the gusA gene fused to promoters from either the cauliflower mosaic virus (CaMV 35S), *Arabidopsis thaliana* phenylalanine ammonia lyase (PAL), or bean (*Phaseolus vulgaris*) chalcone synthase-8 (CHS) genes. Nine days after inoculating transgenic hairy roots with sterile J2 nematodes, CHS-regulated β-glucusonidase (GUS) staining at infection sites increased in hairy roots of resistant Faribault and decreased in susceptible Agassiz. PAL-regulated GUS staining was absent at infection sites in hairy roots of resistant cultivars, but was increased in infection sites in susceptible cultivars. Thirty-five days after inoculation with SCN, the mean number of cysts formed on hairy roots of the resistant cultivars was about 14% of the mean number of cysts formed on hairy roots of the susceptible cultivars, indicating that the SCN resistance phenotypes were preserved in transgenic hairy roots. These results indicated that the transgenic hairy soybean root system will be useful for investigating differential transgene expression during nematode infection and evaluation of candidate SCN resistance genes.

One of the techniques utilized in soybean biotechnological products is the genetic transformation of cultivars expressing double stranded RNAs (dsRNA) in order to drive gene silencing in nematodes. The mechanism of post-transcription gene silencing using dsRNA is known as RNA-mediated interfering (RNAi), or gene silencing (Bosher and Labouesse, 2000; Hunter, 2000; Kuwabara and Coulson, 2000). Gene silencing can be either partial – called knockdown – or total – denominated knockout.

12.3.4. Pineapple reniform nematode, *Rotylenchulus reniformis*

In order to reduce, or even discontinue, its use of pre-plant soil fumigants and post-plant nematicides use in Hawaiian pineapple production (reductions totals to 0.635 million kgs per year in active ingredient with an associated savings of $2.1 million in expenditures), the University of Hawaii, supported in part by the pineapple industry and by federal grants, began a pineapple genetic engineering program in 1995 to produce a nematode resistant Smooth Cayenne cultivar (Rohrbach *et al.*, 2000). Research is underway to improve the efficacy of pineapple transformation via *Agrobacterium*, and to improve plant regeneration techniques.

Pineapple tissue has been transformed by *Agrobacterium* in the laboratory with several different nematode resistant gene constructs, developed by collaborators at the University of Leeds, in the UK. One construct carries genes for the production of cystatin. Cystatin is a proteinase inhibitor that occurs naturally. Cystatin been shown to deter nematode feeding and reduce populations by interfering with their ability to produce digestive enzymes (Atkinson *et al.*, 1996). The cystatin gene found in wild rice has been cloned, amended for improved efficacy against nematodes, and inserted into plants for nematode resistance (Atkinson *et al.*, 1996; Rohrbach *et al.*, 1988). Currently, cystatin is being tested for efficacy against reniform nematodes. Pineapple plants transformed with the modified rice cystatin gene have tested positive for transgene expression, particularly in the root tissue, but plants have not yet been evaluated for reniform nematode resistance.

12.3.5. Potato cyst nematode, *Globodera pallida*

Potato plants were developed that transgenically expressed a disulfide-constrained peptide (nAChRbp) capable of binding to nematode acetylcholine receptors and inhibiting chemoreception of cyst nematodes. A tissue-specific promoter restricted expression of the peptide to the outer cell layers of the root tip. Exudates from the transgenic potato plants inhibited alkaline phosphatase as expected if the peptide was successfully expressed and secreted from the roots. The potato lines tested did display a level of resistance to *G. pallida* in both a containment glasshouse trial and in the field that establishes the potential of this novel approach.

The results validate the root tip-specific promoter of the *Arabidopsis MDK4–20* gene (Lilley *et al.*, 2011) as a means of delivering effective root protection by the peptide under field conditions. This promoter is active in potato in both the zone of elongation and root border cells even after they detach from the root cap, often decorating the zone of elongation. This may enhance protection of this zone from invading nematodes (Lilley *et al.*, 2011). PCN normally invades near root tips which slow root extension, particularly by lateral roots. This reduces the volume of soil from which the plant draws water and nutrients (Trudgill *et al.*, 1998). The peptide's mode of action suppresses this important aspect of the pathology before other defenses such as a cystatin could act as an anti-feedant on just those nematodes that establish in roots. This suggests the resistance conferred on potato roots by expressing these different traits should be additive. If so, this is likely to prevent economic damage by *G. pallida*. Both a cystatin (Lilley *et al.*, 2004) and the peptide have provided >75% resistance, so if fully additive they should provide circa 95% control.

12.3.6. Sugar beet cyst nematode, *Heterodera schachtii*

The sugar beet cyst nematode, *Heterodera schachtii* can attack plants of any age, and seedlings or young beets may be killed *(Fig. 23.4)*, resulting in reduced stands. Young plants infected by the disease have elongated petioles and remain stunted until harvest. Leaves of severely affected plants additionally wilt and have pronounced yellowing. The most easily recognized sign of infection is the white to pale yellow, lemon-shaped adult females (1/32 inch) attached to roots *(Fig. 12.6)*.

 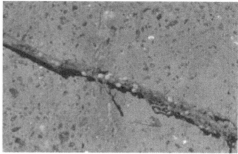

Fig. 12.6. Left - Death of young plant from sugar beet cyst nematode. Right - Lemon-shaped females attached to young tap root early in the season.

The cDNA of the beet cyst nematode was transformed into hairy roots of susceptible sugar beet under the control of the 35S promoter and hairy root clones were inoculated with nematodes. The number of developing females was significantly reduced in 12 out of 15 clones resulting from independent transgenic events suggesting that the gene can be used for inducing cyst nematode resistance in plants.

Hairy roots produced in sugar beet exhibit the resistance phenotypes of the whole plant. Furthermore, when susceptible sugar beet genotypes are co-transformed with *A. rhizogenes* to introduce the (*Hs1.sup.pro-1*) resistance gene, the hairy roots formed exhibit resistance to sugar beet cyst nematode (Cai *et al.*, 1997).

The resistant genes were located on chromosomes 1 and 7 of *Beta procumbens* (designated as *Hs1^{pro1}* and *Hs2^{pro7}*, respectively). DNA markers based on *B. procumbens*-specific repetitive DNA elements or random amplified polymorphic DNA (RAPD) markers linked to the resistance were developed for marker-assisted selection (Jung *et al.*, 1992; Hallden *et al.*, 1997). Studies on pathotypes of *H. schachtii* indicate that the Schach 0-derived population was unable to break the resistance conferred by translocation lines *Hs1^{pro1}*, *Hs2^{pro7}* and *Hs1^{web7}* (Müller, 1998). It was reported that the nematode population decreased by 73% with experimental resistant varieties whereas that increased by 35% with a susceptible variety (Werner *et al.*, 1995). A resistant variety called Nematop was developed which showed acceptable performance in nematode-infested soil (Dewar, 2005). However, it yields less than the highest yielding susceptible varieties in the absence of the pest.

12.3.7. Tomato root-knot nematode, *Meloidogyne incognita*

The *Mi* gene was identified in the wild tomato species, *Solanum peruvianum*, and was introduced into cultivated tomato using embryo culture of an interspecific cross of *S. lycopersicum* and *S. peruvianum* (Smith, 1944), followed by extensive backcrossing to *S. lycopersicum*. Progeny of a single F1 plant are the sole source of nematode resistance in currently available fresh-market and processing tomato cultivars (Medina-Filho and Tanksley, 1983). The observation by Rick and Fobes (1974) of a tight genetic linkage between *Mi* and Aps-1, a gene encoding acid phosphatase-1, greatly facilitated the introduction of the resistance gene into commercial cultivars.

It would be highly preferable to transfer the *Mi* gene as a single unit into tomato cultivars which already have other desirable traits and, thus, eliminate the need for these time-consuming backcrosses. When a clone of the resistance gene becomes available, it can be introduced rapidly into susceptible cultivars using available gene transfer methods.

The *Mi* gene from tomato conferred resistance against a root-knot nematode and an aphid in transgenic potato (Rossi *et al.*, 1998). The gene *Mi* is a true *R* gene, characterized by the presence of NBS and LRR domains. More promising results were obtained when transgenic *Mi-1.2* was expressed in tomato. Transgenic lines sustained significantly lower amounts of *Meloidogyne* reproduction and numbers of egg masses (Fig. 12.7), but aphid performance was not compromised. Although this result is disappointing from the perspective of conferring broad-range pathogen resistance through transgenic expression of *Mi-1.2*, it does provide further evidence supporting the hypothesis that aphids and root-knot nematodes invoke distinct defense cascades downstream of *Mi-1.2*. In order for *R* genes to produce a resistance response following transfer to a susceptible plant, it is a prerequisite that downstream components of the response cascade also be present.

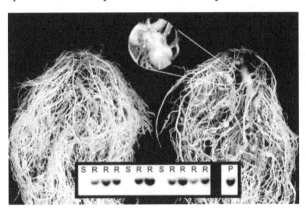

Fig. 12.7. Progeny of tomato line 143-11, which was transformed with *Mi-1.2* free from root-knot galls (*left*) compared to the susceptible check (*right*).

The potential of PIs, a cystatin from the tropical root crop taro (*Colocasia esculenta*) was expressed constitutively in a root-knot nematode-susceptible tomato cultivar. There was a 50% reduction in the number of galls formed by *M. incognita* on the transgenic plants compared to wild-type plants and a larger reduction in the number of egg masses produced per plant (Chan *et al.*, 2010).

12.3.8. Tobacco root-knot nematode, *Meloidogyne incognita*

The performance of the transgenic tobacco lines expressing AtNPR1 was superior compared to the wild type when challenged with the root-knot nematode, *Meloidogyne incognita*. When the growth pattern of different plants was analyzed six weeks after nematode inoculation, the transgenic plants expressing AtNPR1 showed better shoot and root growth when compared to wild type plants. There was significant reduction in the vigor of the plants after infection with *M. incognita* in the case of the wild type plants. The AtNPR1 high expression line 19-1 recorded the highest shoot and root weight, which is about five folds higher, compared to the wild type plants. The number of root galls and egg-masses developed on the infected roots of the transgenic plants were significantly less (up to 50-60% less) compared to the wild type plants. Among the transgenic lines, lowest gall and egg-mass count was recorded in the high expression line 19-1 suggesting a dosage dependant effect of the expression of AtNPR1(Bhanu Priya *et al.*, 2011). This could be correlated with the constitutive expression of the pathogenesis related proteins and the genes for antioxidant enzymes like ascorbate peroxidase and super-oxide dismutase by the NPR1 transgenic plants (Srinivasan *et al.*, 2009). Further, an analysis of the expression of genes for PR1 and PR5 in 19-1 transgenic plants by RT-PCR showed that they had basal constitutive level of expression, which was not observed in wild type plants. The gene expression got further enhanced upon treatment with the nematode. These observations indicate that the resistance to the nematode exhibited by the transgenic plants is associated with enhanced expression of genes for PR proteins (Bhanu Priya *et al.*, 2011).

Tobacco plants were engineered to produce dsRNA of two essential genes of the parasitic nematode *M. incognita*, which infects a wide range of agriculturally important plants. As predicted, the transgenic tobacco plants very effectively resisted *M. incognita* infection (Fig. 12.8). A closer observation of the nematodes in these plants revealed that their development was severely impaired. Further, these nematodes were specifically deficient in the mRNA of targeted genes, indicating that the dsRNA produced in plants did indeed trigger RNAi response in the nematode (Yadav *et al.*, 2006).

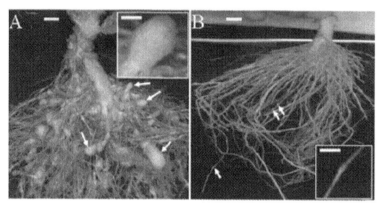

Fig. 12.8. Roots of control (A) and transgenic (B) tobacco plants 45 days after inoculation with 2500 *Meloidogyne incognita* **juveniles.**

12.3.9. Banana nematodes, *Radopholus similis, Pratylenchus coffeae*

The most widespread and damaging nematodes to *Musa* spp. are of the migratory endoparasitic type: the burrowing nematode, *Radopholus similis* and the root-lesion nematode, *Pratylenchus coffeae* (Sarah *et al.*, 1996; Bridge *et al.*, 1997). They cause losses of 20 to 40% of yield. Nematodes in established banana plantations are mainly managed by means of nematicides (Gowen and Quénéhervé, 1990). The most sustainable method of nematode control is the use of resistant plants that suppress nematode reproduction (Starr *et al.*, 2002).

From a biochemical point of view, increased amounts of condensed tannins and flavan-3, 4-diols (Collingborn *et al.*, 2000) and higher levels of vascular lignin and cell-wall bound ferulic acid esters in the cortex (Wuyts *et al.*, 2007) have been observed in nematode resistant cultivars. Cavendish dessert bananas that express the OcIΔD86 engineered variant of rice cystatin under the control of the maize ubiquitin promoter displayed 70±10% resistance to *R. similis* in a glasshouse trial (Atkinson *et al.*, 2004). Plants expressing the same cystatin under the control of a root-specific promoter that is upregulated in giant cells (Green *et al.*, 2002; Lilley *et al.*, 2004) were resistant (83±4%) to *M. incognita*. East African Highland banana plants constitutively expressing a maize cystatin support reduced multiplication of *R. similis* and the plantain cv. Gonja has been transformed (Dr L. Tripathi; IITA, Uganda) to express both a cystatin and a repellent peptide. There could be an additional advantage to cystatin-mediated nematode resistance in banana as cystatin impairs feeding and development of banana weevils (Kiggundu *et al.*, 2010).

12.3.10. Rice nematodes

The most common plant-parasitic nematodes attacking rice are the root-knot nematode *Meloidogyne graminicola*, the cyst nematode *Heterodera oryzae*, the stem nematode *Ditylenchus angustus*, the root rot nematode *Hirschmanniella oryzae* and the white tip nematode *Aphelenchoides besseyi* (Bridge *et al.*, 2005).

To date, the only nematode resistance technology introduced into rice is the cystatin-based defence. Transgenic plants of four elite African rice varieties constitutively expressing the modified rice cystatin OcIΔD86 displayed 55% resistance to *M. incognita* (Vain *et al.*, 1998). Only a low level of cystatin expression was observed, possibly due to a suboptimal CaMV35S promoter or homology dependent silencing of the transgene in combination with the endogenous *OcI* gene. In subsequent work, a maize cystatin has been expressed in the rice variety Nipponbare under the control of a root promoter from Arabidopsis (TUB-1) that is known to be up-regulated in the feeding cells of *M. incognita* parasitizing rice (Green *et al.*, 2002). The construct also included an intron sequence from the maize ubiquitin gene to enhance expression levels and the best transgenic lines displayed 91 ± 7% resistance to *M. incognita*.

12.3.11. Wheat cereal cyst nematode, *Heterodera avenae*

Inhibitory activity of a potato serine proteinase inhibitor (PIN2) expressed in transgenic wheat showed a positive correlation with plant growth and yield following infestation with the cereal cyst nematode, *Heterodera avenae* (Vishnudasan *et al.*, 2005). A protective effect on the plant against nematode infection was inferred; however, the effect of the PI on nematode development was not investigated.

12.4. CONCLUSION

As a consequence of enormous yield losses in crop plants imposed by the PPNs, the understanding of plant–nematode interaction is becoming of utmost importance. The nematode effectors include, for instance, cell wall degrading enzymes, the genes involved in molecular mimicry of plant genes for both compatible and incompatible plant–nematode interactions. Targeted silencing of known nematode effector proteins through *in planta* RNAi technology holds a great potential for plant resistance against various species of nematodes (Gheysen and Vanholme, 2007). A recently developed virus-induced gene silencing (VIGS) method provides a new tool to identify genes involved in soybean–nematode interactions (Kandoth *et al.*, 2013). Similarly, the application of bioinformatics in the form of approaches like OrthoMCL could be very important for computational identification and analysis of effector proteins from PPNs (Williams *et al.*, 1994). During the compatible plant–nematode interaction, nematodes are somehow able to suppress the defense related genes, the overexpression of which has led to enhanced resistance (Ali *et al.*, 2013a). In addition to CaMV-35S promoter, several syncytia specific promoters could be used to overproduce the defense related genes in the feeding sites of the nematode to enhance resistance (Ali *et al.*, 2013a, 2014; Ali and Abbas, 2016). This could be the interesting starting point for further studies to explicate how nematodes are able to suppress systemic plant defense mechanisms. It is concluded that the use of different transgenic strategies has shown good promise for nematode resistance. These have been helpful for reduction of nematode population on the plants on individual basis; however, by stacking all these molecular strategies together in the one plant will result in additive resistance, almost near to immunity against nematodes in crop plants.

Chapter - 13

Genome Editing: New Tool to Combat Phytonematodes

13.1. INTRODUCTION

To date, conventional breeding practices, due to their laborious and time-consuming nature, are being replaced with advanced molecular techniques including genome editing. Recent progress in genome sequencing technologies, together with advances in approaches to genome editing, has opened the door to the development of crops with the desired traits, including nematode resistance. Genome editing is being integrated with plant breeding for the development of crop cultivars with improved resistance against pests and diseases including nematodes. Until now, different genome editing technologies have been practiced to enhance disease resistance in different crop plants. This chapter will briefly focus on recent progress in, and long-term promise of, the use of CRISPR technology for introducing targeted modifications into host genomes with the goal of enhancing resistance against plant parasitic nematodes.

Mutations are the basis of evolution, including all genetic variations. Several mutagenesis tools have been designed to introduce mutations that enable gene functions' characterization and detection of vital genetic mutations for conferring new traits to plants (Shelake *et al.*, 2019a). The advantage of classical mutation breeding is that it allows for the rapid generation of genomic variation that can be used as a source of traits not currently identified in a crop's germplasm. The drawback of these methods is that variation is induced randomly both at desired locations and across the genome. Because of these factors, screening, isolating, and introgressing desired mutations into elite germplasm requires large populations and lengthy breeding development timelines. Thus, classical mutation breeding as a tool to generate variation is of great value but is relatively inefficient.

13.2. GENOME EDITING

Genome editing is a targeted and more exact form of mutation breeding. Genome editing is a process where an organism's genetic code is changed. It is a type of genetic engineering in which DNA is inserted, deleted or replaced in the genome of a living organism using engineered nucleases, or 'molecular scissors' (Fig. 13.1).

These tools, based on the use of Site-Directed Nucleases (SDNs), provide researchers with the ability to modify sequences at specific locations within the genome. SDNs generate targeted Double-Stranded Breaks (DSBs). The cell will die if it does not repair the damage to its genome, so it has two main ways to fix the DNA break. Regardless of the SDN used, cells are repaired by the same native DNA break repair mechanisms, the Non-Homologous End Joining (NHEJ) or Homology-Directed Repair (HDR). NHEJ produces random mutations (gene knockout), while HDR uses additional DNA to create a desired sequence within the genome (gene knock-in). These methods generally result in many fewer mutations elsewhere in the genome relative to earlier mutagenic techniques. SDNs continue to be improved to further reduce the likelihood of non-targeted mutations (Zhao and Wolt, 2017; Hahn and Nekrasov, 2019).

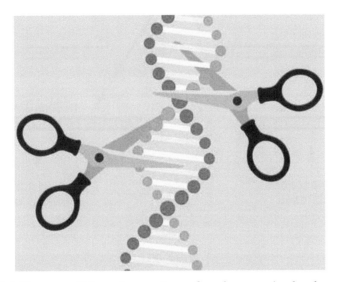

Fig. 13.1. Genome editing using engineered nucleases, or 'molecular scissors'

In the non-homologous end joining (NHEJ) pathway shown on the left side of the diagram (Fig. 13.2), random bits of DNA are inserted or deleted at the cut site before the free DNA ends are reattached. When a gene's sequence is disrupted in this way, it will no longer function, so NHEJ is a great way to stop the effects of a harmful gene.

Another pathway called homology-directed repair (HDR) is shown on the right side of the diagram (Fig. 13.2). Researchers add a piece of DNA containing sequences that match the site of the break, and the cell uses it as a "patch" to repair the cut. In this manner, scientists can deliver new genes with useful functions or replace a mutation with a healthy sequence. Remarkably, genome sequencing of edited plants showed that they were transgene free and indistinguishable from naturally occurring mutations (Nekrasov *et al.*, 2017).

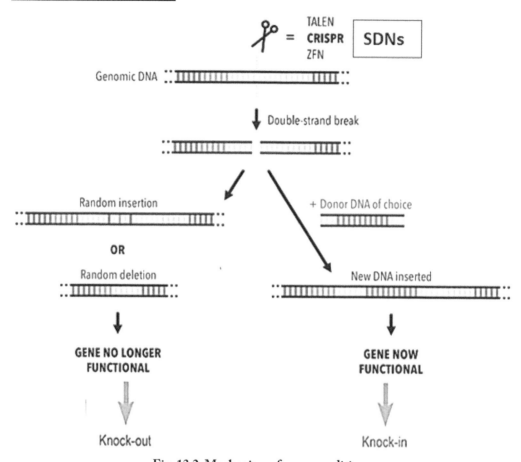

Fig. 13.2. Mechanism of genome editing

13.3. GENOME EDITING TECHNIQUES

Genome editing is accomplished by the four main gene editing techniques that will follow the above-mentioned basic pattern of mechanism of genome editing. Recent genome editing (GE) techniques include (Fig. 13.3):

» Restriction enzymes (Mega nucleases)

» Transcription activator-like effector nucleases (TALENs)

» Zinc finger nucleases (ZFNs),

» Clustered regularly interspaced short palindromic repeats/CRISPR-associated protein (CRISPR/Cas) system.

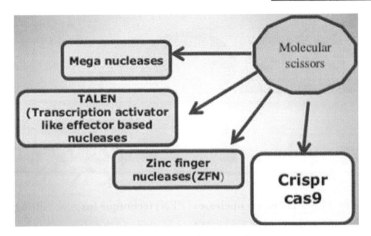

Fig. 13.3. Genome editing techniques

These techniques employ the engineered nucleases to generate a double-strand break (DSB) at the desired loci in the genome. Endogenous DNA repair mechanisms repair the DSBs. Non-homologous end-joining (NHEJ) and homology-directed repair (HDR) are the two major pathways. They may generate various kinds of mutations due to error-prone repair (Shelake *et al.*, 2019b). However, each technique has its own unique advantages and disadvantages.

13.3.1. Restriction enzymes: The original genome editor

The ability to edit genes became a reality with the discovery of restriction enzymes in the 1970s. Restriction enzymes recognize specific patterns of nucleotide sequences and cut at that site, presenting an opportunity to insert new DNA material at that location.

Restriction enzymes are not commonly utilized for gene editing these days, since they are limited by the nucleotide patterns they recognize, but they remain widely used today for molecular cloning. Additionally, certain classes of restriction enzymes play key roles in DNA mapping, epigenome mapping, and constructing DNA libraries.

13.3.2. Zinc finger nucleases (ZFNs): Increased recognition potential

As time went on, the need for precision in genome editing became more evident. Scientists needed a gene editing technique that recognized the site they wanted to edit, as off-target effects could be deleterious. The discovery of zinc finger nucleases (ZFN) in the 1980s addressed this issue.

ZFNs are composed of two parts: an engineered nuclease (Fokl) fused to zinc finger DNA-binding domains (Fig. 13.4). The zinc-finger DNA-binding domain recognizes a 3-base pair site on DNA and can be combined to recognize longer sequences. Additionally, the ZFNs function as dimers, increasing the length of the DNA recognition site and consequently increasing specificity.

ZFNs use about 30 amino acid fingers that fold around a zinc ion to form a compact structure that recognizes a 3-base pair of DNA. Consecutive finger repeats are able to recognize and target a wide area of the target DNA (Segal and Meckler, 2003). However, while specificity increased with ZFNs, it was not perfect.

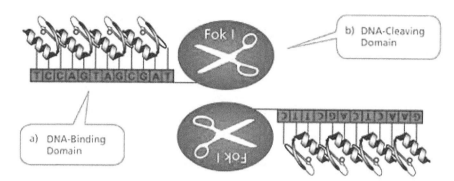

Fig. 13.4. Zinc finger nucleases (ZFN) technique for gene editing.

One of the main hurdles with using ZFNs was that the 3-base pair requirement made the design more challenging. Guanine-rich target sites appeared to be more efficient at editing versus non-guanine-rich sites. Additionally, since the ZFN interaction with DNA is modular (i.e., each ZF interacts with DNA independently); the editing efficiency was also compromised. Therefore, scientists needed to address these issues if they wanted to have more efficiently edited genome.

13.3.3. TALENs Gene editing: Single nucleotide resolution

In 2011, a new gene editing technique emerged, which was an improvement over ZFNs. Transcription Activator-Like Effector Nucleases (TALENs) are structurally similar to ZFNs (Fig. 13.5). Both methods use the FokI nuclease to cut DNA and require dimerization to function, however, the DNA binding domains differ. TALENs use Transcription Activator-Like Effectors (TALEs), tandem arrays of 33-35 amino acid repeats. The amino acid repeats possess single nucleotide recognition, thereby increasing targeting capabilities and specificity compared to ZFNs.

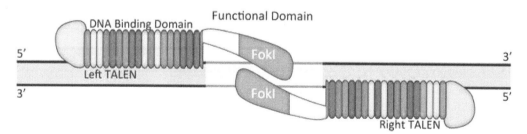

Fig. 13.5. Transcription activator-like effector nucleases technique for gene editing.

When compared with ZFNs, TALENs have better ability in targeting their chromosomal site, have a wide spectrum of sequences that can be targeted and has less chance for off-target since their design is intended to recognize 30-36 base pairs from the target site while ZFNs is designed to recognize 18-24 base pairs (Ran *et al.*, 2013).

Even with the single nucleotide resolution, using TALENs as a gene editing tool was still time and cost intensive and possessed certain design restrictions. The structure of TALENs meant the target site required a 5' thymine and 3' adenine, limiting target customizability. Further, TALENs displayed decreased editing efficiency in heavily methylated regions. Delivery into cells was also challenging, since TALENs are much larger than ZFNs (~6kb vs. ~2kb), increasing the amount of time and money required to create a successful edit. While TALENs showed improvement in genome editing technology, the high labor and monetary cost still hindered its widespread adoption.

Like its predecessor ZFNs, TALENs have been used in the field of agriculture to create pathogen-resistant rice using TALENs. Although ZFNs and TALENs enable genome editing in many species, their wider applications is hindered by their costs in time and labor, have access to the chromatin site but no other part of the genome, unable to have multiplex genome editing and both require engineering of new proteins for each DNA sequences to be targeted (Segal and Meckler, 2003; Gaj *et al.*, 2013).

13.3.4. CRISPR-Cas9 Gene editing

Recently, an alternative genome editing method called CRISPR (Clustered Regularly Interspaced Short Palindromic Repeat)/Cas (CRISPR-associated) system has become a major nuclease-based RNA-guided genome engineering research approach (Shen *et al.*, 2014). The CRISPR/Cas system is widely found in bacteria and archaea and is an adaptive immune response to defend against invading viral and plasmid DNAs (Sorek *et al.*, 2013; Terns and Terns, 2011). When a virus attacks, bacteria integrate the corresponding sequences of the invading DNA or RNA as short fragments/spacers into the cell's genome at the CRISPR locus. Since the DNA sequences on the CRISPR locus are separated by equal length spacer sequences, this 'DNA-repeat'-spacer-'DNA repeat' pattern gave way to the name of 'CRISPR'. When comparing different CRISPR loci, some common repeats were identified (Kunin *et al*, 2007). At the CRISPR loci, spacers interspace a cluster of *Cas* genes and a series of repeat sequences. Following transcription of the spacers to produce short fragments of CRISPR RNA (crRNA), CAS proteins use crRNA to match sequences with foreign invading genetic components of bacteria and destroy foreign DNA (Hsu *et al.*, 2014).

The components of the CRISPR/Cas9 system are less complex than Zinc Finger Nuclease (ZFNs) or Transcription Activator-Like Effector Nuclease (TALENs) making construction of CRISPR/Cas9 constructs less labor intensive and relatively easier to design. Compared to ZFNs or TALENS, CRISPR/Cas9 does not require extra separate 9 diametric proteins for each specific target site (Bortesi and Fischer, 2015). Additionally, by introducing multiple gRNAs one can edit several genes simultaneously (Belhaj *et al.*, 2013; Mao *et al.*, 2013). This allows for precise knocking-out of redundant genes or closely-related pathways. Moreover, by targeting two sites on the gene of interest, large genomic deletions or inversions can be induced by introducing two DSBs (Li *et al*, 2013; Upadhyay *et al.*, 2013; Zhou *et al.*, 2014).

The CRISPR (Clustered Regular Interspaced Palindromic Repeats)/Cas (CRISPR-associated protein)-based tools are the most used programed nucleases compared to other nucleases. Some of the significant features responsible for more popularity of CRISPR/Cas tools include higher efficiency, cheaper, simplicity in designing, diverse applicability, a broad range for target selection, and editing the target region in the genome with higher precision. Genetic modifications can be introduced without foreign gene integration, one of the significant advantages of CRISPR/Cas9 tools.

CRISPR is a two-component system consisting of a guide RNA and a Cas9 nuclease (Fig. 13.6). The Cas9 nuclease cuts the DNA within the ~20 nucleotide region defined by the guide RNA. With CRISPR, scientists can customize their guide RNAs, and algorithms have been developed to assess chances of off-target effects (i.e., does this sequence exist in other places of the genome). However, CRISPR is much more customizable and cost-effective, making it more accessible to the scientists that may have budget and time constraints.

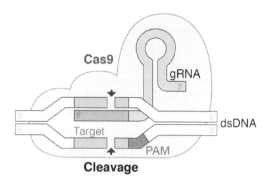

Fig. 13.6. CRISPR-Cas9 gene editing. Treating cells with a well-designed gene editing reagent (including a gRNA homologous to target sequence and adjacent PAM) can create a single SNP or indel at a precise, predetermined location.

The advancements in genome editing techniques have opened up new doors for what genome editing can do to address issues in agriculture. CRISPR has completely revolutionized what genome editing can mean for our future by increasing the speed and breadth of science. We are already feeling the impact of CRISPR in its role in gene drives.

The components of the CRISPR/Cas9 system are less complex than Zinc finger nuclease (ZFNs) or Transcription activator-like effector nuclease (TALENs) making construction of CRISPR/Cas9 constructs less labor intensive and relatively easier to design. Compared to ZFNs or TALENS, CRISPR/Cas9 does not require extra separate diametric proteins for each specific target site (Bortesi and Fischer, 2015). Additionally, by introducing multiple gRNAs one can edit several genes simultaneously (Belhaj *et al.*, 2013; Mao *et al.*, 2013). This allows for precise knocking-out of redundant genes or closely-related pathways. Moreover, by targeting two sites on the gene of interest, large genomic deletions or inversions can be induced by introducing two DSBs (Li *et al*, 2013; Upadhyay *et al.*, 2013; Zhou *et al.*, 2014).

Among the modern GE technologies, the CRISPR/Cas-based tools revolutionize all aspects of life sciences, including agriculture. The CRISPR/Cas-based tools are the most used programed nucleases compared to other nucleases. Some of the significant features responsible for more popularity of CRISPR/Cas tools include higher efficiency, cheaper, simplicity in designing, diverse applicability, a broad range for target selection, and editing the target region in the genome with higher precision. The engineered CRISPR/Cas system for GE applications consists of two major components: nuclease and single-guide RNA (sgRNA) (Jinek *et al.,* 2012). The Cas9 is the most characterized nucleases for GE and beyond. The CRISPR/Cas system is widely adopted to target the genomic DNA in several prokaryotes and eukaryotes for various purposes (Knott and Doudna, 2018; Shelake *et al.*, 2019a, b). The diverse and expanding CRISPR-based GE toolbox allows the engineering of plants at different steps involved in the relay of genetic information across the central dogma of molecular biology.

CRISPR/Cas-based tools were also effectively harnessed to generate stress-tolerance in species, model plant, and commercial crop varieties. Genetic modifications can be introduced without foreign gene integration, one of the significant advantages of CRISPR/Cas tools. The CRISPR-mediated GE technology is applied to edit the DNA and related processes in both the plants and interaction partners, such as pathogens. The rapid research for understanding the fundamental mechanisms and identification of novel genes and regulation-free GE crops can promote improved plant health management for better agriculture.

13.4. CLASSES OF CRISPR-BASED GENOME EDITING

Various adaptations of the CRISPR system (either using the Cas9 endonuclease as is or fused to other 'effector' domains) can be used to generate four classes of CRISPR-based genome editing (Anzalone *et al.*, 2020):

» Nucleases
» Base editors
» Transposases/recombinases
» Prime editors

13.4.1. Nucleases

The most common to date, requires double-stranded cutting of target DNA and repair by either non-homologous end-joining (NEHJ) or homology-directed repair (HDR) to introduce small indels and edits (Fig. 13.7).

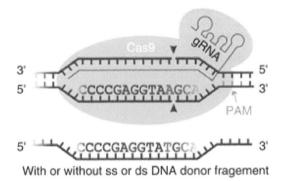

Fig. 13.7. A. Nuclease–NHE/JDR, small indeld/edits (Siddique *et al.*, 2022).

13.4.2. Base editors

Variously fuse cytidine or adenine deaminases to Cas proteins to enzymatically convert endogenous bases (with or without various inhibitors of suboptimal repair pathways and nicking non-deaminated strand to promote repair using the edited strand as template) to introduce a very specific set of edits within a narrow window (Fig. 13.8).

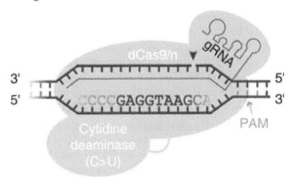

Fig. 13.8. Base edititors-targeted edits in window (Siddique *et al.*, 2022)

13.4.3. Prime editors

Fuse reverse transcriptase to Cas nickase protein and include a longer modified guide to prime reverse transcription using the nicked strand, thereby creating a template for HDR to introduce high fidelity edits (Fig. 13.9).

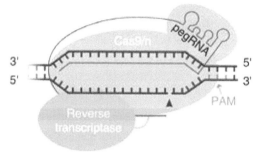

Fig. 13.9. Prime editing-high fidelity targeted edits (Siddique *et al.*, 2022)

13.4.4. Transposases/recombinases

A recent (and not yet deployed in plants) approach to fuse various transposase or recombinase domains to Cas proteins to induce larger indels/structural rearrangements. PAM, protospacer adjacent motif; gRNA, guide RNA; pegRNA, prime editing guide RNA (Fig. 13.10).

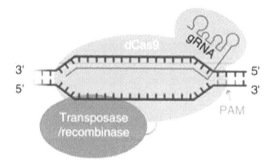

Fig. 13.10. Transpo/recombinase-large indels (Siddique *et al.*, 2022).

The majority of genome editing in plants uses the first class of genome editing: Cas nuclease. The class I CRISPR/Cas nuclease system that is best characterized in a wide range of bacterial and archaeal hosts compared to other class systems (Hsu *et al.*, 2013). The type II CRISRP/Cas9 system is best characterized that consists of a Cas9 nuclease and a single guide RNA (sgRNA) to target a DNA sequence (Belhaj *et al.*, 2013). The Cas9 endonuclease, a large monomeric DNA nuclease compromising RuvC-like domain and HNH nuclease domain, is derived from the *Streptococcus progenies* type II CRISPR/Cas system (Xie and Yang, 2013). Originally, together with two short noncoding RNA molecules-crRNA and trans-activating crRNA (tracrRNA), Cas9 nuclease is guided to a specific DNA sequence and cleaves on both strands to induce a double strand break (DSB) that leads to DNA repair mechanisms through either non-homologous end-joining (NHEJ) or homologous end-joining (HDR). In most cases, plants repair DNA mainly by NHEJ, which results in unfaithful repairs to create small nucleotide deletions or insertions (Indels) (Podevin *et al.*, 2013). Consequently, indels will introduce a frame shift mutation that impact gene function or result in a complete gene knock-out.

To achieve successful Cas9 recognition on the target DNA sequence, a Protospacer-Adjacent Motif (PAM) adjacent to the targeted DNA sequence is necessary. The -NGG is known as the PAM site. In order to target on different DNA sites, the Cas9 nuclease sequences from *Streptococcus progenies* remain the same in a CRISPR/Cas9 plasmid. Only the 20nt sgRNA needs to be re-designed. Based on the target gene sequence, two complementary oligonucleotides are designed to make a new sgRNA. In this way, the new sgRNA will allow the system targeting a different DNA sequence in the form of 5'-N (20)-NGG (Gasiunas *et al.*, 2012).

13.5. APPLICATION OF CRISPR/CAS9 FOR NEMATODE RESISTANCE

As a result of growing demand in boosting plant productivity, robust and versatile genome editing tools became highly demanding. Due to availability of more plant genomes, exploiting the powerful approach of genome editing, CRIPSR/Cas9 in various plant species is becoming realistic (Liang *et al.*, 2014). In recent years, CRISPR/Cas9 has been successfully applied in different model plants such as *Arabidopsis thaliana*, *Nicotiana benthamiana* (Li *et al.*, 2013), and *N. tabacum* (Gao *et al.*, 2015). Also, studies have applied CRISPR/Cas9 to genetically modify a diverse range of crops including wheat (Upadhyay *et al.*, 2013), maize (Liang *et al.*, 2014), rice (Mao *et al.*, 2013), sweet orange, tomato, and sorghum (Fan *et al.* 2015).

While CRISPR-based genome editing has shown considerable potential towards crop disease resistance, its application in nematode management is still in its infancy. A few of the studies are related with plant defense to pathogens including nematodes. In recent years, several studies have reported the successful use of the precise genome-editing tool CRISPR/Cas9 in soybean (Du *et al.*, 2016; Cai *et al.*, 2015). To date, CRISPR genome editing has been used to interrogate plant–nematode interactions, rather than deliver resistance. For example, soybean hydroxymethyltransferase gene (GmSHMT08) has been identified to play a role in resistance to soybean cyst nematodes (Liu *et al.*, 2012). Kang *et al.* (2016) used CRISPR to disrupt the function of GmSHMT08 in nematode resistant soybean. Nematode infection assays found that knocking out GmSHMT08 significantly increases the nematode infection compared to empty vector control roots. The genomes of both dicots and monocots have been edited using the same vector system for CRISPR-Cas9, even though codon optimized versions of Cas9 are available for each plant type (Fan *et al.*, 2015; Jiang *et al.*, 2013).

While CRISPR/Cas9 has been used successfully for genome editing in plants, the biggest concern of CRISPR/Cas9 is often related to mutation efficiency. In Arabidopsis and rice, modification efficiencies up to 90% have been reported (Feng *et al.*, 2014; Liang *et al.*, 2014); however, the mutation frequency is often unpredictable due to different methods of transformation and the uniqueness of various plant species. For example, using PEG-based protoplast transformation in rice and in wheat, the mutagenesis efficiency was reported to be 15%-38% and 3%-8%, respectively (Shan *et al.*, 2013; Xie and Yang, 2013). Agro-infiltration in *N. benthamiana* leaves resulted in mutagenesis efficiency of 2.7%-4.8%, whereas targeting endogenous genes using PEG-based protoplast transformation in *N. benthamiana* showed mutation efficiency up to 38% (Li *et al.*, 2013). For soybean research, CRISPR/Cas9 has been applied to generate mutations in soybean hairy roots (Jacobs *et al.*, 2015). Rates of mutation up to 95% were reported (Jacobs *et al.*, 2015). Nevertheless, in another study applying CRISPR/Cas9 in soybean hairy roots, the modification rate was reported to be ~54% in 170 transgenic hairy roots (Cai *et al.*, 2015). Additionally, targeting of the same gene using the same CRISPR/Cas9 backbones but different sgRNAs could result in variable mutation efficiencies (Cai *et al.*, 2015).

Recently, genome editing using CRISPR/Cas 9 system has been established in free living nematode, *Caenorhabditis elegans* (Dickinson and Goldstein, 2016; Paix *et al.*, 2017). This development would lead to the characterization of several important genes involved in different physiological processes of nematodes. Nonetheless, there are very few reports available on the application of CRISPR/Cas 9 system to study the resistance responses of the plants against nematodes. Kang (2016) recently used soybean hairy roots to study the resistance response knockouts of two serine hydroxymethyltransferase genes, GmSHMT08 and GmSHMT05, generated through this system in soybean-*H. glycines* model organisms.

Several factors influencing mutation efficiency should be considered. For example, the promoters driving sgRNA and components of sgRNA could play important roles in modification efficiency (Belhaj *et al.*, 2015). For example, in soybean hairy roots, use of the Arabidopsis U6 promoter vs. the endogenous soybean U6 promoter to drive the gRNA cassette resulted in different mutation frequencies (Du *et al.*, 2016). Whether or not changing gRNA promoters will improve the capacity of CRISPR/Cas9 in other plants still remains unknown (Cai *et al.*, 2015). As for the sgRNA component, many sources provide the rankings of gRNAs according to various algorithms to evaluate the mutation efficiency, but why some gRNAs proved to have higher mutation rates than others is unknown. Still, researchers are trying to optimize the guidelines of predicting efficiency of gRNAs. Also, plants regenerate during growth and development, therefore, when the mutation occurs often is unpredictable. Consequently, if mutations in different tissues occur independently, a chimeric plant consisting of cells that have different genotypes will form. Therefore, the mixture of wild type, heterozygous, homozygous or biallelic loci could complicate the detection and accuracy of mutagenesis (Belhaj *et al.*, 2015).

Nevertheless, researchers are investing great effort to investigate details of application of CRISPR/Cas9 in plants. Compared to traditional mutagenesis approaches, CRISPR/Cas9 still has relatively high mutation efficiency and low off-targets effects (Podevin *et al.*, 2013). Thus, CRISPR/Cas9 shows promise to accelerate plant breeding through precise genome modification of homologous genes and multiple gene families simultaneously to improve polyploid crops.

13.6. FUTURE OUTLOOK

The recent progress in understanding plant– nematode interactions and the emergence of genome editing tools provide new opportunities for sustainable and broad-spectrum management of nematodes using biotechnology. One strategy to deploy CRISPR-driven nematode resistance would be to generate mutations or deletions in known susceptibility genes as was done for MLO or other plant genes required for successful parasitism. A few susceptibility genes have already been identified in model plants such as Arabidopsis. Mechanisms underlying fundamental biological processes such as nematode susceptibility are often conserved in closely related groups of organisms, making it possible to translate discoveries in one species to other species. Such conservation can be exploited by identifying and editing orthologous genes in crop plants using CRISPR. It is predicted that susceptibility genes will be the prime targets for genome editing over the next 5–10 years. Multiple susceptibility genes can be edited and stacked with classical resistance genes in a single genetic background to help provide durable nematode resistance.

Another promising avenue for future development is establishment of genome editing tools for plant parasitic nematodes. Although what has been achieved without functional genetics is itself remarkable, Siddique *et al.* (2022) propose that arming an able field with new functional genetic tools will accelerate progress in understanding plant– nematode interactions and thereby in achieving solutions for global food security. It is predicted that expanding reverse genetic approaches beyond RNA interference, using low-cost, technically simple and efficient transformation (transient or stable) will be the single most important advance in the field in some years. The impact should be immediate and long-lasting. However, predicting the timescales for development of these technological resources is an uncertain practice. There is a clear pathway for optimizing transient expression of exogenous genes in plant parasitic nematodes to the point of utility, with the potential for expansion to other nematode groups (no obvious biological barriers to deployment in any plant parasitic nematode), over the next 3–5 years. Whether this technology, or others, will lead to stable transgenesis of sedentary endoparasites within or beyond that timeline is unclear, but it is cautiously optimistic. Stable transgenesis is most likely to be first achieved in those plant parasitic nematodes that have the most conducive biology (e.g. *Bursaphelenchus*, *Radopholus*). The long-term effects on integrated nematode management will necessarily follow the development of an efficient protocol for either stable or transient expression, something unlikely before the end of the decade.

13.7. CONCLUSION

Genome editing as an advanced molecular biology technique that can produce precisely targeted modifications in any crop. With the progress already made in the development of genome-editing tools and the development of new breakthroughs, genome editing promises to play a key role in speeding up crop breeding and in meeting the ever-increasing global demand for food. Moreover, the exigencies of climate change call for greater flexibility and innovation in crop resilience and production systems. In addition, we must take into account government regulations and consumer acceptance around the use of these new breeding technologies.

There is current research aimed at delivering dsRNA in spray form (ectopic delivery) rather than by transgenic plants. This strategy requires low cost production of dsRNA sequences, methods to stabilize them for field delivery, uptake of dsRNA by leaves, its systemic basipetal movement through plants to roots, and uptake by nematode on feeding (Fosu-Nyarko and Jones, 2015; Naz *et al.*, 2016). If the technical aspects of ectopic delivery of dsRNA can be overcome in a cost-effective manner, this could bypass the issues of RNAi-based transgenic nematode control. In addition to ectopic delivery of RNAi, gene silencing technology can be used to determine targets for new nematicides. This process involves genome-enabled novel chemical nematicides (Fosu-Nyarko and Jones, 2015). Very recently, it has been reported that the tryptophan decarboxylase 1 (AeVTDC1) gene from a wide relative of wheat *Aegilops variabilis* regulates the resistance against *Heterodera avenae* by altering the downstream secondary metabolite contents rather than auxin synthesis (Huang *et al.*, 2018). This shows the potential use of genetic resources from wide relatives for the enhancement of cereal cyst nematode resistance in cereal crops.

The current generation of genome editing technologies has facilitated the efficient generation of desirable genomic variation and new plant varieties that would have been more challenging to achieve through other breeding or molecular approaches. Genome editing technology continues to be refined, with improved editing specificity, developments in the delivery of editing enzymes, and ever-increasing genomic characterization.

Chapter - 14

Bio-intensive Integrated Nematode Management

14.1. INTRODUCTION

Over-reliance on the use of synthetic pesticides in crop protection programs around the world has resulted in disturbances to the environment, pest resurgence, pest resistance to pesticides, and lethal and sub-lethal effects on non-target organisms, including humans (Prakash and Rao, 1997). These side effects have raised public concern about the routine use and safety of pesticides. At the same time, population increases are placing ever-greater demands upon the "ecological services"—that is, provision of clean air, water and wildlife habitat—of a landscape dominated by farms. Although some pending legislation has recognized the costs to farmers of providing these ecological services, it is clear that farmers and ranchers will be required to manage their land with greater attention to direct and indirect off-farm impacts of various farming practices on water, soil, and wildlife resources. With this likely future in mind, reducing dependence on chemical pesticides in favor of ecosystem manipulations is a good strategy for farmers.

14.2. INTEGRATED NEMATODE MANAGEMENT (INM)

Plant parasitic nematodes cause significant economic losses in crop plants. Root-knot (*Meloidogyne* spp.) and cyst (*Heterodera* spp.) nematodes are among the most significant of all plant parasitic nematodes. To alleviate nematode problems, integrated nematode management (INM) procedures are implemented based on the principles of prevention, population reduction and tolerance (Fig. 14.1).

INM seeks to stabilize populations of target nematodes at acceptable levels resulting in favorable long term socio-economic and environmental consequences. In practice, integration of different methods of nematode control is the rule rather than the exception. In most cases, practical control of a nematode disease involves integration of several diverse control measures.

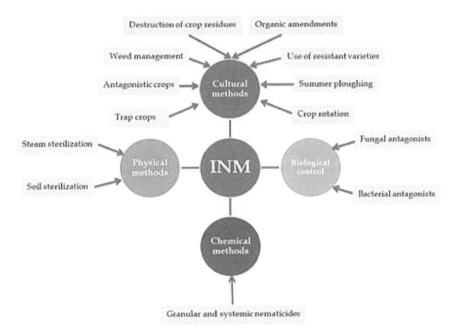

Fig. 14.1. Components of integrated nematode management.

14.3. BIOINTENSIVE INTEGRATED NEMATODE MANAGEMENT

Biointensive integrated nematode management (BINM) is defined as "a systems approach to pest management based on an understanding of nematode ecology. Biointensive IPM incorporates ecological and economic factors into agricultural system design and decision making, and addresses public concerns about environmental quality and food safety. It begins with steps to accurately diagnose the nature and source of nematode problems, and then relies on a range of preventive tactics and biological controls to keep nematode populations within acceptable limits. Reduced-risk nematicides are used if other tactics have not been adequately effective, as a last resort, and with care to minimize risks" (Fig. 14.2).

The objectives of integrated nematode management (INM) include:

- » To minimize environmental and health hazards.
- » Utilization of several compatible control measures.
- » To maximize natural environmental resistance to plant parasitic nematodes.
- » To minimize the use of drastic control measures.
- » To increase reliance on location specific and resource compatible management strategy.
- » To minimize input costs in harmony with potential gains and maximize profit to the concerned grower.

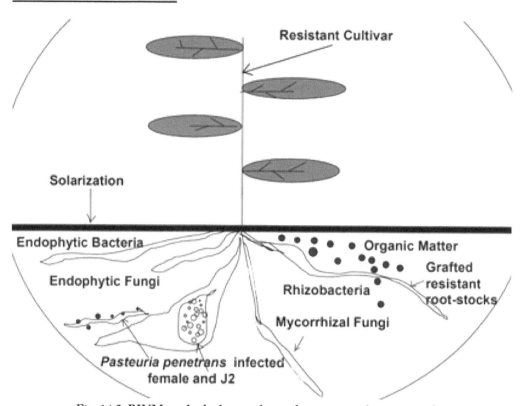

Fig. 14.2. BINM methods that can be used to manage plant nematodes

The primary goal of biointensive IPM is to provide guidelines and options for the effective management of nematode pests and beneficial organisms in an ecological context. The flexibility and environmental compatibility of a biointensive IPM strategy make it useful in all types of cropping systems.

The benefits of implementing biointensive IPM can include reduced chemical input costs, reduced on-farm and off-farm environmental impacts, and more effective and sustainable pest management. An ecology-based IPM has the potential of decreasing inputs of fuel, machinery, and synthetic chemicals—all of which are energy intensive and increasingly costly in terms of financial and environmental impact. Such reductions will benefit the grower and the society.

An important difference between conventional and biointensive INM is that the emphasis of the latter is on proactive measures to redesign the agricultural ecosystem to the disadvantage of a nematode and to the advantage of its parasite and predator complex (Fig. 14.3).

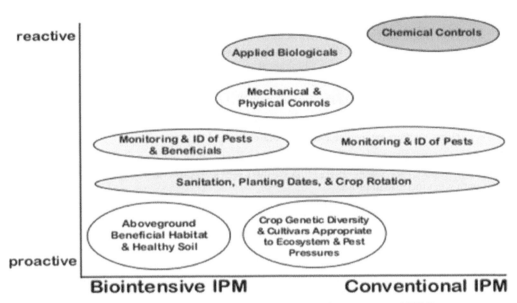

Fig. 14.3. Comparison of biointensive IPM and conventional IPM.

14.4. BIOINTENSIVE INM STRATEGIES

BINM strategies are based on 5 main components/tools:

- » Planning
- » Pest identification
- » Economic injury and action levels
- » Monitoring/observation/regulation
- » Record keeping

14.4.1. Planning

Good planning must precede implementation of any IPM program, but is particularly important in a biointensive program. Planning should be done before planting because many pest strategies require steps or inputs, such as beneficial organism habitat management, that must be considered well in advance. Attempting to jump-start an IPM program in the beginning or middle of a cropping season generally does not work.

When planning a biointensive IPM program, some considerations include:

- » Options for design changes in the agricultural system (beneficial organism habitat, crop rotations).
- » Choice of pest-resistant cultivars.
- » Technical information needs.
- » Monitoring options, record keeping, equipment, etc.

When making a decision about crop rotation, consider the following questions: Is there an economically sustainable crop that can be rotated into the cropping system? Is it compatible? Important considerations when developing a crop rotation are:

>> How might the cropping system be altered to make life more difficult for the pest and easier for its natural controls? What two (or three or several) crops can provide an economic return when considered together as a biological and economic system that includes considerations of sustainable soil management?

>> What are the impacts of this season's cropping practices on subsequent crops?

>> What specialized equipment is necessary for the crops?

>> What markets are available for the rotation crops?

Management factors should also be considered. For example, one crop may provide a lower direct return per hectare than the alternate crop, but may also lower management costs for the alternate crop, with a net increase in profit.

14.4.2. Pest identification

A crucial step in any IPM program is to identify the pest. The effectiveness of both proactive and reactive pest management measures depend on correct identification. Misidentification of the pest may be worse than useless; it may actually be harmful and cost time and money. Help with positive identification of pests may be obtained from university personnel, private consultants, the Cooperative Extension Service, and books and websites.

After a pest is identified, appropriate and effective management depends on knowing answers to a number of questions. These may include:

>> What plants are hosts and non-hosts of this pest?

>> When does the pest emerge or first appear?

>> Where does it lay its eggs? where is the source(s) of inoculum?

>> Where, how, and in what form does the pest overwinter?

>> How might the cropping system be altered to make life more difficult for the pest and easier for its natural controls?

Monitoring (field scouting) and economic injury and action levels are used to help answer these and additional questions (Adams and Clark, 1995).

14.4.3. Monitoring

Monitoring involves systematically checking crop fields for pests and beneficials, at regular intervals and at critical times, to gather information about the crop, pests, and natural enemies. Square-foot or larger grids laid out in a field can provide a basis for comparative nematode counts. Records of rainfall and temperature are sometimes used to predict the likelihood of disease infections.

The more often a crop is monitored, the more information the grower has about what is happening in the fields. Monitoring activity should be balanced against its costs. Frequency may vary with temperature, crop, growth phase of the crop, and pest populations. If a pest population is approaching economically damaging levels, the grower will want to monitor more frequently.

The aim of monitoring/observation/regulation is to determine when and what action to take. Regular monitoring is very crucial for nematode management as it is one of the most important decision making tools. An efficient monitoring program can pay big dividends in lowering nematode management costs. An Agro-Eco-System Analysis (AESA) based on crop health at different stages of growth, population dynamics of nematodes and natural enemies, soil condition, and farmer's past experiences are considered for decision making. Soil sample analysis for nematodes is usually employed as monitoring tool. Economic threshold levels (ETL) are now available to assist in proper timing of INM interventions.

» Quarantine against introduction of exotic nematodes.
» Domestic quarantine against further distribution e.g. potato cyst nematode in Nilgiri Hills of Tamil Nadu.
» Regulation on supply of quality bioagents.

If a harmful species of nematode is not present in an area, legal quarantine is the most effective method for prevention. Numerous attempts have been made to prevent the introduction of nematodes into countries or provinces by means of plant quarantine. Quarantines are established by legislative or other government actions and usually give quarantine authorities power to develop and enforce regulations to accomplish the purpose. Such regulations prohibit bringing infected plants into protected areas where similar crops are vulnerable to nematode infection.

Nematode surveillance and decision making ought to be done either visually or through soil sampling. Similar monitoring of natural enemies is also desirable.

14.4.4. Economic injury and action levels

The economic injury level (EIL) is the pest population that inflicts crop damage greater than the cost of control measures. Because growers will generally want to act before a population reaches EIL, IPM programs use the concept of an economic threshold level (ETL or ET), also known as an action threshold. The ETL is closely related to the EIL, and is the point at which suppression tactics should be applied in order to prevent pest populations from increasing to injurious levels.

ETLs are intimately related to the value of the crop and the part of the crop being attacked. Depending on the severity of the disease, the grower may face a situation where the ETL for a particular pest is zero, i.e., the crop cannot tolerate the presence of a single pest of that particular species because the disease it transmits is so destructive.

14.4.5. Record-keeping: "Past is prologue"

Monitoring goes hand-in-hand with record-keeping, which forms the collective "memory" of the farm. Records should not only provide information about when and where pest problems have occurred, but should also incorporate information about cultural practices (irrigation, cultivation, fertilization, mowing, *etc.*) and their effect on pest and beneficial populations. The effects of non-biotic factors, especially weather, on pest and beneficial populations should also be noted. Record-keeping is simply a systematic approach to learning from experience. A variety of software programs are now available to help growers keep track of and access data on their farm's inputs and outputs.

14.5. BINM OPTIONS

BINM options may be considered as proactive or reactive.

14.5.1. Proactive options

Proactive options, such as crop rotations and creation of habitat for beneficial organisms, permanently lower the carrying capacity of the farm for the nematode. The carrying capacity is determined by the factors like food, shelter, natural enemies complex and weather, which affect the reproduction and survival of a nematode species. Cultural control practices are generally considered to be proactive strategies. Proactive practices include crop rotation, resistant crop cultivars including transgenic plants, disease-free seed and plants, crop sanitation, spacing of plants, altering planting dates, mulches, *etc.*

The proactive strategies (cultural controls) include:

» Healthy, biologically active soils (increasing below-ground diversity).
» Crop diversity with differences in nematode resistance (increasing above-ground biodiversity).
» Resistant crop cultivars.

14.5.2. Reactive options

The reactive options mean that the grower responds to a situation, such as an economically damaging population of nematodes, with some type of short-term suppressive action. Reactive methods generally include inundative releases of biological control agents, mechanical and physical controls, botanical pesticides, and chemical controls.

14.6. PROACTIVE OPTIONS

14.6.1. Healthy, biologically active soils (increasing below-ground diversity)

Various organic amendments like cover crops, animal and green manure, organic wastes, plant residues, composts and peats, *etc.* have been successfully used to increase the soil suppressiveness to different nematode diseases in agricultural and horticultural crops. Application of organic materials to soil can cause a change in soil microflora and microfauna including soil nematodes. The changes in soil nematode fauna can results in an increase in the number of beneficial nematodes such as bacterial or fungal feeders and decrease and/or suppression in the occurrence of economically important plant parasitic nematodes.

The addition of organic matter to soil stimulates natural enemies of nematodes, which in turn attack plant-parasitic nematodes and reduce their numbers (Linford *et al.*, 1938). Linford and Oliveira (1938) noted reduced galling from root-knot nematodes (*Meloidogyne* spp.) when pineapple plant material was added to soil.

Populations of fungal (*e.g.*, *Trichoderma* spp.) and bacterial antagonists of nematodes (*e.g.*, *Pasteuria penetrans*, *Pseudomonas* spp. and chitinolytic bacteria), as well as typical nematode predators (*e.g.*, nematophagous mites and Collembola), are stimulated by the addition of organic matter to the soil (Akhtar and Malik, 2000; Wang *et al.*, 2008). In a field plot experiment, organically managed plots had more species of nematophagous fungi and two species, *Arthrobotrys dactyloides* and *Nematoctonus leiosporus*, were more abundant than in conventionally managed plots (Jaffee *et al.*, 1998). In this 'antagonistic' context, the influence of organic amendments on the degree of root colonization by mycorrhizal fungi, which in return also influences plant-parasitic nematodes, also needs to be considered (Hol and Cook, 2005).

Incorporating mustard (*Sinapis alba*) and oil radish (*Raphanus sativus*) significantly reduced nematode parasitism (Nicolay *et al.*, 1990). Owino *et al.* (1993) showed suppression of egg parasitism by *P. chlamydosporia* when mustard was incorporated into soil. In another study, when *P. chlamydosporia* was applied at the time of cover crop planting, the fungus increased in soil planted to black oats (*Avena strigosa*) and oil radish. In both field and greenhouse studies, amendments containing either composted animal manure or plant material suppressed parasitism of plant-parasitic nematodes by *Hirsutella rhossiliensis* (Jaffee *et al.*, 1994). Organic amendments with a high C/N ratio (in the range of 15 - 20) may be more commonly associated with biological suppression (Rodriguez-Kabana *et al.*, 1987; Stirling, 2011).

The utility of chitin as an amendment for nematode control was demonstrated by several key studies in the early 1980s (Mian *et al.*, 1982; Rodriguez-Kabana *et al.*, 1984). One suggested mode of action was that chitin increased levels of chitinolytic fungi in soil, which then parasitized eggs of plant-parasitic nematodes (Rodriguez-Kabana *et al.*, 1984; 1987). Chitin has a low C: N ratio of 6.4, so it decomposes quickly in soil and releases significant amounts of ammonia (Mian *et al.*, 1982). It is interesting to note that while chitinous amendments resulted in impressive reductions in levels of *M. arenaria* (Mian *et al.*, 1982) or *Heterodera glycines* (Rodriguez-Kabana *et al.*, 1984). The chitinous materials like crushed shells of shrimps and crab should be applied to the soil in order to enhance the population of "nematode-trapping" fungal species. After the complete decomposition of the shell material, these fungi will feed on chitin-containing nematode eggs (Yepsen, 1984).

14.6.2. Crop diversity (increasing above-ground biodiversity)

Maintaining and increasing biological diversity of the farm system is a primary strategy of cultural control. Decreased biodiversity tends to result in agroecosystems that are unstable and prone to recurrent nematode outbreaks and many other problems (Altieri, 1994). Systems high in biodiversity tend to be more "dynamically stable"—that is, the variety of organisms provide more checks and balances on each other, which helps prevent one species (i.e., nematode species) from overwhelming the system.

There are many ways to manage and increase biodiversity on a farm, both above ground and in the soil. In fact, diversity above ground influences diversity below ground. Research has shown that up to half of a plant's photosynthetic production (carbohydrates) is sent to the roots, and half of that (along with various amino acids and other plant products) leaks out from the roots into the surrounding soil, providing a food source for microorganisms. These root exudates vary from plant species to plant species and this variation influences the type of organisms associated with the root exudates (Marschner, 1998).

Factors influencing the health and biodiversity of soils include the amount of soil organic matter; soil pH; nutrient balance; moisture; and parent material of the soil. Healthy soils with a diverse community of organisms support plant health and nutrition better than soils deficient in organic matter and low in species diversity. Research has shown that excess nutrients (e.g., too much nitrogen) as well as relative nutrient balance (i.e., ratios of nutrients for example, twice as much calcium as magnesium, compared to equal amounts of both) in soils affect pest response to plants (Phelan, 1997; Daane *et al.*, 1995). Imbalances in the soil can make a plant more attractive to pests (Phelan, 1997; Daane *et al.*, 1995), less able to recover from pest damage, or more susceptible to secondary infections by plant pathogens (Daane *et al.*, 1995). Soils rich in organic matter tend to suppress plant pathogens (Schneider, 1982). In addition, it is estimated that 90% of all nematode pests infect roots in soil, and many of their natural enemies occur there as well. Overall, a healthy soil with a diversity of beneficial organisms and high organic matter content helps maintain nematode populations below their economic thresholds.

14.6.2.1. Genetic diversity: Genetic diversity of a particular crop may be increased by planting more than one cultivar. Species diversity of the associated plant and animal community can be increased by allowing trees and other native plants to grow in fence rows or along water ways, and by integrating livestock into the farm system. Use of the cropping schemes are additional ways to increase species diversity.

14.6.2.2. Crop rotation: Crop rotations radically alter the environment both above and below ground, usually to the disadvantage of pests of the previous crop. When making a decision about crop rotation, consider the following questions: Is there an economically sustainable crop that can be rotated into the cropping system? Is it compatible? Important considerations when developing a crop rotation are:

>> What two (or three or several) crops can provide an economic return when considered together as a biological and economic system that includes considerations of sustainable soil management?

>> What are the impacts of this season's cropping practices on subsequent crops?

>> What specialized equipment is necessary for the crops?

>> What markets are available for the rotation crops?

An enforced rotation program in the Imperial Valley of California has effectively controlled the sugar beet cyst nematode. Under this program, sugar beets may not be grown more than two years in a row or more than four years out of ten in clean fields (i.e., non-infested fields). In infested fields, every year of a sugar beet crop must be followed by three years of a non-host crop. Other nematode pests commonly controlled with crop rotation methods include the golden nematode of potato, many root-knot nematodes, and the soybean cyst nematode.

14.6.2.3. Multiple cropping: Multiple cropping is the sequential production of more than one crop on the same land in one year. Depending on the type of cropping sequence used, multiple cropping can be useful as a nematode control measure, particularly when the second crop is interplanted into the first.

14.6.2.4. Intercropping: Interplanting is seeding or planting a crop into a growing stand, for example over seeding a cover crop into a grain stand. There may be microclimate advantages (e.g., timing, wind protection, and less radical temperature and humidity changes) as well as disadvantages (competition for light, water, nutrients) to this strategy. By keeping the soil covered, interplanting may also help protect soil against erosion from wind and rain.

Intercropping is the practice of growing two or more crops in the same, alternate, or paired rows in the same area. This technique is particularly appropriate in vegetable production.

14.6.2.5. Strip cropping: Strip cropping is the practice of growing two or more crops in different strips across a field wide enough for independent cultivation. Like intercropping, strip cropping increases the diversity of a cropping area. Another advantage to this system is that one of the crops may act as a reservoir and/or food source for beneficial organisms.

The options described above can be integrated with no-till cultivation schemes and all its variations (strip till, ridge till, etc.) as well as with hedgerows and intercrops designed for beneficial organism habitat. With all the cropping and tillage options available, it is possible, with creative and informed management, to evolve a biologically diverse, nematode -suppressive farming system appropriate to the unique environment of each farm.

14.6.2.6. Disease-free planting material: Disease-free seeds and plants are available from most commercial sources, and are certified as such. Use of disease-free seed and nursery stock is important in preventing the introduction of disease.

14.6.2.7. Sanitation: Following field sanitation helps to reduces the nematode load. Sanitation involves removing and destroying the infected plants, as well as preventing a new nematode from establishing on the farm (e.g., not allowing off-farm soil from farm equipment to spread nematodes or plant pathogens to your land). Sanitation involves removal of infected plants from the soil. As with so many decisions in farming, both the short- and long-term benefits of each action should be considered when tradeoffs like this are involved.

14.6.2.8. Spacing: Spacing of plants heavily influences the development of plant diseases and weed problems. The distance between plants and rows, the shape of beds, and the height of plants influence air flow across the crop, which in turn determines how long the leaves remain damp from rain and morning dew. Generally speaking, better air flow will decrease the incidence of plant disease. However, increased air flow through wider spacing will also allow more sunlight to the ground, which may increase weed problems. This is another instance in which detailed knowledge of the crop ecology is necessary to determine the best pest management strategies. How will the crop react to increased spacing between rows and between plants? Will yields drop because of reduced crop density? Can this be offset by reduced pest management costs or fewer losses from disease?

14.6.2.9. Altered planting dates: Altered planting dates can at times be used to avoid specific nematodes.

14.6.2.10. Optimum growing conditions: Optimum growing conditions are always important. Plants that grow quickly and are healthy can compete with and resist nematode pests better than slow-growing, weak plants. Too often, plants grown outside their natural ecosystem range must rely on pesticides to overcome conditions and pests to which they are not adapted.

14.6.2.11. Mulching: Mulches, living or non-living, are useful for suppression of nematodes. Other researchers have found that living mulches of various crops reduce nematode pest damage to vegetables and orchard crops. Again, this reduction is due to natural predators and parasites provided habitat by the crops. Mulching helps to minimize the spread of soil-borne plant pathogens by preventing their transmission through soil splash. Recent springtime field tests at the Agricultural Research Service in Florence, South Carolina, have indicated that red plastic mulch suppresses root-knot nematode damage in tomatoes by diverting resources away from the roots (and nematodes) and into foliage and fruit (Adams, 1997).

14.6.3. Resistant crop cultivars

Plant resistance is regarded as an extremely feasible method for controlling nematodes. It is an effective, economical and environmentally safe means of reducing losses from nematode pests. Use of resistant plants enables the grower to control the parasitic nematodes without increasing production costs associated with the purchase of expensive chemicals, applicators and in the numerous mechanical operations that go into the production of a crop. Resistance is especially valuable in controlling nematodes in low value crops and minor crops which can increase crop yields equal to that obtained by soil fumigation.

The use of plant varieties resistant to nematodes may be the only method of control available, which is economic on a field scale, for certain nematodes. At present, a few nematode - resistant varieties of crop plants are available to the commercial grower. Resistant varieties are continually being bred by researchers. Growers can also do their own plant breeding simply by collecting non-hybrid seed from healthy plants in the field. The plants from these seeds will have a good chance of being better suited to the local environment and of being more resistant to nematodes. Since natural systems are dynamic rather than static, breeding for resistance must be an ongoing process, especially in the case of plant nematodes, as the nematodes themselves continue to evolve and become resistant to control measures (Elwell and Maas, 1995).

Resistant crops provide an effective and economical method for managing nematodes in both high- and low-cash value cropping systems. In annual cropping systems, resistant crops can reduce nematode populations to levels that are non-damaging to subsequent crops, thereby enabling shortening and modification of rotations. They are environmentally compatible and do not require specialized applications, as opposed to most chemicals and, apart from preference based on agronomic or horticultural desirability, do not require an additional cost input or deficit. In less developed countries and in low-cash crop systems, plant resistance is probably the most viable solution to nematode problems.

14.6.3.1. Biotech crops: Biotechnology has a role to play in incorporation of resistance against nematodes and biological control of plant nematodes. A number of genes that mediate nematode resistance have now been or soon will be cloned from a variety of plant species. Gene transfer technology is being used by several companies to develop cultivars resistant to nematodes. Nematode resistance genes are present in several crop species and are an important component of many breeding programs including those for tomato, potato, soybeans, and cereals (Trudgill, 1991). Several resistance genes have been mapped for chromosomal locations or linkage groups and some of them have been cloned. The first nematode resistance gene to be cloned was *Hs1pro-1*, a gene from a wild relative of sugar beet conferring resistance to *Heterodera schachtii* (Cai *et al.,* 1997). The cDNA, under the control of the CaMV 35S promoter, was able to confer nematode resistance to sugar beets transformed with *Agrobacterium rhizogenes* in an *in vitro* assay (Cai *et al.,* 1997). The *Mi* gene from tomato conferred resistance against a root-knot nematode and an aphid in transgenic potato (Rossi *et al.,* 1998). The gene *Mi* is a true *R* gene, characterized by the presence of NBS and LRR domains. *Gpa2* gene that confers resistance against some isolates of the potato cyst nematode, *Globodera pallida* was identified. This gene shares extensive homology with the Rx1 gene that confers resistance to potato virus X suggesting a similarity in function (Van Der Vossen *et al.,* 2000).

14.7. REACTIVE OPTIONS

14.7.1. Biological control

Biological suppression of plant parasitic nematodes with antagonistic fungi, mycorrhizae, bacteria and predatory nematodes is gaining increasing importance as a result of realization that many environmental and health hazards are linked with the use of chemicals. Standardization of methods for effective utilization of biocontrol agents is very important for evolving ecologically sound integrated nematode management strategies.

Bioagents can be used for seed treatment or soil application with *Trichoderma viride, T. harzianum, Puepureocillium lilacinum, Pochonia chlamydosporia, Bacillus subtilis* and *Pseudomonas fluorescens* for the management of nematodes.

Avermectins are the secondary metabolites of an Actinomycete, *Streptomyces avermictinus,* which are highly effective at a very low concentration against plant parasitic nematodes and mite pests. Scientists at the Indian Institute of Horticultural Research, Bangalore, for the first time in India, have isolated 6 strains of *S. avermictinus* and showed their effectiveness for the management of root-knot nematodes infecting tomato, eggplant, chilli, carnation and gerbera (Parvatha Reddy and Nagesh, 2002).

Biological control is the use of living organisms - parasites, predators, or pathogens -to maintain pest populations below economically damaging levels, and may be either natural or applied. A first step in setting up a biointensive IPM program is to assess the populations of beneficials and their interactions within the local ecosystem. This will help to determine the potential role of natural enemies in the managed agricultural ecosystem.

Natural biological control results when naturally occurring enemies maintain pests at a lower level than would occur without them, and is generally characteristic of biodiverse systems. Fungi, bacteria, and viruses all have a role to play as predators and parasites in an agricultural system. By their very nature, pesticides decrease the biodiversity of a system, creating the potential for instability and future problems.

Creation of habitat to enhance the chances for survival and reproduction of beneficial organisms is a concept included in the definition of natural biocontrol. Farmscaping is a term coined to describe such efforts on farms.

Applied biological control, also known as augmentative biocontrol, involves supplementation of beneficial organism populations, for example through periodic releases of parasites, predators, or pathogens. This can be effective in many situations. Most of the beneficial organisms used in applied biological control today are nematode parasites and predators. They control a wide range of nematodes. The quality of commercially available applied biocontrols is another important consideration.

Inundative releases of beneficials into greenhouses can be particularly effective. In the controlled environment of a greenhouse, nematode infestations can be devastating; there are no natural controls in place to suppress nematode populations once an infestation begins. For this reason, monitoring is very important. If an infestation occurs, it can spread quickly if not detected early and managed. Once introduced, biological control agents cannot escape from a greenhouse and are forced to concentrate predation/parasitism on the nematode(s) at hand. An increasing number of commercially available biocontrol products are made up of microorganisms, including fungi, bacteria, and nematodes.

14.7.2. Mechanical and physical controls

Methods included in this category utilize some physical component of the environment, such as temperature, humidity, or light, to the detriment of the pest. Common examples are tillage, flaming, flooding, soil solarization, and plastic mulches to kill nematodes. Heat or steam sterilization of soil is commonly used in greenhouse operations for control of soil-borne pests. Cold storage reduces post-harvest disease problems on produce.

Heat treatment is one of the oldest methods of nematode control. Plant parasitic nematodes can be killed by heat, desiccation, irradiation, high osmotic pressure, *etc.*, but it is more difficult to employ physical methods in killing nematodes in soil and planting materials. It is virtually impossible to kill nematodes when they are in growing plants. Field scale treatment of soil is also difficult. Nematodes can be killed at 44 to 48 °C. The nematode enzymes are inactivated at short time exposure to about 50 °C *in vitro*. Bare root dip in hot water has to be specifically determined for different nematode species. Solarization is a modern technique to be developed for each situation.

Although generally used in small or localized situations, some methods of mechanical/ physical control are finding wider acceptance because they are generally more- friendly to the environment.

14.7.3. Chemical controls

Nematode monitoring should be done for scheduling the chemical nematicide application. Nematicides should be used in right dosages, time and through appropriate equipment.

The use of chemicals on a field scale for control of plant parasitic nematodes was not possible until early 1940's when effective and economical soil fumigants like D-D and EDB were discovered which made it possible to provide growers with spectacular differences in growth and yield through the effective control of nematodes and other soil pests. Several effective nematicides belonging to organophosphate and carbamate groups were developed and improvements in methods of application provided more economic control. In many parts of the world today, soil treatment for nematode control is an established farm practice.

The primary advantage of chemical control over other methods is the quick reduction of nematode population within days after the chemical is applied. The grower is able to plant a crop soon after treatment or, in some cases, at the time of treatment. Most crops are especially vulnerable to nematode attack during the seedling stage when the young root system is becoming established. Crops planted in treated soil develop extensive root systems, and the crop, especially annuals, mature before residual population of nematodes can increase to a damaging level.

Included in this category are both synthetic pesticides and botanical pesticides. Synthetic pesticides comprise a wide range of man-made chemicals used to control nematode pests. These powerful chemicals are fast acting and relatively inexpensive to purchase. nematicides are the option of last resort in INM programs because of their potential negative impacts on the environment, which result from the manufacturing process as well as from their application on the farm. Nematicides should be used only when other measures, such as biological or cultural controls, have failed to keep nematode populations from approaching economically damaging levels.

If chemical nematicides must be used, it is to the grower's advantage to choose the least-toxic nematicide that will control the nematode but not harm non-target organisms such as birds, fish, and mammals. Nematicides that are short-lived or act on one or a few specific organisms are in this class.

Biorational nematicides are generally considered to be derived from naturally occurring compounds or are formulations of microorganisms. Biorationals have a narrow target range and are environmentally benign. Formulations of *Bacillus subtilis, Pseudomonas fluorescens, Trichoderma harzianum, Pochonia chlamodosporia* and *Purpureocillium lilacinum*, are perhaps the best known biorational pesticides.

Botanical pesticides are prepared in various ways. They can be as simple as pureed plant leaves, extracts of plant parts, or chemicals purified from plants. Botanicals are generally less harmful in the environment than synthetic pesticides because they degrade quickly, but they can be just as deadly to beneficials as synthetic pesticides. However, they are less hazardous to transport and in some cases can be formulated on-farm. The manufacture of botanicals generally results in fewer toxic by-products.

Botanicals and their products play an important role in the reduction of nematode population in soil. A large number of reports have been published showing that growing of antagonistic plants (marigold, mustard, sesame, asparagus, *etc.*) or addition of organic materials (green manures, oil cakes, crop residues, cellulosic soil amendments, *etc.*) to the nematode infested soils results in definite reduction of several plant parasitic nematodes.

If the monitoring program indicates that the pest outbreak is isolated to a particular location, spot treatment of only the infested area will not only save time and money, but will conserve natural enemies located in other parts of the field. The grower should also time treatments to be least disruptive of other organisms. This is yet another example where knowledge about the agroecosystem is important.

One way to increase application efficiency and decrease costs of nematicide use is through seed treatment, nursery bed treatment, bare root-dip treatment, row application, and spot application. This puts the nematicide only where it is needed.

Microencapsulation of nematicides is promising technology. This technique reduces the nematicide used in fields by up to 80%, releases the chemical slowly, long lasting, and conserves beneficials.

Some of the approaches that can be utilized for the development of bio-intensive integrated nematode management strategies are presented in Table 14.1.

Table 14.1. Approaches that can be utilized for the development of bio-intensive integrated nematode management

S. No	Approaches	Bio-intensive INM practices
1	Regulatory methods	Seed certification (disease-free seeds and plants)
2	Physical methods	Hot water treatment of planting material, soil solarization
3	Cultural methods	Crop rotation, intercropping, strip cropping, fallowing, flooding, summer plowing, green manuring, trap crops, organic soil amendments, field sanitation, spacing of plants, altered planting dates, organic mulches
4	Chemical methods	Reduced-risk pesticides, site specific treatment, nanotechnology
5	Biological methods	Nematophagous fungi, arbuscular mycorrhizal fungi, nematophagous bacteria, predatory nematodes
6	Host plant resistance	Resistant varieties, genetically modified crops, genome edited crops

14.8. LOW INPUT SUSTAINABLE BINM STRATEGY

14.8.1. Summer

» Two or three deep summer plowings with a soil turning harrow at fortnightly intervals, preferably with a light irrigation between two plowings.

» Soil solarization of nursery beds or pit soil using clear thin polyethylene mulch for 3-6 weeks before sowing.

» Rabbing of nursery beds (burning of crop residues over the soil surface), preferably after turning of soil, before sowing.

» Application of FYM/compost enriched with *Trichoderma harzianum/ Purpureocillium lilacinum/Pochonia chlamydosporia* to nursery beds.

» Growing of non-host green manure/antagonistic crops/trap crops and incorporating in the soil at the time of flowering.

» Plowing back non-commercial crop residues into the soil before monsoon.

14.8.2. Monsoon

» Growing of non-host green manure/antagonistic crops/trap crops and incorporating in the soil at the time of flowering.

» Application of organic amendments (FYM/compost/oil cakes/crop residues) to the soil while preparing the land.

14.8.3. *Kharif*

» Growing of non-host or nematode-resistant crop cultivars.

» Application of FYM/compost enriched with *Trichoderma harzianum/ Purpureocillium lilacinum /Pochonia chlamydosporia* to nursery beds.

» Uprooting and burning of roots of host crops and weeds after harvest.

14.8.4. *Rabi*

» Delaying sowing/planting to mid-November (when soil temperature falls to 15-18 °C).

» Restricting of growing susceptible or tolerant crops in *rabi* season with a suitable period of rotation.

» Removal and burning of non-commercial but disease-free crop residues.

14.9. CASE STUDIES

14.9.1. Rice white tip nematode

Farmers saved seeds + brine solution (20% NaCl) + hot water treatment (55 ^{0}C for 10 minutes) + application of Carbofuran (Furadan 3G) in the main field at the time of final land preparation would help greatly to minimize the loss of yield caused by seed borne nematode, *Aphelenchoides besseyi* in BR11 and Nizersial rice cvs. commonly cultivated in the country (Islam *et al.*, 2015).

14.9.2. Black pepper foot rot and nematodes disease complex

Integrated management of foot rot (*Phytophthora capsicii*) and nematodes (*Meloidogyne incognita* and *Radopholus similis*) on black pepper was achieved by:

» Mixing VAM and *Trichoderma harzianum* in solarized nursery mixture to raise healthy and robust seedlings.

» Application of *T. harzianum* and FYM in planting pit.

» Field application of Neem cake at 1 kg/vine mixed with 50 g of *T. harzianum* during August (Sarma, 2003).

14.9.3. Banana burrowing nematode

Integration of neem cake at 200 g/plant with *Glomus mosseae* at 100 g/plant (containing 25-30 chlamydospores/g of inoculum) was most effective in reducing the *Radopholus similis* population both in soil and roots, while karanj cake with *G. mosseae* gave maximum increase in fruit yield of banana. Mycorrhizal root colonization and number of chlamydospores of *G. mosseae* were highest in neem cake amended soil (Table 14.2) (Parvatha Reddy *et al.*, 2002).

Table 14.2. Effect of *Glomus mosseae* **and oil cakes on population of** *Radopholus similis* **and yield of banana.**

Treatment	Dose (g)/ plant	Popn. of *R. similis*		Yield(kg)/ plant
		Roots (10 g)	Soil (250 ml)	
G. mosseae	200	112	122	8.64
Castor cake	400	146	132	8.18
Karanj cake	400	118	128	10.34
Neem cake	400	118	112	8.91
G. mosseae + Castor cake	100 + 200	90	108	12.68
G. mosseae + Karanj cake	100 + 200	76	80	16.61
G. mosseae + Neem cake	100 + 200	48	62	14.80
Control	---	218	184	5.45
CD (P = 0.05)		11.97	8.31	0.84

14.9.4. Potato nematodes

The integrated control of the potato cyst nematode, *Globodera rostochiensis* can be achieved by a rational combination of crop rotation, resistant cultivars and nematicides (Jones, 1969) (Table 14.3).

Table 14.3. Integration of control methods for *Globodera rostochiensis*

Control method (s)	Resulting population (% initial population)	Kill (%)	Population after growing and harvesting a susceptible cultivar, calculated at two assumed nematode multiplication rates* (% initial population)	
			30 x	70x
4 years without potatoes	3	97	90	<100
1 year with resistant potatoes	20	80	>100	>100
Nematicide treatment (fumigant)	25	75	>100	>100
1 and 2	0.6	99	18	42
1 and 3	0.75	99	22.5	52.5
2 and 3	5	95	>100	>100
All three methods	0.15	100	4.5	10.5

*The observed maximum reproductive rate is thought to lie between 25 and 75 times but could be greater.

The control of potato rot nematode, *Ditylenchus destructor* was achieved by the combination of disease escape; hygiene, in the form of seed certification and crop rotation (Winslow and Willis, 1972).

» Healthy seed, planted late, harvested early and stored as cool and dry as possible.

» Proper rotation-potatoes not more frequently than once in 3 or 4 years and avoid other susceptible crops.

» Field hygiene (removal of old infested tubers and weed control).

14.9.5. Tomato root-knot nematode

Proper site selection and preparation for transplant productions are essential components of INM. Fallowing, crop rotation, cover crops and nematicides can be used. The transplants are inspected at the site of origin for root-knot nematode galls. Certification and inspection should be followed. *Meloidogyne* spp. infections avoiding detection are usually below the damage thresholds (Bird, 1981).

New variations on these involve combinations of cultural (deep plowing to a depth of 20 cm) and chemical methods (nursery bed treatment with Carbofuran at 0.4 g a.i. per m^2 and main field treatment with Carbofuran at 1 kg a.i. per ha) for the control of root-knot nematodes in tomato which also registered maximum yield.

Red gram (*Cajanus cajan*) and *Crotalaria spectabilis* were used in conjunction with clean fallow and a nematicide (Fensulfothion) for managing root-knot nematode populations in tomato transplants.

In nursery, integration of *Pasteuria penetrans* (at 28 x 10^4 spores/m^2), *Purpureocillium lilacinum* (at 10 g/ m^2 with 19 x 10^9 spores/g) and neem cake (at 0.5 kg/m^2) gave maximum increase in plant growth and number of seedlings/bed. Parasitization of *M. incognita* females was highest when neem cake was integrated with *P. penetrans*, while parasitization of eggs was highest when neem cake was integrated with *P. lilacinum*. In field, planting of tomato seedlings (raised in nursery beds amended with neem cake + *P. penetrans*) in pits incorporated with *P. lilacinum* (at 0.5 g/plant) gave least root galling and nematode multiplication rate and increased fruit weight and yield of tomato (Table 14.4) (Parvatha Reddy *et al.*, 1997).

Table 14.4. Effect of Neem cake, *Pasteuria penetrans* **and** *Purpureocillium lilacinum* **on root galling, nematode multiplication rate and yield of tomato**

Treatment		Root-knot index	Yield (kg) /6 m^2
Nursery (m^2)	Main field (per plant)		
Neem cake -1 kg	*P. lilacinum* – 0.5 g	3.4	9.168
Neem cake -1 kg	*P. penetrans* (28 x 10^4 spores)	3.2	9.312
P. lilacinum – 20 g	*P. penetrans* (28 x 10^4 spores)	3.0	9.504
P. penetrans (28 x 10^7 spores)	*P. lilacinum* – 0.5 g	2.9	9.624
Neem cake – 0.5 kg + *P. lilacinum* – 10 g	*P. penetrans* (28 x 10^4 spores)	2.5	9.672
Neem cake – 0.5 kg + *P. penetrans* (28 x 10^4 spores)	*P. lilacinum* – 0.5 g	2.0	9.984
Neem cake – 0.5 kg + *P. lilacinum* – 10 g + *P. penetrans* (28 x 10^4 spores)	---	2.6	9.600
Control	---	4.6	8.352
CD (P = 0.05)		0.14	0.100

14.9.6. Brinjal root-knot nematode

Deep plowing (20 cm) together with spot application of Carbofuran at 1.0 kg a.i. per ha increased growth of brinjal plants and significantly reduced the root-knot nematodes.

Nursery treatment with Metham sodium at 25 ml per sq. m. along with deep plowing (20 cm) in the main field gave significant reduction in root-knot nematode population and better brinjal yields.

14.9.7. Okra root-knot and reniform nematode

Deep plowing (20 cm) together with seed treatment (Carbofuran at 3% w/w) was found to be effective for the control of root-knot nematodes in okra.

Deep plowing (20 cm) followed by fallowing for one month, fallowing for one month after weeding, integration of Carbofuran application at 1 kg a i per ha at sowing after either of the cultural practices or deep plowing (20 cm) together with Carbofuran seed treatment resulted in the control of the reniform nematode (*Rotylenchulus reniformis*) and better yield of okra (Lakshmanan and Sivakumar, 1981).

14.9.8. Sweet potato root-knot nematode

Control of root-knot damage to sweet potatoes involves integration of at least three methods:

» The selection of seed roots which are either free of nematodes or freed of nematodes by hot water treatment.

» Planting of the seed roots into beds of sand or coarse textured soil that is either nematode - free or, if infested, is pre-plant fumigated.

» Transplanting of the clean `slips' into nematode-free soil or pre-plant fumigated soil.

14.9.9. Carrot root-knot nematode

Meloidogyne hapla is a severe pathogen of carrots. The damage threshold is very low and infection by one *M. hapla* can make a carrot unsuitable for the fresh market. A highly commercial INM system has been evolved using a 5 year rotation in conjunction with a cover crop, soil fumigation and tolerant cultivars. A soil fumigant is applied the fall before the first of two consecutive carrot crops, followed by 2 years of onion production and 1 year of the cover crop sudex. Onions are relatively poor hosts for *M. hapla* and have a high economic threshold. In fields where nematode problems are detected after the first growing season, a low rate of a soil fumigant may be applied before the second carrot crop, or the rotation design may be altered to delete the second carrot crop. Tolerance to *M. hapla* exists in commercial cultivars of carrots (Bird, 1981).

14.9.10. Tobacco root-knot nematode

The development of an integrated program for the control of *Meloidogyne incognita* in tobacco has been of immense value to the state of North Carolina (Taylor and Sasser, 1978). On problem fields where the average yearly income without control is $ 1,047 per hectare:

» Sanitary procedures, such as plowing up tobacco roots at the end of the season, add $ 711.

» Rotation with resistant crops, such as groundnut or cotton, adds $ 1,104.

» Use of resistant tobacco varieties, such as NC 95, adds $ 939.

» Use of nematicides adds another $ 1,378.

The total increase is $5,179 or almost five times as much as without nematode and disease control.

14.9.11. Strawberry lesion nematode

When a soil is heavily infested with *Pratylenchus penetrans*, intelligent combination of nematicides and crop rotation is desirable. Soil application of nematicides may reduce the nematode population to a very low level. An advisable cropping sequence could then be:

» Strawberry (which is susceptible but does not cause a large increase of the residual nematode population)

» Potato (moderately susceptible and causing moderate nematode reproduction)

» Oats or barley (not susceptible but causing nematode reproduction)

» Beet (not susceptible and suppressing the population)

After the above cropping sequence, moderately susceptible crops can be grown again successfully.

It is seen from the above discussion that the integration of the different methods offers effective control possibilities which are of great practical value in the suppression of harmful nematode population densities as well in preventing the buildup of measurable populations of nematodes.

14.10. FUTURE STRATEGIES

» Soil treatment before planting or between crops

• Soil drying

• Soil solarization in hot season

• Rotation with non-host crops

» Biological enhancement during seedling production
  • Arbuscular mycorrhizal fungi
  • Mutualistic fungal or bacterial endophytes
  • *Pasteuria penetrans*
  • Resistant cultivars

» Biological enhancement of seedlings prior to transplanting
  • Plant health-promoting rhizobacteria e.g. *Bacillus cereus* S18 and *Rhizobium etli* G12
  • Grafting resistant/tolerant rootstocks

» Stimulation of natural antagonistic potential
  • Incorporation of organic composts e.g. chicken manure
  • Planting and incorporation of green manure e.g. castor, sesame

» Management methods during crop growth
  • Systemic nematicides
  • Optimum plant fertilization and irrigation

» Control methods at harvest
  • Trap cropping using hosts and degree day monitoring
  • Antagonistic cropping with e.g. Tagetes or Crotalaria
  • Removal or composting roots to kill nematodes

» Biofumigation with plant residues or green manure crops

In conclusion, the logical use of the methods listed above can lead to effective INM of nematodes over time in the field. However, research both on basic and applied aspects is still needed to make such approaches available to the farmer under practical conditions. Government support for small scale industry may be needed in order to produce biocontrol agents on a large scale for market use by resource poor farmers. In addition, training of extension experts is required to teach them how to optimize INM combined with biological control for nematode management.

14.11. CONCLUSION

Research opportunities remain in characterizing nematode communities and in the assessment of some management tactics as well as in developing related damage functions/thresholds and the integration of compatible strategies/tactics. Components for this integration include crop rotation/cropping systems, resistant cultivars, fallow/destruction of residual infected roots, solarization, nematicides, and soil amendments/biological control. Carefully developed research-extension priorities, and the use of some type of data/decision-support system, should facilitate the development and deployment of effective INM programs. Long-term IPM goals and priorities might focus on interdisciplinary IPM (all pathogen and pest groups) and sustainable crop-production systems with critical inputs from experts in technology development/transfer, institutional resources/training, and economic/ social/ cultural components.

Promising new technologies for nematode diagnosis and identification, critical to IPM, are on the horizon. In addition, novel management weapons, including transgenic plants with nematode resistance and/or biological control genes, new safer nematicides and more effective biological control agents should be forthcoming. Improved data management and decision-support systems should facilitate the integration of new and traditional INM strategies and tactics.

Chapter - 15

Integrated Nematode Management under Protected Cultivation

15.1. INTRODUCTION

Due to climate change, population explosion, shrinkage of arable land and reduced water availability, a paradigm shift in global crop production is occurring towards protected agriculture. This input-intensive farming provides year-round production under controlled environment and by occasionally utilizing vertical spaces for soilless cultivation. Among the productivity enhancing technologies, protected cultivation has a tremendous potential to increase the yields several folds. Protected agriculture allows growing crops under extended favorable climatic conditions (enabled with sophisticated heating, cooling and lighting systems) in order to realize greater crop yield potential (than open field) by safeguarding the crop from meteorological adversities, diseases and pests (Maynard and O'Donnell, 2018).

The structures that are mainly adopted for covering the crops are (1) low tunnels: row cover with less than 1 m height, cultural practices performed from outside, mostly made up of conventional polyethylene or ethylene vinyl acetate or polyvinylchloride; (2) high tunnels: walk-in tunnels with covers alike of low tunnel, cultural practices can be performed from inside, and moderately tall crops can be grown; (3) greenhouses: high and large protection structure made up of glass or plastic, provide improved environmental conditions to the crops against cold, rain, hail, wind *etc.* (Albajes *et al.*, 1999; Maynard and O'Donnell, 2018). On a very brief account of working principle, the incident radiation from the sun passes through the transparent or translucent covering/wall material, falls on the opaque surface, i.e., leaves, floor, soil *etc.*, and the heat energy cannot escape *via* convection resulting in inside temperature rise (Vleeschouwer, 2001).

Growing of vegetable and floriculture crops under protected cultivation is receiving utmost attention and gaining popularity among farming community across the country. Government of India launched Horticultural Technology Mission Project, and under this program large number of polyhouses in the States of Kerala, Himachal Pradesh, Karnataka, Maharashtra, North Eastern hill region, Haryana, Gujarat, Tamil Nadu, Andhra Pradesh and NCR region have come up in a big way. The high-value crops such as vegetables (cucumber, capsicum, tomato), fruits (strawberry), ornamentals and flowers (carnation, roses, gerbera, chrysanthemum) that are labor-intensive and energy demanding during cooler season are mostly preferred for protected agriculture to optimize the production cost, maximize yield, secure the quality and nutritional value, and simultaneously to minimize abiotic and biotic stresses by exploiting the modern production technologies. The protected cultivation has shown promise in respect of higher crop productivity both in terms of quality and quantity of the produce.

15.2. ADVANTAGES AND LIMITATIONS

15.2.1. Advantages

- » Higher productivity and higher income
- » Quality produce
- » Off season or round the year cultivation
- » Hardening of tissue culture plants
- » Better management of insect pest
- » Less use of chemicals
- » Efficient use of resources

15.2.2. Limitations

- » High cost of initial infrastructure (capital cost).
- » Non-availability of skilled human power and their replacement locally.
- » Lack of technical knowledge of growing crops under protected structures. All the operations are very intensive and require constant effort.
- » Requires close supervision and monitoring.
- » A few pests and soil-borne pathogens are difficult to manage.
- » Repair and maintenance are major hurdles.
- » Requires assured marketing, since the investment of resources like time, effort and finances, is expected to be very high.

15.3. AREA UNDER PROTECTED CULTIVATION

15.3.1. International scenario

Presently, there are more than 90 countries in the world that have adopted indoor farming of different crops including vegetables, fruits and flowers with vegetables contributing the maximum share (Hickman, 2018). Crops in order of their importance include tomato, cucumber, capsicum, eggplant, and strawberry which correspond to an annual worldwide production of 910, 400, 370, 53 and 41.5 million kg, respectively (FAO, 2017). Countries using protected agriculture extensively include China, Spain, South Korea, Turkey, Japan, Mexico, Brazil, Italy, Morocco and Israel, which cumulatively account for 70% of world's total protected area (5.63 million ha) (Fig. 15.1). Among few interesting facts, Netherlands contributes 70% of world's flower export *via* protected agriculture. China contributes 90% of world's protected vegetable output. Israel grows quality fruits, vegetables and flowers in water deficit desert areas using indoor agriculture (Chang *et al.*, 2013; Hickman, 2018).

15.3.2. Indian scenario

Presently total space covered under protected cultivation in our country is approx. 30,000 ha. Leading states space underneath protected cultivation square measure Maharashtra (Carnation, gerbera, rose, Capsicum), Karnataka (Roses, gerbera, carnation, vegetable seed production and nursery raising of vegetables), Himachal Pradesh (Capsicum, carnation, gerbera, tuberose), North-Eastern States (Floricultural and vegetable crops), Uttarakhand (Gerbera, Capsicum), Tamil Nadu (Floricultural crops), and Punjab (Vegetable crops). The most important crop full-grown within the protected cultivation square measure tomato, capsicum, melons, rose, gerbera, carnation, and chrysanthemum.

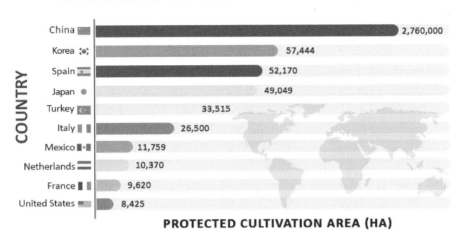

TOTAL AREA OF PROTECTED CULTIVATION WORLDWIDE

China — 2,760,000
Korea — 57,444
Spain — 52,170
Japan — 49,049
Turkey — 33,515
Italy — 26,500
Mexico — 11,759
Netherlands — 10,370
France — 9,620
United States — 8,425

COUNTRY / PROTECTED CULTIVATION AREA (HA)

Fig. 15.1. Area under protected cultivation worldwide.

15.4. NEMATODE PROBLEMS

Demand for high quality, export-oriented horticultural products and need for the availability of horticultural crop produce round the year especially in off season, compelled the growers to cultivate select crops under protected cultivation in 1980s. As a result, people started cultivation of crops under protected conditions in all the states of India. These covered structures largely maintain monoculture with extended diurnal temperature variation, high humidity and less ventilation (Dasaeger and Csinos, 2006; Engindeniz and Engindeniz, 2006), aiding in population build-up of humidity loving microbes (such as fungi and bacteria), soil-borne pathogens, insects and plant-parasitic nematodes (PPNs) as well. Unlike open field agriculture, soil-borne plant-parasitic nematodes proliferate heavily inside the protected structure due to continuous monoculture, maintenance of stable microclimate, recycling of nematode-infected growing medium and planting materials by unaware growers. The greenhouses and polyhouses mostly have drip/sprinkler irrigation facility, which ensures optimum moisture level in rhizosphere throughout the crop. In the current decade, nematodes have emerged as a major yield reducing factor in protected agriculture, especially after phasing out of methyl bromide. Nematode damage has skyrocketed under the protected structures in several parts of the world accounting a heavy toll, and the growers can hardly get rid of this menace. Nematode incidence under protected cultivation, particularly the root-knot nematodes, became severe and led to complete crop losses because of congenial conditions of higher temperature, humidity and use of high agronomic inputs like fertilizers and plant growth promoters in polyhouses.

Symptoms such as chlorosis and stunting appear after sufficient damage is inflicted. The proliferation rates of nematodes in polyhouse cultivation reached up to 10 to 30 folds more than in the open field cultivation. As yet, farmers continue to incur losses in crops under protected cultivation without appropriate solutions to problems posed by nematodes. The population build up is very rapid in the polyhouses and nematode population reaches 5 - 6 times of threshold levels within 18 - 24 months, making the polyhouse cultivation a wasteful exercise. In tomato, dynamics of root-knot nematode showed enhanced population build up from 1 to 30 juveniles J_2/cc soil within a period of 6-12 months, which is comparatively higher in contrast to the open cultivation.

Crops such as capsicum (bell-pepper), tomato, colored capsicum, okra, gherkins, muskmelon, watermelon, carnations, roses, gerbera and anthuriums are being grown under protected cultivation (in poly houses/ greenhouses/ shade nets). The ideal conditions provided by protected cultivation by way of continuous availability of the host plant round the year often result in high population build-up of soil-borne pathogens including plant parasitic nematodes. These crops grown in some states of India are seriously infested with nematodes such as *Meloidogyne incognita*, *M. javanica* (root-knot nematodes) and *Rotylenchulus reniformis* (reniform nematode). Nematode problems on all these crops under protected conditions have assumed alarming proportions leading to huge crop losses (up to 80%) in select crops. The nematode infestations exacerbate severity of fungal diseases leading to complete crop losses.

The most common and prevalent source of nematode infection under greenhouse and polyhouse condition arise from the contaminated soil or the soil mixture used as a component of growing medium. Irrigation water and manures also serve as source of inoculum.

Besides causing direct damage, nematodes are also responsible for causing disease complexes in association with soil-borne pathogens (bacteria and fungi). *M. incognita* infection makes the plants highly susceptible for the attack of *Fusarium oxsporum* f. sp. *dianthi*. *Phytophthora parasitica* + *M. incognita* interact to produce a disease complex in gerbera leading to reduction in the yield to the extent of 40 to 60 %. In capsicum, a pathogenic bacteria *Ralstonia solanacearum* gets entry into the roots infested by root-knot nematode and together produce wilting disease that reduces yield by 60-70%.

Management practices adopted by farmers include continual use of chemical nematicides, often at higher than recommended rates resulting in build-up of resistance. In addition, biomagnification and environment deterioration due to hazardous chemicals has rendered several cultivated ecosystems unstable and non-profitable. Hence, there is an urgent need to develop prophylactic as well as curative measures to check the build-up of nematode populations under polyhouse conditions.

15.5. REASONS FOR NEMATODE FLARE UP UNDER PROTECTED CULTIVATION

Nematodes basically require three essential conditions for survival and multiplication which include moisture, temperature and monocropping (Fig. 15.2).

15.5.1. Moisture

Drip irrigation in polyhouses ensures availability of optimum moisture around the root zones continuously and this factor ensures their rapid movement favoring infection, as compared to open field conditions where irrigations are given after 15-20 days and moisture levels in rhizosphere vary from saturation to almost dry.

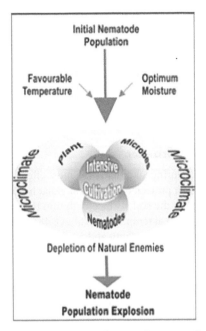

Fig. 15.2. Reasons for flare-up of nematode population under protected cultivation

15.5.2. Temperature

Nematodes multiply optimally from 25-35°C, though they can reproduce from 15-40 °C. Below 15 °C is not lethal for nematodes, but multiplication is temporarily arrested, and they can survive through cold spells. But temperatures higher than 45 °C are lethal for nematodes. Compared to open field conditions, particularly in north India, the night temperature during winter remains high inside the polyhouses, so the nematode multiplication continues, while it is arrested in open field conditions. Nematodes are able to complete their life cycles within the shortest possible time (25- 30 days) inside the polyhouses compared to open-field conditions.

15.5.3. Continuous cultivation of susceptible hosts

Intense and continuous cultivation of most susceptible hosts in the polyhouses ensures uninterrupted availability of food for nematodes. There is little choice for crop rotation with non-host crops in protected cultivation systems considering the market compulsions. Besides this, major polyhouse vegetable crops like tomato have longer duration as compared to open field which leads to higher number of nematode generations. On the other hand, crops like cucumber are grown up to three times in a year in the same greenhouse, which again leads to nematode population explosion due to continuity of host crop. While perennial crops like gerbera (3 to 5-year life-span) not only increase the nematode population considerably but also give little options for nematode management on standing crop.

Protected cultivation conditions favor nematode multiplication e.g., reduced sunlight, and depletion of natural enemies due to continuous and injudicious use of chemicals and pesticides. Therefore, all these crucial factors contribute to explosion of nematode populations in polyhouses.

15.6. NEMATODE MANAGEMENT

Nematode problems arise from contaminated soil or soil mixture used as a component of the growing medium, monocropping and infested planting materials. Nematode management here must be considered primarily as exclusion or avoidance. Once nematodes are introduced it is difficult to manage them.

Nematode management can be achieved by following physical, cultural, biological, chemical and integrated methods.

15.6.1. Physical methods - Soil solarization

Soil solarization for 4 to 6 weeks during summer is the most effective, economical and eco-friendly way of nematode management under protected cultivation. Solar heating involves a process of trapping solar energy in the soil by covering the soil surface with thin transparent polyethylene films (0.05-0.06 mm). Most nematodes are killed at temperatures above 48 °C. The temperature of solarized soil in green houses may reach 60 °C at 10 cm depth and 53 °C at 20 cm depth. Even *M. incognita* eggs are killed at 43 °C for 13 hrs.

Tomato fruit yields in solarized greenhouses were four-fold higher than those obtained in untreated control. The yield of lettuce was about two-fold higher in solarized soil as compared with untreated control. The soil solarization treatment reduced root-knot nematode populations by more than 95% in tomato and 90% in lettuce.

15.6.2. Cultural methods

- » Growing cowpea (trap crop for root-knot nematode) closely and uprooting the plants 45 days after sowing.
- » Raising marigold (antagonistic to nematodes) and incorporating *in situ* after plucking flowers.
- » Adding well decomposed farm yard manure at 20 t/ha to enrich soil.
- » Soil incorporation of neem cake at 2.5 t/ha two weeks before sowing/planting.
- » Rotation of cucumber which is most susceptible to root-knot nematodes with bell pepper which is relatively more tolerant to nematode infection. Besides, some tomato varieties like Pusa Cherry-I (an indeterminate variety), moderately resistant to nematodes may also be grown in rotation with cucumber.

15.6.3. Biological methods

- » Seed treatment with biopesticide - *Pseudomonas fluorescens* @ 10 g/kg seed.
- » Soil application of *Pseudomonas fluorescens, Purpureocillium lilacinum, Trichoderma harzianum* and *Pochonia chlamydosporia.*
- » Soil application *P. fluorescens* and *P. lilacinum* (at 2.5 kg/ha and 10 g/m², respectively) effectively reduced nematodes number of *R. reniformis* and *H. dihystera* on gerbera.

15.6.4. Chemical methods

Pre-plant soil application of dazomet/carbofuran at 1 kg a.i./ha at planting and repeating the treatment at 45 days after planting.

15.6.5. Grafting on nematode-resistant rootstocks

Grafting as a technique is gaining wide attention throughout the world, especially for greenhouse cultivation of vegetable crops, mainly the Solanaceous and Cucurbitaceous ones, from the view point of resistance against nematodes in addition to obtaining better yield and quality.

Using *Cucumis metuliferus* as a rootstock to graft RKN-susceptible melons, led to lower levels of root galling and nematode numbers at harvest (Siguenza *et al.*, 2005). Cucumbers grafted on the bur cucumber (*Sicyos angulatus*) rootstock exhibited increased RKN resistance (Zhang *et al.*, 2006). Commercial watermelon 'Fiesta' (diploid seeded) (*Citrullus lanatus* var. *lanatus*) plants grafted on rootstock from the wild watermelon germplasm line RKVL 318 (*C. lanatus* var. *citroides*) had significantly less root galling (11%) due to *M. incognita* than non-grafted 'Fiesta' watermelon (36%) plants.

Tomato varieties Kashi Aman, Kashi Vishesh and Hisar Lalit grafted on wild brinjal (*Solanum torvum*) plants were compatible between root stock and scion and also showed significant resistance against root-knot nematode by reducing soil population, reproduction and gall index (Gowda *et al.*, 2019). Tomato plants grafted onto 'Survivor' rootstock (*Solanum lycopersicum*) gave significant root galling reduction, while tomato plants grafted onto 'Multifort' rootstock (*S. lycopersicum* x *S. habrochaites*) resulted in significantly higher (P<0.05) yields than the non- and self-grafted treatments (Barrett *et al.*, 2012).

15.6.6. Integrated methods (Fig. 15.3)

Fig. 15.3. Integrated nematode management under protected conditions.

» Pre-plant treatment of beds with dazomet followed by the application of neem cake enriched with *P. lilacinum* or *P. chlamydosporia* (1 kg/m², 15 days later) significantly reduced populations of *M. incognita* and the mortality of carnation and gerbera plants, and suppressed the nematode infection for nearly 2 years.

» Soil application of *P. fluorescens/T. harzianum* @ 2.5 kg/ha mixed with 50 kg farmyard manure 10 days before sowing/planting.

» Raising the seedlings in the coco-peat or any other substrate treated with *P. fluorescens* @

1kg + *T. harzianum* @ 1kg + neem or pongamia cake @ 50 kg/ton.

» Treating the beds with neem or pongamia cake enriched with *T. harzianum* (50g) or *P. chlamydosporia* (25 g) + *P. lilacinum* @ 25 g/sq. m at an interval of 2 months.

» Application of 2 tons of farmyard manure enriched with *T. harzianum* per ha before planting, along with 100-200 kg of neem or pongamia cake.

15.6.7. IIHR Schedule for INM (Rao *et al.*, 2015)

15.6.7.1. In nursery beds

i) Soil application

» Before preparation of the beds in the poly-house, incorporate 20 tons of FYM enriched with the bio-pesticides in the soil.

» Also add carbofuran or phorate @ 50 g/sq. m. + 200 g neem/ pongamia/ mahua cake enriched by bio-pesticides per sq. m. of nursery bed.

» Further, incorporate bio-pesticide enriched FYM @ 2 kg/sq. m. or biopesticides enriched vermicompost @ 500 g/sq. m. in top 18 cm of soil in the beds.

ii) Spraying

» The organic formulation containing *Pseudomonas fluorescens* and *Trichoderma harzianum* has to be sprayed on the plants at regular intervals of 20 days at a dosage of 5 g/ lit. or 5 ml/lit.

» Alternately, take 20 kg of neem/ pongamia/ mahua cake enriched in the above mentioned manner and mix it in 200 lit. of water, leave it for a period of 2-3 days. Filter this suspension and use it for spraying by mixing 250 ml of suspension in 1 lit. of water at regular intervals of 20 days.

iii) Drenching or application through drip irrigation system

» The IIHR patented organic formulation has to be given through drip/ by drenching @ 5 g/ lit. or 5 ml/ lit. at regular intervals of 20 days.

» Alternately, take 20 kg of neem/ pongamia/ mahua cake enriched in the above mentioned manner and mix it in 200 lit. of water, leave it for a period of 2-3 days. Filter this suspension and use it for drenching at regular intervals of 20 days.

15.6.7.2. In standing crop

i) Soil application

» Apply 100 g of neem / pongamia / mahua cake or 250 g of vermicompost enriched with *Pseudomonas fluorescens* + *Trichoderma harzianum* + *Purpureocillium lilacinum* on 1 sq. m. beds or around the rhizosphere of the plants.

ii) Spraying

Same as in nursery beds.

iii) Drenching or application through drip irrigation system

Same as in nursery beds.

By following all these methods farmers can get significant increase in the yield of the crops and the benefit: cost ratio will be above 3.

15.7. NEMATODE MANAGEMENT IN VEGETABLE CROPS

15.7.1. Tomato root-knot nematodes, *Meloidogyne* spp.

Above ground symptoms are stunting, yellowing, wilting, reduced yield, and premature death of plants. Below ground symptoms are swollen or knotted roots (root galls) (Fig. 15.4) or a stubby root system. Root galls vary in size and shape depending on the nematode population levels, and species of root-knot nematode present in the soil.

Fig. 15.4. Heavy galling of tomato roots infected with *Meloidogyne incognita*

15.7.1.1. Soil solarization: Tomato fruit yields in solarized greenhouses were four-fold higher than those obtained in untreated control. The soil solarization treatment reduced root-knot nematode populations by more than 95% in tomato.

15.7.1.2. Chemical methods: While nursery bed preparation, addition of Dazomet at 1 kg a.i. /ha to the nursery beds gave effective control of root-knot nematodes.

15.7.1.3. Grafting: The interspecific tomato hybrid rootstock 'Multifort' (*S. lycopersicum* x *S. habrochaites*) and tomato hybrid rootstock 'Survivor' (*S. lycopersicum*) significantly reduced root galling compared with the non-grafted and self-grafted scions by ~ 80.8%. The root galling reduction by 'Survivor' (97.1%) was significantly greater than that by 'Multifort' (57.6%) compared with non-grafted scions (Table 15.1). 'Brandywine' grafted to 'Multifort' resulted in significantly higher (P<0.05) yields than the non- and self-grafted 'Brandy-wine' treatments (Fig. 15.5) (Barrett *et al.*, 2012).

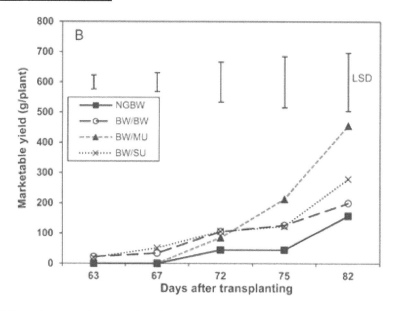

Fig. 15.5. Effect of rootstocks on marketable yield on tomato cultivar Brandywine (NGBW-Non-grafted Brandywine, BW/BW- Brandywine grafted on Brandywine, BW/MU- Brandywine grafted on Multifort, BW/SU- Brandywine grafted on Survivor).

Table 15.1. Effect of grafting treatments on root-knot nematode galling ratings on tomato cultivars.

Treatment	Root galling (1-10 scale) in cultivar	
	Brandywine	Flamme
Non-grafted	9.30 a	8.06
Self-grafted	7.30 b	6.12
Grafted on 'Multifort' rootstock	3.88 c	3.48
Grafted on 'Survivor' rootstock	0.54 d	0.00

The Asian Vegetable Research and Development Centre (AVRDC) recommends eggplant accessions EG195 and EG203 [resistant to flooding, bacterial wilt, root-knot nematode (*Meloidogyne incognita*), *Fusarium* wilt (*Fusarium oxysporum* f. sp. *lycopersici*), and southern blight (*Sclerotium rolfsii*)] as rootstocks for tomato (Black *et al.*, 2003).

Tomato plants grafted on "Big Beef," "Celebrity," and "Jetsetter" rootstocks (resistant to *Verticillium* wilt, *Fusarium* wilt, nematodes, and tobacco mosaic virus) had the least nematode populations in the greenhouse. In field experiments, nematode population levels were lower in "Big Power" that had been grafted on Celebrity, Jetsetter, and Big Beef rootstocks, compared to self-grafted or non-grafted "Big Power" (Fig. 15.6) (Owusu *et al.*, 2016).

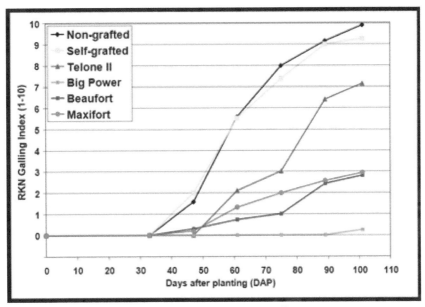

Fig. 15.6. Effect of rootstocks on root galling on tomato

The root system of tomato plants grafted on wild brinjal (*S. sisymbriifolium*) was free from gall formations; however, non-grafted plants had an average of 7.5 gall index (GI). Fruit yields significantly (P > 0.05) increased by 34.64% in the grafted plants compared with the non-grafted plants (Table 15.2) (Baidya *et al.*, 2017).

Table 15.2. Effect of tomato grafted on 'Wild Brinjal' rootstock *Solanum sisymbriifolium* **on root galling and fruit yield.**

Treatment	Yield (Kg/plant)	RKI
Non-grafted	8.980	8
Grafted on *Solanum sisymbriifolium*	12.064	0
Yield increase over non- grafted	3.104	---
% yield increase	34.64	---

Tomato plants grafted on 'Aloha' and 'Multifort' rootstocks reduced root galling (Fig. 15.7).

15.7.1.4. Integrated methods: Addition of enriched FYM/compost/vermicompost (with *Trichoderma harzianum/T. viride* + *Purpureocillium lilacinum* + *Pseudomonas fluorescens* each at 2 kg/ton of organic amendment) at 5 t/ha helps in the management of root-knot nematodes in nursery.

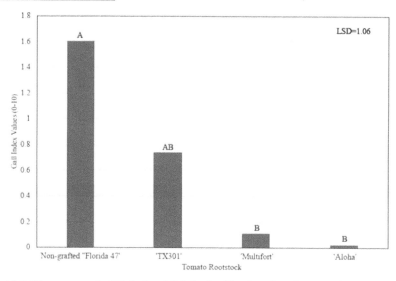

Fig. 15.7. The rootstocks 'Aloha' and 'Multifort' both reduced root galling in tomato.

After 15 to 30 days of planting, addition of enriched neem cake (with *Trichoderma harzianum/T. viride* + *Purpureocillium lilacinum* + *Pseudomonas fluorescens* each at 2 kg/ton) at 50 g/m² gave effective management of nematodes under polyhouse conditions. Further, drip application of neem cake suspension enriched with bioagents (20 kg neem cake + *Trichoderma harzianum/T. viride* + *Purpureocillium lilacinum* + *Pseudomonas fluorescens* each at 2 kg) helps in keeping the polyhouse beds free from root-knot nematodes.

15.7.2. Bell pepper root-knot nematodes, *Meloidogyne* spp.

Root galls caused by root-knot nematodes on sweet pepper are frequently small (Fig. 15.8).

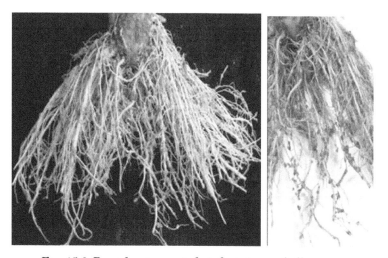

Fig. 15.8. Root-knot nematode infestation on bell pepper

15.7.2.1. Physical methods: Burning of paddy husk or saw dust on infested nursery proved to be reasonably effective. Use of plastic sheets for covering soil in greenhouse further enhances nematode reduction.

15.7.2.2. Cultural methods: Several cultural methods like selection of nematode-free nursery sites, destruction of infested roots after crop harvest, crop rotation with antagonistic crop (marigold) or intercropping with marigold, either alone or in combination proved to be reasonably effective and economical to check multiplication of root-knot nematodes on bell pepper.

15.7.2.3. Chemical methods: Bare-root dip treatment of bell pepper seedlings in 0.1% Carbosulfan/ Monocrotophos for 6 hr eliminates root-knot infection.

15.7.2.4. Biological methods

i) Antagonistic bacteria: Application of a nematicidal *B. thuringiensis* strain to bell pepper in Puerto Rico reduced galling in roots due to *M. incognita* and increased yield significantly (Zuckermann *et al.*, 1993).

ii) Antagonistic fungi: Application of 'Royal 350' (*Arthrobotrys irregularis* cultured on oat seed medium) at 140 g/m^2 a month before transplantation of bell pepper resulted in good protection against root-knot nematodes. Among the egg parasites, efficacy of *Purpureocillium lilacinum* (commercially formulated as 'Biocon' in Philippines) has been found to be comparatively higher in suppressing the population of *Meloidogyne* spp. on bell pepper.

Treatment of the greenhouse bed with the formulation of *Pochonia chlamydosporia* at 50 g/m^2 was significantly effective in reducing the galling index, number of nematodes in roots and soil, increasing the percent parasitization of eggs by bioagent and also yields. Seed treatment with *P. fluorescens* alone and the nursery bed treatment with *P. chlamydosporia* alone were effective (Naik, 2004).

15.7.2.5. Host resistance: Bell pepper cvs. Mississipi-68, Santanka, Anaheim Chile and Italian Pickling are reported to be resistant to root-knot nematodes. Black Indica, Naharia and Pant C1 were reported to be resistant to *M. incognita*, while All Big as moderately resistant. California Wonder and Naharia were resistant to *M. javanica*, while Early California Wonder as moderately resistant. Naharia was reported to be resistant to *M. arenaria*.

The southern root-knot nematode (*M. incognita*) resistant 'Carolina Wonder' and 'Charleston Belle' are recommended for use by greenhouse growers.

Sweet pepper (*Capsicum annuum*) cultivars possessing the *N* gene, which controls resistance to RKNs (*M. incognita, M. arenaria*, and *M. javanica*), have been effective as rootstocks to control RKNs in pepper (Table 15.3) (Oka *et al.*, 2004; Thies and Fery, 1998, 2000).

Table 15.3. Rootstock for root-knot resistance in sweet pepper

Scion/rootstock	Gall index (1-10 scale)	Fruit yield (kg/6m^2) 20 plants
Celia non-grafted	7.8a	20.2b
Celia grafted on AR-96023 (*C. annuum*)	0.7b	43.8a

15.7.2.6. Integrated methods: Integration of *P. fluorescens* and *P. chlamydosporia* in the greenhouse bed has proved significantly effective in reducing the root-galling index (*M. incognita*), number of nematodes in the roots and soil and increasing the yield of the bell pepper crop (Naik, 2004).

In capsicum, the seedling stand was good where the combination of neem-based *P. fluorescens* and *T. harzianum* was used (4.4), followed by seed treatment with *P. fluorescens* + soil application of *T. harzianum* (Naik, 2004).

15.7.3. Cucumber root-knot nematodes, *Meloidogyne incognita, M. javanica*

The infected plants exhibit stunted plant growth, leaf yellowing and reduced yield. Large number of galls on root system and multiple galling is common. In case of heavy attacks, galls can become very large, the root system being reduced to a swollen stump without hairs (Fig. 15.9). It restricts the uptake of nutrients from the root system to the foliage, resulting in a leaf yellowing and stunted plant growth. Ultimately plants wilt and die.

Fig. 15.9. Symptoms of root-knot nematode infestation on cucumber plants (*Left*) and roots (*Right*).

15.7.3.1. Cultural methods: Application of Neem cake @ 30 g/plant 10 days before sowing is effective in cucumber (Anon, 2012).

In polyhouse grown cucumber, Neem cake at 320 g/m² proved most effective and resulted in highest yield followed by Mustard cake (Hema, 2014). The benefit: cost ratio with Neem cake was found to be 3.07 in cucumber.

15.7.3.2. Biological methods: The lowest rate of root galling was observed in cucumber treated plants with bionematicide BioAct WG (*Purpureocillium lilacinum* strain 251) along with and without compost as compared to bacterial strain (*Bacillus thuringiensis* strain Bt1 + *B. amyloliquefaciens* strain 2/7A). Plant growth parameters were relatively higher in plants grown with compost and treated with bioproducts (Fig. 15.10) (Yankova *et al.*, 2016).

The lowest root galling (1.7 as compared to 3.1 in control) and highest yield (54.190 t/ha as compared to 46.040 t/ha in control) were obtained in the combined application of BioAct WG (*Purpureocillium lilacinum* strain 251) + *Trichoderma viride* strain T6 (Yankova *et al.*, 2014).

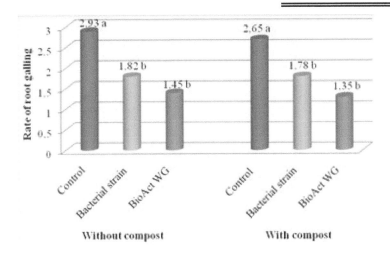

Fig. 15.10. Rate of root galling by *Meloidogyne incognita* **in cucumber treated with bacterial strain** (*B. thuringiensis* **strain Bt1** + *B. amyloliquefaciens* **strain 2/7A) and BioAct WG** (*Purpureocillium lilacinum* **strain 251).**

The soil drenched with *Amphora* or *Spirulina* algal extracts at 2g/L had significant increments in vegetative growth, yield, and fruit quality of cucumber. *Amphora* (sprayed with soil drenched) treatment gave 2.5 and 2.69 folds the control in marketable yield in 2016 and 2017 seasons, respectively. The combination of sprayed and soil drenched with *Amphora* was more effective in nematode's control and reduced nematode reproduction factor (RF), which reached 0.42 and 0.45 in both seasons, respectively. It had insignificant differences compared with the nematicide (10% Ebufos, at the rate of 5 g/m^2) . Therefore, using algae for the biological control of root-knot nematodes is recommended, especially in sustainable agriculture for maintaining the soil and improve fertility (El-Eslamboly *et al.*, 2019).

15.7.3.3. Chemical methods: Spot application of Carbofuran 3G @ 10 g/plant at sowing is effective in cucumber (Anon, 2012).

Metham sodium (40 ml/m^2) and Dazomet (40 g/m^2) treatments proved promising in polyhouse grown cucumber. The benefit: cost ratios with Metham sodium and Dazomet were found to be 3.84 and 3.76, respectively (Anon, 2015; Chandel *et al.*, 2014).

15.7.3.4. Host resistance: Cucumbers grafted on the bur cucumber (*Sicyos angulatus*) rootstock (good compatibility with cucumbers) exhibited increased root-knot nematode resistance (Zhang *et al.*, 2006).

Cucumber cv. 'Adrian' plants grafted onto interspecific hybrid rootstock 'Strong Tosa' (*Cucurbita maxima* × *C. moschata*) had higher plant growth and total number of fruits (19.9/plant as compared to 10.1 in non-grafted plant) and yield (5.38 kg/plant as compared to 2.56 in non-grafted plant) in first season, and the same result was found in the second season also (6.96 kg/plant and more than 28.9 fruits/ plant as compared to 4.5 kg/plant and 20.4 fruits/ plant in non-grafted plant) (Ban *et al.*, 2014).

Wild rootstock, *Cucumis pustulatus* with simultaneous resistance to root-knot nematode (*M. incognita*) and Fusarium wilt (*Fusarium oxysporum* f. sp. *melonis*) was found suitable for cucumber, melon, and watermelon (Fig. 15.11) (Liu *et al.* 2015). *Cucumis pustulatus* gave effective control of root-knot nematodes on cucumber and increased fruit yield (Table 15.4) (Liu *et al.* 2015).

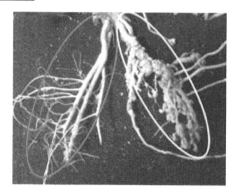

Fig. 15.11. Effect of resistant rootstock on root-knot nematode incidence in cucumber (Left red loop -Resistant, Right yellow loop -Susceptible

Table 15.4. Cucumber root-knot control using wild cucumber rootstock *Cucumis pustulatus*

Treatment	Gall index	Yield in kg/plant
Cucumber grafted on *C. pustulatus*	1	2.032
Cucumber non-grafted	5	1.234

15.7.3.5. Integrated methods: Pre-plant treatment with *Purpureocillium lilacinum* (20 kg/ha) in combination with organic amendments effectively reduced root-knot infection in gherkin.

Combined inoculation of AMF and *Pseudomonas fluorescens* had positive effect on root-knot nematode control on cucumber.

Enhancement of cucumber fruit yield, reduction in root galling, and second stage larvae of *M. javanica* in soil was achieved by soil amendments (using alfalfa, cauliflower, olive-cake residues, poultry manure, and tomato) or soil solarisation. Organic amendments reduced densities of *Fusarium* spp., generally increased *Aspergillus* spp., while *Trichoderma* spp. was not affected. Combinations of solarization and addition of organic amendments substantially augmented each other, particularly with poultry manure, alfalfa hay, and to a lesser extent with cauliflower and tomato residues.

Bello (1998) reported that initial very high levels of *M. incognita* were reduced to nil in susceptible cucumber crop by integrating the cultivation of short-cycle vegetables acting as trap crops and biofumigation with mushroom residue and sheep manure in commercial greenhouse trials in Spain.

In a cucumber crop in glasshouse trial, the use of solarization and *Pasteuria penetrans* had an additive detrimental effect on *M. javanica* populations (Tzortzakakis and Gowen, 1994).

Oxamyl increased the efficacy of *P. penetrans* in trials against *M. javanica* infection of cucumber crop and the effects on nematode control were additive (Tzortzakakis and Gowen, 1994).

In field study conducted by Tempta (2017), application of Neem cake @ 50 g/m^2 + Carbofuran 3G @ 10 g/m^2 was found most effective in reducing nematode population and root gall index as well as in increasing the plant growth and yield in cucumber.

15.7.4. Cruciferous vegetable crops root-knot nematodes, *Meloidogyne* spp.

Symptoms of root-knot nematode infection include galls on roots which can be up to 3.3 cm in diameter but are usually smaller; reduction in plant vigor; yellowing of plants which wilt in hot weather.

15.7.4.1. Physical methods: Solarizing soil can reduce nematode populations in the soil and levels of inoculum of many other pathogens.

15.7.4.2, Host resistance: Planting resistant varieties if nematodes are known to be present in the soil.

15.7.5. Lettuce root-knot nematodes, *Meloidogyne* spp.

Root-knot nematodes (*Meloidogyne incognita, M. hapla, M. arenaria, M. javanica*) are polyphagous nematodes causing characteristic knots (galls), swellings and other malformations on the roots of lettuce (Fig. 15.12), which results in poor growth, occasional wilting leading to poor yields. The smaller discrete galls with adventitious root proliferation formed by *M. hapla* are distinctive. Although *M. hapla* prefers coarse-textured soils, it also occurs in organic soils, and it is a major pest of lettuce which is commonly grown in this kind of soil.

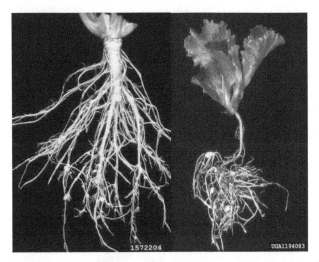

Fig. 15.12. Root-knot nematodes, *Meloidogyne incognita* **and** *M. hapla* **on lettuce roots**

15.7.5.1. Physical methods: The yield of lettuce was about two-fold higher in solarized soil as compared with untreated control. The soil solarization treatment reduced root-knot nematode populations by more than 90% in lettuce.

Steam sterilization of the soil is an effective curative measure.

15.7.5.2. Cultural methods: Use of nematode-free planting material in non-infested soil is normally sufficient to keep lettuce free from these nematodes. Weeds should be thoroughly controlled throughout the areas where lettuces may be grown.

15.7.5.3. Chemical methods: It is not good practice to treat soil systematically with nematicides. Such treatments should be limited to what is strictly necessary, and may be subjected to official limitations. Pre-planting soil treatment, allowing for the above considerations (that nematicide use should be restricted rather than recommended): Dazomet, Oxamyl.

15.8. NEMATODE MANAGEMENT IN FLOWER CROPS

15.8.1. Carnation and gerbera root-knot nematode, *Meloidogyne incognita*

Root knot nematode is a serious pest of both carnation and gerbera. Infected carnation plants usually appear stunted and tend to wilt on warmer days. When such plants are dug, the root galls are generally conspicuous and easily identified (Fig. 15.13). Symptoms on gerbera include stunting of plants, yellowing of leaves (Fig. 15.14), wilting of plants, and heavy galling on roots.

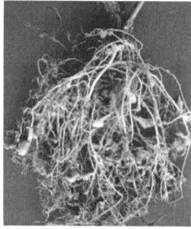

Fig. 15.13. Carnation plants and roots infected by root-knot nematodes

Fig. 15.14. Symptoms of root-knot nematode infection on gerbera

15.8.1.1. Chemical methods: Chlorpyriphos and Carbofuran (each applied twice in 6 months) significantly reduced nematode populations in roots and soil. However, there was a build-up of nematode populations in beds treated with these two chemicals after 1 year.

15.8.1.2. Integrated methods: Pre-plant treatment of beds with Dazomet followed by the application of neem cake (1 kg/m², 15 days later) along with *Purpureocillium lilacinum* or *Pochonia chlamydosporia* significantly reduced populations of *M. incognita* and the mortality of plants, and suppressed the nematode infection for nearly 2 years. The antagonistic fungi established themselves better in the beds treated with Dazomet than in untreated beds. On a long-term basis, soil management with pre-plant treatment of Dazomet, followed by the application of oil cakes plus antagonistic fungi, was more effective against *M. incognita* than post-plant treatment with Carbofuran, Carbosulfan and

Chlorpyriphos on carnation and gerbera grown in polyhouses.

15.8.2. Carnation and gerbera spiral nematode, *Helicotylenchus dihystera*

Helicotylenchus dihystera was found to be the most predominant nematode as revealed by the community analysis of carnation and gerbera rhizosphere samples. An initial inoculum level of 1000 nematodes of *H. dihystera* per plant was found to be pathogenic. *H. dihystera* was responsible for the reduction in growth parameters as well as for delay in flowering. Staining of the affected root system showed the presence of this nematode pathogen in the cortical tissues.

Pseudomonas fluorescens and *P. lilacinum* (at 2.5 g and 10 g/m², respectively) effectively reduced the population of *Rotylenchulus reniformis* and *H. dihystera*. These treatments were followed by farm yard manure, Neem powder and Carbofuran.

15.8.3. Chrysanthemum foliar nematode, *Aphelenchoides ritzemabosi*

The symptoms on chrysanthemum include characteristic brown spots limited to the veins, and a progressive yellowing of the whole leaf. These symptoms are due to combined action of the nematode and other organisms (Cayrol and Combettes, 1972). Leaf symptoms on infested Chrysanthemum include reddish-yellow lesions on the lower leaves of young plants; in older plants these leaves are markedly chlorotic and a large area of the leaf surface becomes necrotic (Fig. 15.15).

Fig. 15.15. Foliar nematode symptoms on chrysanthemum leaves

The foliage is scanty and the flowers are few and deformed. Leaves in the upper part of plants have shown slightly higher resistance than those in the lower part (Cid del Prado and Sosa-Moss, 1978). Direct effects are mechanical damage caused by the stylet, and damage due to growth and division hormones (Cayrol and Combettes, 1972).

Suggested control measures include cleaning and burning infested leaves, submerging infected cuttings in hot water, spraying of foliage with Chlorpyrifos (Gill, 1981).

Treatment of chrysanthemum nursery soil with an organophosphorus nematicide was very effective in control of this nematode (Fukazawa and Kobayashi, 1971).

15.8.4. Gladiolus root-knot nematodes, *Meloidogyne* spp.

Severe galling on roots results in yellowing of leaves (Fig. 15.16) which subsequently leads to stunted growth. The nematode invades roots, daughter corms and cormels which develop after flowering.

Fig. 15.16. Root-knot nematode symptoms on gladiolus

15.8.4.1. Physical methods: Hot water treatment of corms at 58 °C for 30 min. eliminates root-knot nematode infection.

15.8.4.2. Chemical methods: Dipping of gladiolus corms in Thionazin or Fensulfothion solution (0.5 g a.i. per litre) gave reduced root-knot nematode infestation (Overman, 1970). Application of Vorlex (35 gal/acre) by broadcast method also resulted in better flower yield.

15.8.4.3. Integrated methods: Soil application of neem cake enriched with *Trichoderma harzianum* at 1 t/ha is effective.

Application of 5 tons of FYM enriched with *Pseudomonas fluorescens* (with 1×10^9 cfu/g) per ha significantly reduced *M. incognita* by 61-76%, *Rotylenchulus reniformis* by 65-70% in the roots of gladiolus and increased the flower yield by 19-22%.

15.8.5. Lilies root-knot, lesion and foliar nematodes

Nematodes penetrate root tissues, killing cells as they move. They move inside the root, feeding, laying eggs, and destroying additional cells. The roots become soft and flabby, eventually succumbing to infection that moves into the basal plate, turning it into mush. River water often carries nematodes, which can then enter croplands through irrigation. These pests also host bacteria; some species even carry virus diseases. Nematode infestation causes stunting of growth and can severely reduce commercial production. Crops parasitized by nematodes are seldom uniformly affected.

Foliar nematodes (*Aphelenchoides* spp.) move through the stem in a surface film of moisture to invade the leaves and flowers. They produce discolored streaks in lily foliage (Fig. 15.17) and responsible for decline of bulbs.

Fig. 15.17. Foliar nematode damage on Easter lily leaves

The stem and bulb nematode (*Ditylenchus dipsaci*) causes individual bulb scales to rot, and eventually can kill the bulb.

15.8.5.1. Physical methods: Soils are solarized mainly to kill fungi, nematodes, perennial weeds, and weed seeds. The treatment is applied during the summer when soil temperatures are high and moisture levels relatively low; however, the soil must be moist for optimal results.

Treat bulbs with hot water at 44 °C for one hour. Steam sterilization of greenhouse soils is very important.

15.8.5.2. Cultural methods: Keep the foliage as dry as possible to control foliar nematodes by preventing movement of the organisms. Avoid planting lilies continuously in the same site; this prevents harmful nematode populations from building up in the soil.

15.8.5.3. Chemical methods: Systemic insecticides are very effective. Apply a granular nematicide such as Fenamiphos (e.g., Nemacur) when planting bulbs. Fumigate soil with Metham Sodium (e.g., Vapam). This is a highly successful technique in commercial plantings.

15.8.5.4. Biological methods: Try adding carnivorous nematodes to the soil.

15.8.6. Anthuriums burrowing nematode, *Radopholus similis*

Anthurium root rots caused by *R. similis* are brown or dark brown to black. The rots develop relatively slowly. Initially, although older roots are infected and rot, new roots are produced and the plant often continues to grow well. But compared to healthy plants, the number of functional roots in diseased plants is greatly reduced. With time, fewer new roots are produced, and gradually the entire root mass is destroyed (Fig. 15.18). The steady and progressive destruction of the root system usually causes plant decline in the second to fourth years. The leaves turn yellow and may have other symptoms of nutrient deficiency. The plants become smaller and lose vigor, producing fewer and smaller flowers.

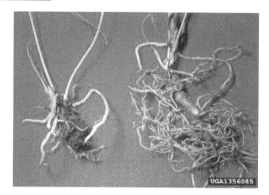

Fig. 15.18. Burrowing nematode infected (*left*) and healthy (*right*) roots of anthurium

In moist environments, the nematodes are able to migrate short distances outside the plant, above the soil line. They maybe splashed onto or migrate up stems to aerial roots, where they penetrate the soft, fleshy root tips. Invasion of the root tips frequently cause cessation of root growth. Infected tips are brown (Fig. 15.19) and may be missing. When the environment is dry, nematodes either migrate back into the soil or desiccate, and many rotted tips are devoid of nematodes. The tips produce a callus, stop growing, and have a rounded form that suggests the description "stubby root." When the environment becomes favorable for root growth, a new lateral root is produced at or near the tip of the original root. This new tip can be infected and become stubby also. Repetition of this cycle results in a short aerial root with several stubby tips (Fig. 15.19). Stubby roots and branched aerial roots are not normal in anthurium. While this condition is not always caused by nematodes, these abnormal roots should be a clue to the possible presence of burrowing nematodes.

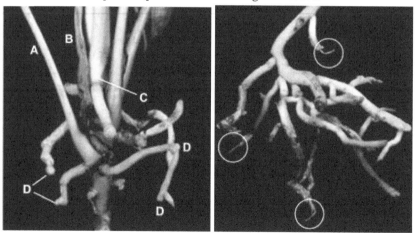

Fig. 15.19. *Left* - **Typical stubby roots on an anthurium plant: A, petiole; B, sheath; C, node; D, stubby roots,** *Right* - **Early stages of root rot on anthurium. Note the missing root tips with exposed vascular strands.**

Drenching potted anthuriums with hot water maintained at 49-50 °C for 15 to 20 min eliminated 95–100% of burrowing nematodes infecting the roots and stems of plants to ≤1g⁻¹ in infested roots of potted anthuriums.

15.9. NEMATODE MANAGEMENT IN FRUIT CROP

15.9.1. Strawberry root-knot nematode, *Meloidogyne hapla*

M. hapla is a serious pathogen of strawberry. Above-ground symptoms include wilting during hot days, stunting, chlorosis, and reduction of yields. Feeding by root-knot nematodes results in the production of characteristic symptoms on roots called galls (15.20). *M. hapla* feeding results in very small galls (swellings) and these are often inconspicuous on some hosts. Root-knot nematodes have very high reproductive potentials as single females can produce up to 500 eggs.

Fig. 15.20. *Left* - **Stunted strawberry plants and discoloration of leaves.** *Right* - **Galls on strawberry roots caused by the northern root-knot nematode.**

15.9.1.1. Chemical methods: In the field trials, the most effective treatments for reducing *M. hapla* soil densities were dazomet and metam-sodium, whose efficacies ranged from 78% to 87% (Talavera *et al.*, 2019).

15.9.1.2. Integrated methods: Soil biosolarization (Dried chicken manure was uniformly distributed onto the surface of the soil and then incorporated into the top 20-cm layer. Plots were then drip irrigated until the soil reached field capacity. Solarization was carried out under a low-density transparent polyethylene film (0.03 mm thick) during July and August for about 6 weeks) with chicken manure was more effective in reducing *M. hapla* soil densities when applied at a rate of 25 (86%) than at 20 tons ha⁻¹ (67%) (p < 0.05). Soil biosolarization reached efficacies of 73% reduction for *M. hapla* and 67% for *P. penetrans*, which seems to provide sufficient nematode control in most situations (Talavera *et al.*, 2019).

15.10. CONCLUSION

Nematodes assume new proposition in protected cultivation of crops due to moderate climate and intensive cultivation. With this technology, the problems of nematodes had cropped up. Therefore, a fresh look into the dynamics of soil-borne pathogens like root-knot nematodes has to be intensified. As the accrued loss due to nematodes is tangible, proper attention is must for their management.

Various chemicals for the management of nematodes have been used under protected cultivation but fumigants, dazomet and metham sodium used as pre-plant under tarp, gave effective nematode management. There is further need to develop viable options for nematode management, including use of crop rotation, grafting on resistant rootstocks, biocontrol agents and antagonistic crops like crotalaria (a green manure crop) and marigold. Since the choice of crops and nematode incidence in protected agriculture differ according to locations, futuristic management studies should run in a holistic approach considering the interactions between nematode communities, host plants and biotic/abiotic environmental factors.

Chapter - 16

Climate Change Adaptation
and Mitigation Strategies
for Phytonematodes

16.1. INTRODUCTION

Plant pathogenic nematodes are one of the important biotic constraints in crop production. They will be affected by climate changes in a complex way. Beneficial or negative effects are expected, depending on species, regional situations, crops, climate and geography. Significant changes induced by variations in climate extremes (mainly, but not uniquely, related to temperature and moisture regimes) are expected within the next decades. Among other environments and natural systems, they will also affect crops as well as plant parasites and biological control agents, with different outcomes, depending on geographic regions and agricultural systems (Olfert and Weiss, 2006). Nematodes by virtue of their trophic diversity act directly as pests and indirectly as vectors of other plant pathogens, and play a significant role in regulating plant growth in agro-ecosystems.

Scientific research on climate change and its impact on herbivorous nematodes are very limited. However, based upon their environmental requirements some assumptions are possible. Severe droughts resulting in a reduction of soil water will most likely negatively affect soil nematodes. Higher average temperatures will probably have little effect, since thermal conductivity of soils is low.

Apparently, a prediction of how climate change will affect herbivorous soil nematodes and thus yields cannot be made. There is some evidence that population dynamics may change, but so far no trend is clear. Basically, and most likely true for all ecological research, the impacts of climate change are specific to crop/plant, region and interacting species.

Majority of plant-pathogenic nematodes spend part of their lives in soil, and therefore, soil is the source of primary inoculum. Life cycle of a nematode can be completed within 2–4 weeks under optimum environmental conditions. Temperature is the most important factor, and development is slower with cooler soil temperatures. Warmer soil temperatures are expected to accelerate nematode development, perhaps resulting in additional generations per season. The drier temperatures are expected to increase symptoms of water stress in plants infected with nematodes such as the soybean cyst nematode. Overwintering of nematodes is not expected to be significantly affected by changes in climate, although for some, such as the soybean cyst nematode, egg viability may be reduced in mild winters.

Climate change due to increased emission of greenhouse gases is posing a serious challenge to sustainability of crop production by interfering with biotic and abiotic components and their interactions with each other. Global warming resulting in elevated carbon dioxide (CO_2) and temperature in the atmosphere may influence plant pathogenic nematodes directly by interfering with their developmental rate and survival strategies and indirectly by altering host crop physiology. Available information on effect of global warming on plant pathogenic nematodes though limited, indicate that nematodes show a neutral or positive response to CO_2 enrichment effects with some species showing the potential to build-up rapidly and interfere with plant's response to global warming. Studies have also demonstrated that the geographical distribution range of plant pathogenic nematodes may expand with global warming spreading nematode problems to newer areas. Besides plant parasites, other trophic groups (microbial feeders, predators and insect parasites) of soil nematodes also shown to influence the plant productivity indirectly by regulating the key ecosystem processes including decomposition, nutrient mineralization, biological pest suppression and energy transfer in food webs. These findings underline the importance of understanding the impact of climate change on soil nematodes and its implications to crop production while developing mitigation and adaptation strategies to address impact of climate change on agriculture.

16.2. EFFECTS OF CLIMATE CHANGE

16.2.1. Elevated temperature

Increasing temperatures are expected to enhance plant growth rates and yields, providing a greater food source for nematode pests but also increasing the whole ecosystem complexity. Temperatures will also affect plant's phenology (earlier germination of seeds, plant flowering or ripening). Both factors will be responsible for a higher carrying capacity of plants, an earlier emergence of diseases/vectors and crop attacks, longer life-cycles and reduced pest/disease generation times.

The fact that plants become efficient hosts at high temperatures is probably due to three factors:

» High temperatures are optimum conditions for nematode activity.
» Stress caused by high temperature makes the plants more vulnerable to nematode attack.
» Chemicals responsible for cell necrosis may not be produced or may be neutralized or counter-acted at high temperature.

As Goudriaan and Zadoks (1995) pointed out for plant diseases that also applies to nematodes that will thrive where host plants grow best. However, the range of genetic variability and adaptability is still unknown thus far, for several nematode species. The identification of these adaptive boundaries is important, since they represent the basis for future expansions and colonization of new areas, following isothermal shifts. The reduction or elimination of natural barriers confining a species in a particular geographic or climatic area should always represent a source of concern: for example, changes are expected in the distribution of insects, due to movements towards higher altitudes or latitudes, as a response to higher temperatures (Walther *et al.*, 2002). Reaching a higher altitude may allow a species to overcome a natural barrier, thus colonizing new geographic regions from which it was previously excluded. Although this factor is a major concern for nematode phoresis or vectoring capacity by insects, i.e. *Monochamus* sp., provide a link to the potential spread of nematode pests (i.e. pine wilt nematodes, *Bursaphelenchus* spp.).

Temperature is the most important factor influencing the biology of nematodes. Nematode developmental rate is directly influenced by the temperature with slower development at cooler and faster growth rate at warmer soil temperatures. In plants under environmental stress (at high temperatures), nematode reproduction is higher. Hatching and embryogenesis is faster and migration and penetration by the nematode is favored by high temperatures. The nematode life cycle is also completed faster at high temperature; therefore, more generations are produced. Moreover, at high temperatures, fewer males develop. Therefore, increase in atmospheric temperature due to global warming is expected to result in more number of generations per season and expansion of their geographical distribution range. Other potential effects of elevated temperature on parasitic nematodes include altered sex ratio, host defence responses and interference in their survival strategies like dauer juveniles or egg diapauses in extreme environments.

Drier temperatures are expected to increase symptoms of water stress in plants infected with nematodes such as the soybean cyst nematode. Overwintering of nematodes is not expected to be significantly affected by changes in climate, although for some, such as the soybean cyst nematode, egg viability may be reduced in mild winters.

Plantain (*Musa* spp. AAB) is both an important staple and cash crop throughout the West/ Central African humid forest zone. Major yield constraints are root nematodes, particularly *Radopholus similis*. Data from lab and field experiments demonstrated that higher nematode population densities and greater plantain root damage occurs to banana at the projected temperature increases. *R. similis*, currently absent from cooler, higher altitude areas is likely to expand its range.

Climate change may also influence the plant nematode interactions by interfering with host defence mechanisms. Rebetez and Dobbertin (2004) reported that strong climate warming that has occurred in recent years favored pine wood nematode (*Bursaphelenchus mucronatus*) and bark beetles and increased drought stress reduced tree resistance against these pests. This resulted in rapid tree mortality in pine forests in Switzerland.

The alfalfa stem nematode appeared in Yolo County due to climate change. The stem nematode parasitizes alfalfa crops and can cause severe crop losses. The nematode disperses through several vectors, including waterways and irrigation runoff, contaminated farm equipment, and other anthropogenic means similar to other plant diseases. As a new and increasing threat in Yolo County, stem nematode populations are thought to have appeared due to warmer minimum temperatures moving closer to the reproductive threshold for the nematode.

Warming will generally cause a pole-ward shift of the risk of damage from potato nematodes which would increase in all regions and may become a serious problem with additional generations per year.

16.2.2. Enrichment of CO_2 levels

Herbivorous nematodes showed neutral or positive response to CO_2 enrichment effects with some species showing the potential to build up rapidly and interfere with plant's response to global warming. A recent publication presents results of a long-term agricultural experiment conducted in winter wheat and sugar beets in Germany. Winter wheat and sugar beet were grown in rotation under 550 ppm atmospheric CO_2 compared to ambient (380 ppm) atmospheric CO_2. The number of herbivore, bacterivore and fungivore nematodes was significantly higher under wheat and sugar beets grown under elevated CO_2, while the number of carnivore was not changed. The total numbers of herbivore, bacterivore and fungivore nematodes were higher under elevated CO_2 wheat than under elevated CO_2 sugar beet, most likely due to the very different root system of both plant species (Sticht *et al.*, 2009). However, impacts on yields were not determined.

CO_2 concentration plays a crucial role in various aspects - biology of plant and insect parasitic nematodes including host recognition, recovery from dauer stage or diapause, *etc.* Elevated CO_2 levels may also influence these nematodes indirectly by altering host physiology (defence mechanisms such as production of secondary metabolites and nutrient status such as C: N ratio, *etc.*). It may also influence microbial feeding nematodes due to changes in quality and availability of food under enriched CO_2 conditions in soil. The impacts of climate change can be positive, negative or neutral, since these changes can decrease, increase or have no impact on nematode abundance, depending on each region or period.

Available information on effect of global warming on soil nematodes though limited, indicate that abundance of soil nematodes in general is either increased or unaffected by elevated CO_2 levels while individual species and trophic groups differ considerably in their response to climate change. Herbivorous nematodes showed neutral or positive response to CO_2 enrichment effects with some species showing the potential to build up rapidly and interfere with plant's response to global warming.

Effects of elevated CO_2 on nematode densities as mediated by the host plant are "plant species-specific" and include negative effects (Niklaus *et al.*, 2003), positive effects (Yeates *et al.*, 2003), and no significant effects (Ayres *et al.*, 2008). Most of these studies proposed that changes in root biomass and C/N ratio were the main factors responsible for the effects of elevated CO_2 on nematode abundance (Niklaus *et al.*, 2003; Yeates *et al.*, 2003; Ayres *et al.*, 2008). However, the mechanisms underlying the effect of elevated CO_2 on the interaction between plant-parasitic nematodes and their host plants are poorly understood.

Systemic acquired resistance (SAR) is considered to be the major induced plant defence that confers long-lasting protection against nematodes (Durrant and Dong, 2004). SAR depends on the salicylic acid (SA) pathway and is associated with accumulation of pathogenesis related proteins, which are considered to contribute to resistance. Researchers have recently suggested, however, that the jasmonic acid (JA) pathway is also an indispensable component of plant resistance to nematodes (Howe and Jander, 2007; Bhattarai *et al.*, 2008). The JA pathway is associated with expression of proteins (including proteinase inhibitors, phenylalanine ammonialyase, and lipoxygenase), up-regulation of secondary metabolites, and induction of plant volatile organic compounds (VOC). Cooper *et al.* (2005) reported that the artificial induction of JA-pathway defences reduced reproduction of root-knot nematodes on tomato plants (Cooper *et al.*, 2005). A tomato genotype (35S::Prosystemin) in which induced defence was dominated by the JA pathway (hereafter referred to as a "JA defence-dominated genotype") has stronger resistance to nematodes than a JA defence-recessive genotype (spr2, a jasmonate-deficient mutant) and that the specific responses of these isogenic tomato genotypes to elevated CO_2 requires further investigation (Sun *et al.*, 2010).

Based on several works referring to plant induced defence under elevated CO_2 (Zavala *et al.*, 2008; 2009; Sun *et al.*, 2010), it has been hypothesized that elevated CO_2 would reduce the resistance of a JA defence dominated genotype against *M. incognita* by altering the JA pathway but enhance the SA-pathway defence of a JA-defence-recessive genotype infected by *M. incognita*. In this study, it is determined whether elevated CO_2 affects the regulation of genes and the production of secondary metabolites and the emission of VOC associated with the JA pathway of isogenic tomato genotypes. It has also been determined whether elevated CO_2 affects the regulation of genes associated with the SA pathway. Finally, it is determined whether the changes in these pathways and genes are associated with the performance of *M. incognita* under elevated CO_2.

Similar to other organisms which feed on plants, increased CO_2 levels are believed to have an impact on herbivorous nematodes (Ayres *et al.*, 2008) and several studies have been conducted, where the above ground plant community was exposed to elevated CO_2. Almost all of these studies were done in different grasslands and forests, and thus results have been variable and contradictory. Research results regarding nematodes, from experiments conducted on agricultural crops in arable soils, are very limited. Basically, all kinds of results were determined: increase, decrease and no change of nematodes populations (Sticht *et al.*, 2009).

The observations that elevated CO_2 levels often induces increased root production, it can be presumed that herbivorous nematode communities will be relatively more affected by increases in atmospheric CO_2 concentration. Positive effects of CO_2 enrichment on the abundance of herbivorous nematodes have been reported in some studies (Yeates *et al.*, 2003). The abundance of *Tylenchus* and *Longidorus* increased after 5 years of CO_2 enrichment, but there was no effect on the abundance of *Paratylenchus, Trichodorus* and members of Hoplolaimidae in pasture plots (Yeates *et al.*, 2003). Yeates *et al.* (1997) reported increase in abundance of *Meloidogyne* in response to CO_2 enrichment in grassland turfs while 7 other herbivorous nematode taxa were not affected. The abundance of *Pratylenchus* was positively associated with CO_2 concentration in gley, but not in organic soil around a natural CO_2 vent in New Zealand (Yeates *et al.*, 1999).

Although root biomass increased under elevated CO_2, the damage due to root-feeding nematodes was more under elevated CO_2 compared to the ambient levels in a grass species (Wilsey, 2001). Similarly, neutral responses of herbivorous nematodes to CO_2 enrichment were observed despite increase in root production by 3-32% in different locations (Ayres *et al.*, 2008). This may be due to decrease in root quality (low nitrogen content) or increase in nematode antagonists.

The interaction of elevated CO_2 with Nitrogen fertilization or residue addition significantly affected the soil nematode community indices. The residue addition stimulated structure index and inhibited plant-parasites response to the elevated CO_2 in a wheat field (Li *et al.*, 2009).

Experiments with rice have showed no adverse effects of elevated CO_2 levels up to 700 ppm on the abundance of soil nematodes and penetration of rice root-knot nematode, *M. graminicola* (Fig. 16.1) (Somasekhar and Prasad, 2010).

Positive effects of increased CO_2 levels alone are expected on plant's productivity (Chakraborty *et al.*, 2000). They will also increase the numbers of herbivores or efficiency in water, N use and conversion of radiation (Olesen and Bindi, 2002). However, nematodes reaction to rising CO_2 levels is complex, depending on trophic groups. No effect was reported on nematodes from prairie soil (Freckman *et al.*, 1991), but numbers decreased in cotton rhizosphere (Runion *et al.*, 1994), or increased in forest (Hoeksema *et al.*, 2000), grassland (Hungate *et al.*, 2000) and pasture soils (Yeates *et al.*, 1997; 2003). Higher CO_2 levels lowered the numbers of bacterial feeders, increasing fungal feeders and predators in forest (Neher *et al.*, 2004) or prairie soils (Yeates *et al.*, 2003).

Available information on effect of global warming on soil nematodes though limited, indicate that abundance of soil nematodes in general is either increased or unaffected by elevated CO_2 levels while individual species and trophic groups differ considerably in their response to climate change (Table 16.1). Nematode's reaction to rising CO_2 levels is complex, depending on trophic groups: population increased in forest (Hoeksema *et al.*, 2000), grassland (Hungate *et al.*, 2000) and pasture soils (Yeates *et al.*, 1997; 2003).

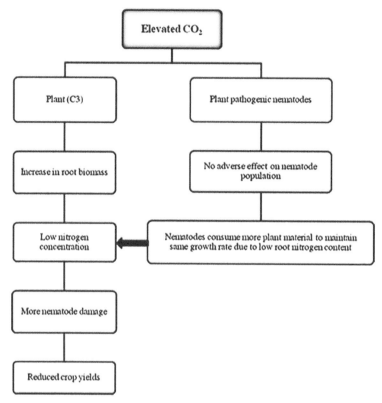

Fig. 16.1. Consequences of elevated CO_2 levels on plant-nematode interaction and crop productivity (Somasekhar and Prasad, 2012)

Table 16.1. Response of herbivorous nematodes to CO_2 enrichment

Cropping system	Experimental arena[*]	Nematode response[**]	Location	Reference
Grassland	CER	+/N	New Zealand	Yeates *et al.* (1997)
Grassland	Vent	N	New Zealand	Yeates *et al.* (1999)
Grassland	FACE	+/N	New Zealand	Yeates *et al.* (2003)
Grassland	OTC	+/N	California, USA	Hungate *et al.* (2000)
Grassland	OTC	N	California, USA	Ayres *et al.* (2008)
Grassland	OTC	N	California, USA	Ayres *et al.* (2008)
Grassland	FACE	+/N	Germany	Sonnemann and Wolters (2005)
Sugar beet & wheat rotation	FACE	+	Germany	Sticht *et al.* (2009)
Grassland	SACC	N	Switzerland	Niklaus *et al.* (2003)

Grassland	CER	N	Montpellier, France	Ayres *et al.* (2008)
Rice- wheat rotation	FACE	+/N	China	Li *et al.* (2007, 2009)
Rice	OTC	N	India	Somasekhar and Prasad (2010)

*OTC = Open top chambers, SACC = Screen aided CO_2 control, FACE = Free air CO_2 enrichment, CER = Controlled environment room, Vent = Natural CO_2 vent.

** + = Positive or increase in abundance, N = Neutral or not affected

16.2.3. Fluctuating precipitation patterns

Changes in the water regime will influence the water availability for plants and will affect yields, increasing densities and levels of nematode parasitism, due to positive effects on total plant biomass.

Increased flood rates are expected to raise the frequencies of several water-borne (aquatic fungi) or moisture dependent (powdery mildew) diseases, indirectly affecting plants productivity and nematode densities. Availability of water affects soil nematodes, whereas changes in plants communities are known to indirectly affect their abundance and community structure (Kardol *et al.*, 2010). Changes in water regimes may also influence the duration of a nematode parasitic event, i.e. the infection phase or the progeny produced in a season, with final outcomes on total numbers. Similarly, higher/lower incidence of droughts will affect either survival of field inocula, frequency of parasitism by endemic aquatic or nematophagous fungi, or the time spent in soil by juveniles seeking for an available root penetration sites.

Factors affecting a function's response (i.e. crop productivity, pest or disease prevalence) to a given climatic change are key elements and must be correctly identified, in order to yield reliable information for modelling or preventive measures. They include also physical properties of soil, i.e. texture or water retention capacity which may affect, at different extents, soil nematodes responses to changes of the hydrologic cycle (Wessolek and Asseng, 2006).

The effect of increasing rainfall regimes may vary: in arid climates higher water availability may yield positive consequences on agricultural practices, increasing crops productivity or cultivated surfaces. Increasing rainfalls may also induce changes in the selection of varieties or cultivated species that may increase the incidence of nematodes, switching either their species composition (replacement) or even increasing natural antagonists and prevalence levels, due to higher moisture, affecting i.e. the spreading of antagonists or predators (i.e. *Pasteuria* spp. parasites). On the opposite, in an already moist climate, an increasing rainfall regime of the same magnitude may yield floods, with soil losses due to erosion, forcing the adoption of other plants/cultivars or sowing periods. Higher amounts of water in soil will either increase the incidence of nematodes dispersion and the probability of a local extinction. Catastrophic rainfalls may also affect some management practices, like soil labour, solarisation, sowing and fertilization, or vanish the effects of chemicals.

16.2.4. Combined changes in temperature and moisture regimes

Combined changes in temperature and moisture regimes will also affect biological control agents (BCA). As for nematodes, increased numbers due to higher yields and plants carrying capacities will affect BCA densities, an effect also expected to produce earlier emergence times and outbreaks due to a corresponding earlier host nematode emergence. Other effects will be the longer life cycle and/or the reduced generation time, an increased spatial spread to newly colonized areas following nematodes spreading, or the increased spread of water related parasites (i.e. *Catenaria*).

Other indirect mechanisms will favour nematodes insurgence or spreading of invasive species (Fuhrer, 2003). Among them, the increased (reduced) leaf moisture in wet (dry) conditions (affecting i.e. foliar nematode species), the changes in the survival of BCA or propagules of endophytes (spores, bacterial cells, mycorrhizae) in soil or other host tissues, the reduced (increased) plants resistance due to physiological adaptations to temperature changes (i.e. prolonged vegetation, lignification, longer lasting resistance-breaking temperatures, insurgence of virulent nematode populations), changes in the nutritive value of host plant tissues, higher densities of alternated or secondary hosts (weeds).

16.2.5. Severe droughts

Soil nematodes are dependent on the continuity of soil water films for movement. Their activities are largely controlled by soil biological and physical conditions (Yeates and Bongers, 1999). Increased water stress due to climate change diminishes plant vigor and alters C: N ratios, lowering plant resistance to nematodes. Severe droughts resulting in a reduction of soil water will negatively affect soil nematodes.

Infection with plant-parasitic nematodes can exacerbate or counteract the effects of abiotic stress on plants, as their parasitism in roots severely disrupts plant water relations (Smit and Vamerali, 1998). Fig.16.2 shows an example of the positive interactive effect of drought stress and nematode infection on rice plants. In this case, the addition of nematodes ameliorated the severity of the drought stress. Several studies have been carried out to examine the effect of combined nematode and drought stress on plant growth and development. In the Ivory Coast, the nematode *Heterodera sacchari* increased drought-related losses in upland rice by contributing to reduced leaf water potential, stomatal conductance, and leaf dry weight (Audebert *et al.*, 2000). A similar study investigated the effect of drought and the cyst nematode *Globodera pallida* on water use efficiency in potato (Haverkort *et al.*, 1991). Both factors were found to affect growth negatively, although the simultaneous effect of both stresses was non-additive, perhaps because the infected plants used less water, thus reducing drought stress. *Globodera pallida* has also been shown to cause retardation in potato root development, which in turn had the effect of reducing drought tolerance (Smit and Vamerali, 1998).

Fig. 16.2. Growth of rice under combined biotic and abiotic stress conditions. Plants are shown after exposure to drought stress, infection with the plant-parasitic nematode *Meloidogyne graminicola*, **or the two stresses in combination.**

Increased drought stress reduced pine tree resistance against pine wood nematode (*Bursaphelenchus mucronatus*) and bark beetles. This resulted in rapid tree mortality in pine forests in Switzerland (Robetez and Dobbertin, 2004).

16.2.6. Air pollutants

Air pollutants affect phytoparasitic nematodes through changes in the physiology of the host plant. Synergistic interactions between ozone (O_3) or sulphur dioxide (SO_2) and *Meloidogyne incognita* were observed on tomato, as higher levels of foliar damage were found on nematode infested plants. Galls on roots were higher in nematodes infested plants exposed to 100 ppb ozone at 5 hours intervals every third day. Ozone, however, reduced the reproductive performance of *M. incognita*, with lower numbers of eggs and egg masses at 50 and 100 ppb (Khan and Khan, 1997).

The root galling on eggplants caused by *M. incognita* was severe at SO_2 and NO_2 polluted sites but egg laying and egg mass production was significantly reduced (Khan and Khan, 2007).

The okra plants grown at SO_2 and NO_2 polluted sites exhibited suppressions in plant growth, yield and photosynthetic pigments. The joint effects of the nematode (*M. incognita* race 1) and coal-smoke were synergistic. Egg laying was reduced by 13-27% (Khan and Khan, 1994).

The effects of the increase in UV-B radiation were highly significant for the occurrence of disease epidemics.

16.3. IMPACTS OF CLIMATE CHANGE

The climate change can influence host plant growth and susceptibility; pathogen reproduction, dispersal, survival and activity; as well as host-pathogen interaction. Once climate and nematodes are closely related, climate change will probably alter the geographical and temporal distribution of phytosanitary problems. New nematodes may arise in certain regions, and other nematodes may be economically important, especially if the host plant migrates into new areas (Coakley *et al.*, 1999). The host range changes can be extended due to crop migration. A new geographical distribution of potato cyst nematode (*Globodera rostochiensis*) is predicted, with northward expansion of potato in Finland and a higher number of generations per year. The geographical distribution of the virus-vector nematodes (*Xiphinema* and *Longidorus*) in Great Britain and Continental Europe were associated with increase in temperature. Since the nematodes geographical distribution is directly related to temperature, there could be more problems with these microorganisms in northern Europe, caused by the increase in existing populations and to the dissemination of these species from the southern region.

Nematode management strategies depend on climate conditions. Climate change will cause alterations in the nematode geographical and temporal distributions and consequently the control methods will have to be adapted to this new reality. Changes in temperature and precipitation can alter the nematicide residue dynamics, and the degradation of products can be modified. Alterations in plant morphology or physiology, resulting from growth in a CO_2-enriched atmosphere or from different temperature and precipitation conditions, can affect the penetration, translocation and mode of action of systemic nematicides.

Global warming and climate changes will result in:

- » Extension of geographical range of nematode pathogens.
- » In cooler latitudes, global warming brings new nematode species.
- » Increased risk of invasion by migrant nematodes.

» Reduced effectiveness of crop protection technologies.

» Increase in pesticide use.

» Increased probability of nematodes developing faster resistance to nematicides.

» Warmer winter temperatures would reduce winter kill, favoring the increase of nematode populations.

» Rising temperatures extend the crop growing season.

» Overall temperature increases may influence crop nematode interactions by speeding up nematode growth rates which increases reproductive generations per crop cycle.

16.3.1. Expansion of geographical distribution

Boag *et al.* (1991) used data from soil samples collected during the European plant parasitic nematode survey to assess the possible impacts of climate warming on the geographical range of virus-vector nematodes. Initial analyses of nematode presence-absence data suggested a close association between mean July soil temperature and nematode distribution. Based on this result, the authors predicted that climate change could result in increased nematode and virus problems in northern Europe; a 1°C warming would allow the species in study to migrate northward by 160 to 200 km (Neilson and Boag, 1996). Although nematodes migrate very slowly, humans are credited with efficiently disseminating them. Hence, nematode spread into new regions could put a wide range of crops at risk; additionally, introduction of new crops into a region could also expose them to infestation by nematode species already present. Changes in precipitation could influence nematode distribution on a large scale; although previous findings had suggested that soil moisture would not affect nematode distribution in most agricultural soils in northern Europe (Neilson and Boag, 1996).

Studies have also demonstrated that the geographical distribution range of plant pathogenic nematodes may expand with global warming spreading nematode problems to newer areas. The soybean cyst nematode (*Heterodera glycines*) is the cause of great economic losses to soybean producers in the US. The pest has been expanding since the early 1950s, but the increase has been more dramatic since the early 1970s. Before 1970, the soybean cyst nematode was mainly distributed in the Mississippi River Delta area, northern Arkansas, southern Missouri, southern Illinois, and western Kentucky. It is now distributed throughout the main soybean production area and has become the number one soybean pest in the US (Rosenzweig *et al.*, 2000). In Iowa alone, it caused an estimated yield loss of 201 million bushels (worth about $1.2 billion) during the 1998 growing season (USDA, 1999). In the northern production region, the nematode has up to three generations per year, depending on planting and weather conditions during the growing season. A longer growing season, associated with a warmer climate, would result in an increased risk of losses similar to the ones reported during the year 1998. This pest has been monitored and mapped since the 1950s.

Analysis of the distribution of soybean cyst nematode between 1971 and 1998 in the United States showed rapid dispersal and a very high establishment rate in the central and northern US producing regions (Chakraborty and Datta, 2003). This phenomenon was probably accentuated by the adoption of new production zones linked to climate fluctuations (Somasekhar and Prasad, 2012). In addition, the modeling of soybean phenology within the agricultural potential area of Quebec using climatic parameters of the recent past (1971-2000) and the near future (2041-2070) has shown that more northern latitudes offer new areas for growing soybeans (Gendron St-Marseille, 2013). Similarly, modeling the life cycle of *H. glycines* based on climatic parameters has demonstrated that it can already be established within the agricultural regions where soybean is currently grown

(Koenning, 2004; Niblack, 2005) and that it has the capacity to follow the move of its host to the north (Gendron St-Marseille, 2013).

Warming will generally cause a pole-ward shift of the risk of damage from potato nematodes which may become a serious problem. Using simulated climate change, Carter *et al.* (1996) predicted that warming will expand distribution of the potato cyst nematode, *Globodera rostochiensis* by 2050 in Finland. The predicted effects of climate change on diseases of selected major agricultural and forestry species in Ontario showed that the cyst nematode, *Heterodera glycines*; root lesion nematodes, *Pratylenchus* spp.; and pine wood nematode, *Bursaphelenchus xylophilus* severity increases due to increase in rate of disease development and potential duration of epidemic due to climate change (Boland *et al.*, 2004).

In India, rice root-knot nematode, *Meloidogyne graminicola*; once considered to be serious pest only in upland rice, has made its importance felt in almost all rice growing areas and in all types of rice culture including hill ecosystems in recent years (Prasad and Somasekhar 2009; Pankaj *et al.*, 2010).

Neilson and Boag (1996) assessed the possible effect of climate change on the distribution of some common virus-transmitting *Xiphinema* and *Longidorus* species within Great Britain. They observed that theoretically an increase in 1 °C in mean temperature would result in the northward extension of these nematode species by about 160-200 km. Colonization of new areas by virus-vector nematodes has serious implications for agriculture.

Using mathematical models, Carter *et al.* (1996) simulated climate change in Finland and concluded that warming will expand the cropping area for cereals by 2050 (100 to 150 linear km per 1 °C increase in mean annual temperature). Furthermore, higher yields are expected with higher CO_2 concentration. In this scenario, potato cropping will also be benefited with an estimated 20 to 30% increase in yield. However, a new distribution of the potato cyst nematode (*Globodera rostochiensis*) is also predicted, with northward expansion in the country.

In coffee crop, the potential impacts of climate change on the spatial distribution of the coffee nematode (races of *Meloidogyne incognita*) in Brazil were determined using a geographic information system (Ghini *et al.*, 2008). Maps obtained considering the A2 scenario indicated that there could be an increase in infestation of the nematode when compared to the climatological normal from 1961-1990. For the B2 scenario, the infestation level will also be higher as compared to the current scenario, but it will be lower than for the A2 scenario for both.

16.3.2. Breakdown of host resistance

Genetic resistance to *Meloidogyne* spp. is sensitive to soil temperatures above 28 °C. Tomato, bean, and sweet potato lose resistance at elevated soil temperatures (Jatala and Russell, 1972). High soil temperatures appear to be the main reason that root-knot nematode resistance is not effective in Florida, USA (Walter, 1967), and in many tropical countries. Results by Araujo *et al.* (1983) indicate that race 4 of *M. incognita* reproduces better on resistant tomato genotypes than race 1.

The resistance to root-knot nematode in tomato (cv. 'Sanibel') has often failed as a result of the heat instability or apparent temperature sensitivity of the resistant Mi gene (Fig. 16.3). For example, previous research has demonstrated threshold soil temperatures and incremental reductions in nematode resistance with each degree above 78 °F, such that at 91 °F tomato plants are fully susceptible. This would suggest that in Florida use of these varieties may have to be restricted to spring plantings when cooler soil temperatures prevail.

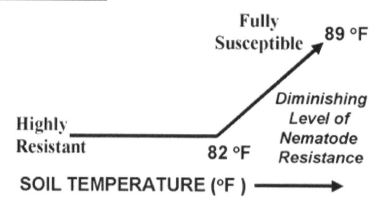

Fig. 16.3. Complete loss of root-knot nematode resistance conferred by the Mi gene in tomato due to increasing soil temperature.

Susceptibility of bell pepper (Charleston Belle, Carolina Wonder, Keystone Resistant Giant, and Yolo Wonder B) to *M. incognita* increased as temperature increased from 24 to 32 °C. Reproduction factor of *M. incognita* and root galling increased ($P < 0.05$) for all cultivars as temperature increased (Table 16.2). Overall, reproduction of *M. incognita* and severity of root galling on the resistant isolines Charleston Belle and Carolina Wonder were less ($P < 0.05$) than on susceptible Keystone Resistant Giant and Yolo Wonder B, and the two cultivars within each group did not differ. However, temperature x cultivar interaction was found ($P < 0.05$) for reproduction index and root galling (Thies and Fery, 1998). Further research is necessary to characterize the usefulness of these varieties under the high soil temperature conditions of Florida. Like tomato, use of these varieties may have to be restricted to spring plantings when cooler soil temperatures prevail.

Table 16.2. Effect of bell pepper cultivars differing in resistance to *Meloidogyne incognita* grown at different temperatures on reproductive index and root galling (eight weeks after inoculation)

Temperature/cultivar	Reproductive index[*]			Gall index[**]		
	24°C	**28°C**	**32°C**	**24°C**	**28°C**	**32°C**
Charleston Belle	0.3 a	4.9 b	24.7 c	1.53 a	2.46 b	3.61 c
Keystone Resist. Giant	6.1 c	32.2 de	29.9 f	4.79 d	7.62 e	8.74 f
Carolina Wonder	0.1 a	4.0 b	22.9 c	1.57 ab	2.37 ab	3.55 c
Yolo Wonder B	4.7 c	29.4 de	125.8 ef	4.37 cd	7.82 ef	8.61 ef

Mean separation within columns by Duncan's multiple range test at $P \leq 0.05$. Means were compared across all temperature and cultivar combinations.

[*] Reproduction index (final population/initial population) of *M. incognita*

[**] Gall index: 1=no galls, 2=1-3%, 3=4-16%, 4=17-25%, 5=26-35%, 6=36%-50%; 7=51-65%, 8=66-80%, and 9=greater than 80% of root system galled.

In grapevine, the number of galls and egg sacs of *M. javanica* increased with time and temperature on both rootstocks, although not significantly in all cases (Table 16.3). This increasing infestation with time can be ascribed to normal population increase with increasing degree days (Loubser, 1988). Furthermore, pathogenicity as measured by the degree of galling, also appeared to increase with increasing temperature. This was more evident on the moderately resistant rootstock 143 B Mgt (*Vitis vinifera* x *V. riparia*) [compared to the susceptible rootstock Jacquez (*Vitis aestivalis* x *V. cinerea* x *V. vinifera*)] where the number of eggs increased significantly (and apparently galling also) between treatments B and C, irrespective of the number of degree days which remained the same. The reason for this was seen as a breakdown in resistance at the higher temperature. Chitambar and Raski (1984) also found that the grapevine rootstock cultivars Harmony and Couderc 1613 lost their resistance at 36°C.

Table 16.3. Effect of temperature on root growth and infestation of grapevine rootstocks by *Meloidogyne javanica*

Treatment	Jacquez			143 B Mgt		
	Galling[1]	Egg sacs[2]	Eggs[3]	Galling[1]	Egg sacs[2]	Eggs[3]
A. 23 °C/44 days ($572DD_{10}$)	16.2 a	5.2 a	460 a	1.2 a	0.0 a	0 a
B. 23 °C/178 days ($1014DD_{10}$)	62.6 ab	22.2 a	560 a	2.0 a	2.0 a	75 a
C. 33 °C/44 days ($1012DD_{10}$)	41.2 ab	14.6 a	464 a	66.6 ab	16.6 a	396 b
D. 33 °C178 days ($1794DD_{10}$)	613.2 b	613.2 b	855 b	230.6 b	220.6 b	825 b

Treatments which differ significantly (P ≤ 0.05) are marked vertically with different letters.

[1.] Galling is expressed by the number of galls visible per 5 g of new roots under 20x magnification.

[2.] Egg sacs are expressed by the number visible per 5 g new roots under 20 x magnification.

[3.] Eggs are the average number calculated per egg sac.

DD_{10}: Physiological time expressed as degree-days above a pre-determined threshold of 10 °C.

Various stages in the soybean cyst nematode (SCN) disease cycle are affected differentially by temperature and moisture. The highest winter survival of SCN eggs occurs in the colder areas of the continent. Thus, spring inoculum levels may be highest in the northern range of soybean culture. Optimal soil temperatures for egg hatch, root penetration, and juvenile and adult development are 24 °C, 28 °C and 28–32 °C, respectively. Temperatures below 15 °C and above 35 °C, little nematode development occurs. Thus, temperature can affect the number of SCN generations per growing season. In theory, with fewer generations, new races will develop less quickly. The more moderate winter temperatures will reduce egg survival, while the higher temperatures in the growing season will increase egg hatch, the rate of nematode development, and perhaps, the number of generations per season. Soil water is important for SCN movement and development, but water is unlikely to be a limiting factor early in the season. More importantly, the drier growing conditions of summer will increase the yield loss due to SCN because of reduced root surface.

Down-regulation of genes for defence signalling (Iox7, Iox8 and acc-s) reduced the production of cysteine proteinase inhibitors (CystPI), which are specific deterrents of nematodes resulting in increased activity by the nematodes (Fig. 16.4).

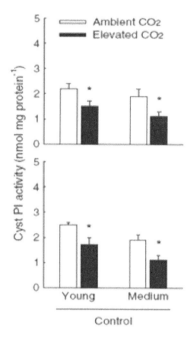

Fig. 16.4. Effect of down-regulation of defence signalling genes on reduced production of cysteine proteinase inhibitors (deterrents of nematodes)

While the exact cause of the recent alfalfa stem nematode outbreak in Yolo County is unclear, an increase in winter temperatures is likely to be an important contributing factor. Stem nematodes do not actively reproduce below 41 °F. In Yolo County, average minimum winter temperatures have increased 3 °F since 1983 and are currently approaching the lower reproductive threshold (Fig. 16.5). Higher temperatures allow the nematode to complete a larger number of breeding cycles during the winter and thus impact the severity of the infestation. If climate change causes winter temperatures to rise further, outbreaks of alfalfa stem nematode may become more frequent in the region. Additionally, the use of organophosphates and carbamates in alfalfa crops has decreased 50% since 2005. These pesticides are known to suppress stem nematode populations, but are being replaced by pyrethroids, which do not affect the stem nematodes. Consequently, the decreased use of these pesticides may have also played a role in the recent outbreak.

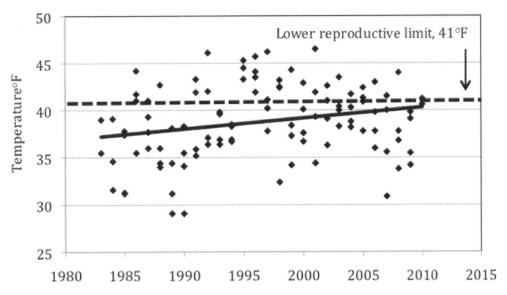

Fig. 16.5. Average minimum monthly ambient temperatures in Davis, California, from November to February 1983-2010. The lower reproductive limit of stem nematode females is 41°F.

16.3.3. Additional generations per season

The expected increase in temperature would positively favour the development of the soybean cyst nematode, *Heterodera glycines*, which would result in the addition of one or two additional generations to reach up to two generations per growing season in the northernmost part of Quebec and up to six generations in the extreme south (Gendron St-Marseille, 2013). This scenario would further weaken the soybean, as higher numbers of nematode infections increase the susceptibility of roots to other pathogens and, on the other hand, restrict the delivery of nutrients and the water to the aerial parts (Gendron St-Marseille, 2013).

Ghini *et al.* (2008) compared climatological norms from 1961-1990 with future scenarios (A2 and B2) of the decades of 2020's, 2050's and 2080's from five General Circulation Models (IPCC, 2001) to predict the changes in spatial distribution of infestation levels of the coffee root-knot nematode (races of *Meloidogyne incognita*) based on number of generations per month in Brazil. They predicted that the nematode infestation will increase in future due to greater number of generations per month. The number of generations of nematodes will increase in both scenarios, but, it will be lower in B2 than A2 scenario.

Warming will generally cause the risk of damage from potato cyst nematode (*Globodera rostochiensis*) that may become a serious problem with additional generations per year. Using simulated climate change, Carter *et al.* (1996) predicted that warming will increase number of potato cyst nematode generations per year by 2050 in Finland.

16.3.4. Impact on overwintering

Overwintering of nematodes is not expected to be significantly affected by changes in climate, although for some, such as soybean cyst nematodes, egg viability may be reduced in mild winters.

16.3.5. Reduced effectiveness of nematicides

Reduced effectiveness of nematicides is expected in view of increased use of nematicides.

16.3.6. Reduced effectiveness of bioagents

Nematode, bacterial and fungal based bio-pesticides are highly vulnerable to environmental stress. Increase in temperature, UV radiation and a decrease in relative humidity may reduce the efficacy of these bio-agents.

16.4. PREDICTION MODELING

Predictive models may help in evaluating a number of different outcomes and situations, i.e. providing a basis to identify the best among different options, to select in case of invasive species (eradication, suppression, no action) (Fraser *et al.*, 2006). In general, an insight on the possible outcomes of climate changes may be facilitated by the application of simulation models. Their sensitivity should be checked, however, with real scenarios, to verify the level of uncertainty and affordability, as well as their effectiveness as informative tools (Kickert *et al.*, 1999).

One key issue in modeling is the level of resolution achieved and the forecasting precision. The asymmetric increase of temperatures (with different shifts for minima and maxima), as well as the non-linearity in growth or development responses, require sub-daily resolutions when modeling climate changes affect fungal plant pathogens, due to a bias introduced when only mean temperatures are used, without considering the amplitude of the fluctuations (Scherm and Van Bruggen, 1994). Nematodes, however, due to a longer time required for completion of the different life-stages (eggs, juveniles, adults), may be modeled with different scenarios of daily mean temperature changes, aiming at providing a first rough insight about possible field situations and outcomes.

Although modeling necessarily represents a simplification of the more complex, non-linear relationships occurring in soil among different organisms, it provides an insight on dynamics otherwise difficult to trace, due to the amount of experimental data and time required to investigate these relationships. At small scales, generic dynamic crop modeling may result informative about the effects of weather, variety, pests, soil, and management practices on crop growth and yield, as well as on soil N and organic carbon dynamics in aerobic as well as anaerobic conditions. Data generated include pest induced yield losses, allowing a comparative analysis with different greenhouse gas emission situations (Aggarwal *et al.*, 2006). Weather variables introduced in modeling result is fundamental to explore the different outcomes of climate changes and hence estimate the spatial and seasonal dynamics of pests, in the mid-term (Yonow *et al.*, 2004).

A further tool is given by probability distribution maps (PDM). Their use allows the identification of potential risks related to the distribution of nematode pests, vectors or plant diseases. PDMs are calculated on the basis of the combined use of local records of pest occurrence, climate data and subsequent statistical analysis. They show the areas susceptible for colonization or of endemism, for a given organism (Morales and Jones, 2004). This tool has several practical advantages, including the possibility of early identification of invasion areas by a pest or the possibility to anticipate the insurgence of epidemics for secondary pests, already present in areas with sub-optimal conditions for life-cycle. PDMs require the monitoring of different climatic variables at the regional scale and a given resolution level, as well as the implementation of an early monitoring and detection support system.

16.5. ADAPTATION AND MITIGATION STRATEGIES

Development and application of new diagnostic protocols for accurate identification of nematode species is beneficial and necessary for national and international regulatory and quarantine agencies relative to free trade and economics. Understanding genetic variability and adaptation in nematodes is important for effective nematode management strategies that involve rotation, host plant resistance, cultural manipulations and biological control, because they are influenced directly by genetic variability existing in target nematode field populations. Results from current work indicate that genetic variability and adaptation potential in nematode populations are responsible for the aberrant results of many experiments assessing resistance, crop rotations, host ranges, cover/trap cropping, and biological control.

Implementation of new nematode management tactics that reduce pesticide usage, and thereby benefit human health and the environment; promotion of sustainable farm management practices through new nematode management strategies; increased knowledge base in plant-nematode biology and genetics for use in identifying novel targets for nematode control; economic benefits to producers and consumers through reduction in nematode management costs and food production; and more efficient and effective response and mitigation capabilities for invasive and trade product contamination issues are the needs of the hour.

Bridge (1996) outlines and discusses four different strategies for specific nematodes and some other pests in sustainable and subsistence agricultural systems:

» Preventing the introduction and spread of nematodes by the use of nematode-free planting materials.

» Using non-chemical, cultural, and physical control methods, particularly crop rotation and soil cultivation.

» Encouraging naturally occurring biological control agents by understanding of cultivation methods and appropriate use of soil amendments.

» Maintaining or enhancing the biodiversity inherent in traditional farming systems that use multiple cropping and multiple cultivars to increase the available resistance or tolerance to nematodes.

16.5.1. Alternative management strategies

Scientific evidence and public awareness have heightened concerns about environment quality, food quality, and human health and safety relative to pest management in agricultural production. The need for alternative, integrated nematode management has been propelled by the actions triggered by the Montreal Protocol and the Food Quality Protection Act (FQPA) of the 1990's. The agricultural production community (our stakeholders) is scrambling to find viable and agro-ecologically sustainable alternatives to chemical-based soil pathogen and nematode control.

A major shift in nematode-management strategies is occurring, from almost exclusive reliance on soil-applied nematicides, to the use of combinations of alternative strategies such as crop rotation, host plant resistance, cultural manipulations and biological control (Ferris *et al.*, 1992). Greater understanding of nematode genetic response and adaptation to abiotic factors will be important in optimizing the design of cultural management tactics such as manipulations of planting and harvest times, wet or dry fallow, and soil solarization. Wet or dry fallowing starve nematodes while active (wet fallow) or die from extreme moisture stress (dry fallow). Soil solarization involves natural heating of soil under plastic cover to attain the thermal death point of nematodes. Avoidance may include changes in planting and harvest dates, such as delaying planting in the fall to avoid infection activity, and early planting or late harvest of crops to avoid additional nematode generations. Soil amendments such as green manures and various bio-products show promise for nematode suppression in some systems. Some potential biological control agents are known to give effective management of nematodes. Current research is addressing many areas of nematode management, including: development and deployment of nematode-resistant plants; crop rotation to reduce population densities of these pathogens; cover crops, trap crops and soil amendments including green manures to reduce population densities; the role of weed hosts in bringing about phenotypic changes in nematode populations; characterization of resistance genes and resistance responses; the development of molecular diagnostic protocols for nematode identification and the reference databases necessary for their implementation. Efforts to develop improved techniques for nematode diagnostics, understanding of nematode diversity and genetic variability, processes of nematode fitness and adaptation, and the incorporation of this knowledge into the design and analysis of improved nematode management strategies needs to be intensified.

Climate change will cause alterations in the spatial and temporal distribution of nematodes and consequently the control methods will have to be altered to suit these new situations. Assessments of the impact of climate change on nematode infestations and in crops provide a basis for revising management practices to minimize crop losses as climate conditions change (Ghini *et al.*, 2008).

Recent observations suggest that nematode pressure on plants may increase with climate change (Ghini *et al.*, 2008). As a result there may be substantial rise in the use of nematicides in both temperate and tropical regions to control them. Non-chemical nematode management methods (green manuring, crop rotation, mulching, application of organic manures, etc.) assume greater significance under changing climate scenario.

16.6. PHYSICAL METHODS

16.6.1. Soil solarization

Solarization-heating soil under clear plastic tarps that trap and increase the sun's heat can be an effective means of controlling nematodes in the soil. The soil needs to be moist, well tilled, and heated to at least 50 °C for several days, preferably several weeks. This method can be practical for nursery sites and home gardens, but it should be done during the hot months and long days of mid-summer.

Similarly, other heat and steam-based pasteurization methods can be used to prepare potting soil. Healthy plants grown in nematode-free media have a better chance to survive after being transplanted to the field.

16.7. CULTURAL METHODS

16.7.1. Crop rotation

Crop rotation often reduces nematode populations by reducing reproductive potential and survival, eliminating or reducing population densities by increasing the abundance or activity of beneficial soil micro-flora. Among the organisms that are most likely favored by cover crops are fungal egg-parasites, nematode-trapping fungi, endoparasitic fungi, arbuscular mycorrhizal fungi, plant growth promoting rhizobacteria, and obligate bacterial parasites (Sikora, 1992). The use of legume cover or rotation crops provides organic nitrogen while also suppressing plant-parasitic nematodes. A number of legume crops have been identified that are highly resistant to various species and/or races of RKN. Some of these, cowpea (*Vigna unguiculata*), velvet bean (*Mucuna deeringiana*), and sun hemp (*Crotalaria juncea*) have potential in cropping systems (McSorley, 1999; Sipes and Arakaki, 1997).

Several nematode-antagonistic crops including rotation and cover crops such as rapeseed (*Brassica napus*), marigold (*Tagetes* spp.), forage pearl millet (*Pennisetum typhoides*), black-eyed Susan (*Rudbeckia hirta*), Sudan grass (*Sorghum bicolor*) and sorghum-Sudan grass were identified (LaMondia and Halbrendt, 2003). Native prairie plants that are resistant to *M. hapla* (LaMondia, 1996) may be useful as cover crops alone or in mixtures for managing this and other plant-parasitic nematode species.

Selected Brassica species, including rapeseed and mustard, may suppress nematode populations, soil-borne pathogens, and weeds in crop rotations (Halbrendt, 1996; Koch and Gray, 1997). These plants produce glucosinolates, and their decomposition products are toxic to nematodes. Nematode-resistant radish is very effective in suppressing *Heterodera schachtii* on sugar beet (Koch and Gray, 1997).

The utilization of certain Sudan grass hybrids as a green manure provides excellent control of *Meloidogyne chitwoodi* on potato by producing a higher concentration of toxin, dhurrin (Mojtahedi *et al.*, 1993).

The use of legumes including velvet bean, *Mucuna deeringiana* in a soybean rotation enhances the activity of rhizosphere bacteria antagonistic to the soybean cyst nematode, *H. glycines*, and the southern root-knot nematode, *M. incognita* (Kloepper *et al.*, 1992).

16.7.2. Fallowing

Keeping the soil free of plants (fallow) deprives plant-parasitic nematodes of a host, which, over time, reduces their populations. Maintaining good weed control is a critical component of fallowing for nematode control because weeds are hosts of many species of plant-parasitic nematodes.

16.7.3. Multicropping

Multicropping (intercropping) with plants that either are not good nematode hosts or are antagonistic to the nematodes also reduces nematode numbers.

16.7.4. Green manuring

Green manuring—tilling under a crop that grows rapidly and produces a large quantity of biomass-adds organic matter and, depending on the green manure crop used, may add substances that repel or kill nematodes. Sudan grass and corn are excellent green manure crops that provide good nematode control.

16.7.5. Time of planting

Another potential nematode-management strategy involves adjusting the schedules for susceptible crop production to limit nematode reproduction. For example, delayed planting of soybean, which occurs in wheat-soybean double cropping systems, allows for greater nematode attrition in the absence of a host and results in lower at-planting population densities of *Pratylenchus brachyurus* and *H. glycines* (Koenning and Anand, 1991; Koenning *et al.*, 1985). The wheat-soybean double-cropping system was shown to be economically superior to a rotation with grain sorghum in Arkansas (Dillon *et al.*, 1996). This approach also has been tried for root-knot nematode in carrot. Shifts in the planting and harvest dates of carrot to minimize root-knot development caused by *M. incognita* have produced striking results (Roberts, 1993). Delaying planting to late autumn or early winter clearly restricts root-gall development on carrot. In some regions, early harvest of peanut is critical to limiting damage to seeds by *Ditylenchus africanus* (Venter *et al.*, 1992).

16.7.6. Weed control

Wardle *et al.* (1995) found that cultivation for weed control was an important factor influencing the species diversity of the nematode community. Increased numbers of fungivore nematodes were found in one study in Georgia in reduced tillage plots compared to conventional tillage (Parmelee and Alston, 1986) during the summer, but the reverse was true at other times of year. Similarly, higher numbers of the plant parasites *Helicotylenchus dihystera, Tylenchus, Aphelenchoides* spp., dorylamids and mononchids were associated with conventional till systems in North Carolina (Mannion, 1991). Numbers of bacterial feeding and total nematode numbers were greatest in a no-till system in Spain (Lopez-Fando and Bello, 1995). The ratio of fungivores to bacterivores can be regarded as an indicator of the decomposition pathway in detrital food webs (Freckman and Ettema, 1993). The decrease in this ratio associated with no-till may indicate a shift from a bacteria-based food web to a fungus-based food web.

16.8. BIOLOGICAL METHODS

Nematode, bacterial and fungal based biopesticides are highly vulnerable to environmental stress. Increase in temperature, UV radiation and a decrease in relative humidity may reduce the efficacy of these bio-agents.

A more realistic strategy for biological control of nematodes is to incorporate soil amendments such as manure (particularly poultry manure) and compost. Such additions of organic matter contribute to biological activity in the soil and enhance the natural activity of organisms antagonistic to nematodes.

In addition to suppressing nematodes, certain plant-growth–promoting rhizobacteria may induce systemic resistance to foliage pathogens such as *Pseudomonas syringae* pv. *lacrymans* and *Colletotrichum orbiculare* on cucumber (Wei *et al.*, 1996). They suggested that these rhizobacteria may control a spectrum of plant pathogens/pests, including fungi, bacteria, nematodes, and insects. In a split-root system, treatments with *Bacillus sphaericus* B43 or *Agrobacterium radiobacter* G12 also induced a significant degree of resistance in potato to *Globodera pallida* (Hoffmann-Hergarten *et al.*, 1997). These rhizobacteria suppressed infection of potato roots by the juveniles, but had no effect on egg production.

Pasteuria spp. are obligate, mycelial, endospore forming bacterial parasites of endoparasitic nematodes, such as RKN (*Meloidogyne* spp.), and several ectoparasitic nematodes, such as lesion (*Pratylenchus*) and sting nematodes (*Belonolaimus*) (Chen and Dickson, 1998; Giblin-Davis, 1990). These parasites are promising biological control agents for management of important agricultural species of *Meloidogyne* and *Pratylenchus* (Chen and Dickson, 1998). *Pasteuria penetrans* and *P. thornei* are widespread in soils in the south-eastern USA where RKN and lesion nematode populations occur. However, *Pasteuria* occurs in northern climates because it has been reported on stunt nematodes from turf soil samples taken from Massachusetts (Wick and Dicklow, 2001), on lesion nematodes from potato fields in Minnesota, on soybean cyst nematodes in Illinois (Atibalentja *et al.*, 1998), and it also has been observed on *M. hapla* in Connecticut and New York and Michigan.

16.9. HOST RESISTANCE

Any research program aiming at introducing new varieties should take into account the effects of climate and environmental changes. Particular attention must be paid in the management of nematodes through the introduction of resistance genes in commonly used varieties, since an evaluation is needed about the persistence of the genetic pool introduced, on a scale of decades or years. Hence, protection of biodiversity of plant genetic pools has a practical and immediate justification. The term "adaptive strategy" reflects, in this sense, the need for a quick and flexible response, since some climate changes are expected to occur on a short temporal scale. Some changes are already in action and newly introduced varieties may display in a few years unsuitable agronomic traits, i.e. higher susceptibility due to increased air moisture or rainfall, or may become susceptible to newly colonizing or substitution pests, thus vanishing long term research investments. Programs on the introduction of resistance genes in plant varieties should consider their durability in a changing environment, and how much they will remain useful.

Climate change mediated changes in physiology can alter the expression of resistance genes. The most serious threat to genetic resistance to nematodes may be posed by the increased selection pressure resulting from acceleration of nematode developmental rate and increase in number of generations per season due to global warming (Ghini *et al.*, 2008).

Therefore, there is a need to develop appropriate strategies for nematode management that will be effective under situations of global warming in future.

16.9.1. Pre-emptive plant breeding

Investment is necessary in pre-emptive crop pest resistance breeding against future high-risk pests based on current and future geographic distributions in order to avoid the consequences of shock pest episodes. Predictions of what pests are likely to be a future threat must structure plant breeding programs more than is currently the case. That a pest is absent from an area only by virtue of not having been introduced, or that a particular pest under a changed cropping system may emerge as a major threat can be anticipated. Pre-emptive breeding, as opposed to the current reactive approach, on crop pests must be seen as a sound investment if the consequences of shock pest episodes are to be avoided.

16.10. INTEGRATED NEMATODE MANAGEMENT

More comprehensive integrated management farming systems that include more restricted tillage, fertilization, pesticide use, the addition of organic manure, and under sowing with clover greatly alter the soil fauna and micro-flora (El Titi and Ipach, 1989). For example, the numbers and biomass of earthworms were six times greater in the integrated plot with limited tillage than in the conventionally managed plot. Predatory mites and microbivorous nematodes (bacterivores and fungivores) also are often greatly increased through this type of integrated management (El Titi and Ipach, 1989). Population densities of *Heterodera avenae* and *Ditylenchus dipsaci* were lower in integrated systems with minimal tillage than in conventional systems with standard tillage practices (El Titi and Ipach, 1989).

16.11. INNOVATIVE TECHNOLOGIES

As sustainable nematode management becomes increasingly based on soil biology–health, new complementary technologies are developing. These new tools undoubtedly will improve the accuracy of nematode diagnoses and assessments of potential problems, and will result in more effective management, reduced pesticides, pesticide usage, and less contamination of groundwater with agricultural chemicals such as nematicides, nitrogen, and fertilizers.

16.11.1. Precision nematode management

Modern computerized harvest-management and data systems offer new opportunities for more precise management of nematodes and general crop production. This technology has the potential to improve water use and limit fertilizer and pesticide application on a spatial and temporal basis as dictated by soil fertility and, more important, differential spatial crop yields (Daberkaw and Christensen, 1996; Evans *et al.*, 1996). Based on early results, this management tool should allow specially prescribed nematode control in high-intensive crop production such as *Radopholus similis* on banana and root-knot nematodes on potato in the north-western United States (Evans *et al.*, 1996). Approaches that focus on a harvest index to locate environmental stress (Copeland *et al.*, 1996) should be able to relate nematode kinds and numbers to poor yield and other stress factors. This approach is now being used in some banana operations in which fruit is harvested in small subunits and yield data are recorded and analysed by computer. Poor-yielding sections can be examined for nematode densities and other potential problems.

16.11.2. Genetically engineered and traditional host resistance

There has been considerable progress made in engineering host resistance to nematodes, genetic mapping, and diagnostics (Opperman and Conkling, 1998). However, genetically engineered resistance to nematodes is still at the developmental stage in contrast to the recently deployed herbicide- and insect-resistant cultivars of cotton, soybean, and other crops. One strategy involves transformation of plants with a transgene(s) encoding a product detrimental to the target nematode or that suppresses the expression of key plant genes involved in the nematode-host interaction (Opperman and Conkling, 1998). Candidate genes for this strategy include collagenase, genes expressed in the development of specialized feeding cells induced by species of *Globodera* or *Heterodera* (syncytia) and *Meloidogyne* (giant cells). Constructs of the root-specific TobRB7 gene in tobacco have been used to develop promising root-knot nematode–resistant genotypes (Opperman and Conkling, 1998). Linking this gene with a BARNASE gene resulted in root knot–resistant plants. Transformed plants with an antisense TobRB7 construct also exhibited root-knot resistance; root-gall development was about 70% less than in susceptible plants (Opperman and Conkling, 1998).

A second approach for engineering nematode-resistant plants involves identifying, cloning, and introducing natural plant-resistance genes into susceptible crop plants. Exciting results with this strategy were recently reported with *Heterodera schachtii* on sugar beet (Cai *et al.*, 1997). In one major development, Cai *et al.* (1997) cloned the cyst-resistant gene in wild *Beta* species. A transformed, normally susceptible sugar beet line exhibited the typical incompatible resistant reaction. Similar progress is being made with the Mi gene, which confers resistance to the common *Meloidogyne* species and populations attacking tomato. With the wide host range of these nematodes, the transfer of the Mi gene to numerous crop species, for which root-knot nematodes affect major crop yields, has great economic promise.

New molecular techniques and markers also have positively affected traditional plant-breeding programs related to the development of host-resistance to nematodes. Recently, two markers for parasitism in *H. glycines* were identified (Dong and Opperman, 1997) and molecular markers for crop resistance for various cyst nematodes are being investigated. These resistance markers included soybean (*H. glycines*) (Cook, 1991), potato (*G. rostochiensis*) (Pineda *et al.*, 1993), and wheat (*H. avenae*) (Williams *et al.*, 1996). Markers for *M. incognita* races 1 and 3 resistance in tobacco also have been described (Yi, 1997). Undoubtedly, combining markers for parasitism (virulence) within different nematode populations and host-resistance genes should spur advances through traditional plant breeding.

16.11.3. Advisory programs

Advisory programs have successfully contributed to lower pesticide usage and greater farm profits. For peanut alone, growers in Virginia were able to reduce their nematicide use by 35% after a predictive nematode assay program was established (Phipps, 1993). Savings in production costs for 1989 were estimated at $800,000, primarily through fewer nematicide applications. Currently, about one half of the states in the United States offer their farmers some type of nematode advisory program, usually through the Extension Service, State Departments of Agriculture, or private consultants. Many growers monitor the relative magnitude of nematode problems in given fields by observing root symptoms and signs of nematodes and through field histories.

Where detailed data on production and nematode populations are maintained, more precise approaches in decision-making are becoming available. Burt and Ferris (Burt and Ferris, 1996) developed a sequential decision rule to aid in choosing a rotation crop versus host crop where this practice is the management tactic rather than using a nematicide. The static model used by Ferris (1978) is unsuitable for quantifying the optimal dynamic threshold that would be characterized by population densities lower than where returns from the nematode host and non-host are equal. A dynamic model for this type of crop-nematode management system was recently developed (Burt and Ferris, 1996). Application of this model should allow better economic management of nematodes, but data will still be needed on annual nematode population change under host and non-host crops and the relationships between nematode numbers and crop yields. More comprehensive pest-host simulators and expert systems (Saracino *et al.*, 1995) have bolstered research in recent years (McSorley, 1994).

16.12. FUTURE PROSPECTS

Water-limiting environments, nematode diseases, declining fertility, availability and degradation of the soil resource are among key constraints to increasing production and quality of food. Climate change adds an extra layer of complexity to an already complex agro-ecological system. Crop protection professionals routinely develop and deploy strategies and tools based on well-established principles to manage plant nematodes and many may also be applicable under climate change when projected changes, processes and interactions are factored in. Therefore, research to improve adaptive capacity of crops by increasing their resilience to nematodes may not involve a totally new approach, although managing plant nematodes may have the added advantage of mitigating rising CO_2 levels (Mahmuti *et al.*, 2009). The bulk of any new investment to nematode-proof food crops, therefore, need only to accelerate progress of new and existing promising strategies and approaches and not to 're-invent the wheel' under the guise of climate change research. Such an investment model will ensure that nematode management solutions span the entire range of uncertainties associated with climate change including the 'business as usual' scenario.

There has only been limited empirical research on plant nematodes under realistic field conditions mimicking climate change and this severely restricts development of options to enhance crop adaptation or nematodes management under climate change. In addition, a relatively large body of knowledge has been gathered on potential effects of global climate change using models. First-pass assessments are now available for some countries, regions, crops and particular nematode pathogens. From a food security perspective, emphasis must now shift from impact assessment to developing adaptation and mitigation strategies and options. Two broad areas of empirical investigation will be essential; first, to evaluate the efficacy of current physical, chemical and biological control tactics including nematode resistant varieties under climate change, and second, to factor in future climate scenario in all research aimed at developing new tools and tactics. Transgenic solutions (Huang *et al.*, 2006) must receive serious consideration in integrated nematode management strategies to improve food security.

16.13. CONCLUSION

Research on impact of climate change on soil nematodes has been limited, with most work concentrating on the effects of a single atmospheric constituent under controlled conditions. A few recent experiments have also reported the response of herbivorous nematodes to elevated CO_2 beyond the trophic group level. Nevertheless, findings of these studies give an insight into the response of nematode trophic groups to climate change and its consequences to agricultural production. Responses of herbivorous nematodes to CO_2 enrichment were observed to be either neutral or positive but not negative. Further, studies predicting changes in geographical distribution of plant parasitic nematodes using simulation models give a fairly good idea about future scenarios of nematode diseases of plants. More long-term studies in varied agro-ecosystems under different cropping systems of particularly tropical regions are needed to critically assess the impacts of climate change on soil nematodes. This knowledge is vital for developing appropriate adaptation and mitigation strategies to minimize effect of climate change on agriculture.

Chapter - 17

A Vision of the
Future Outlook

17.1. INTRODUCTION

Plant-parasitic nematodes must be addressed in crop production and in integrated pest management (IPM) systems if a sustainable agriculture is to meet world demands for increasing food and fiber production. Crop losses due to nematode damage have been estimated to be US$ 157 billion annually to the world agriculture (Abad *et al.*, 2008). The nematodes diseases are responsible for 21.3% crop loss in India amounting to Rs. 102 billion annually. Economic damage of 40 to 50 per cent can occur for the most damaging nematodes with sometimes complete crop losses. The main nematode problems throughout the world are caused by root-knot nematodes, *Meloidogyne* spp. Others include *Rotylenchulus reniformis, Radopholus similis, Tylenchulus semipenetrans, Helicotylenchus multicinctus, Pratylenchus* spp. *and Ditylenchus dipsaci*, and cyst nematodes (*Globodera/Heterodera* spp.) are important in the cooler areas.

The specific concept of integrated nematode management (INM) is a relatively recent development. INM consists of the development and deployment of nematode management strategies and tactics that result in favorable socio-economic and environmental consequences. Theoretically, INM is a holistic systems approach to limit nematode damage to tolerable levels through a combination of tactics and techniques, including parasites and predators, host resistance, cultural practices, environmental modification, and nematicides where appropriate. A number of national and international IPM programs have resulted in many research and extension activities becoming more interdisciplinary.

17.2. NEW PARADIGMS FOR NEMATODE MANAGEMENT

Biological control amplifies biotic imbalances, because the economic pressures compel drastic methods based on the physical elimination of antagonists. High grade results achieved by crop protection researches carried out on crop cultural practices, plant resistance to nematodes or biocontrol, are very diverse. But, all these control practices, included or not in integrated pest management strategies, seem not to be sufficient: they all target some nematode species (population approach), and then involve changes in nematode communities but do not necessarily modify their overall pathogenicity. So, binary researches focused on plant-nematode or nematode-parasite relationships should be extended to ecological investigations on nematode communities.

These reasons support eco-epidemiological approaches, based on the ecology of communities (interspecific competitions, biological, and edaphic constraints) and focused on the management of the natural enemy biodiversity, appearing as a new research alternative for crop protection (Fig. 17.1).

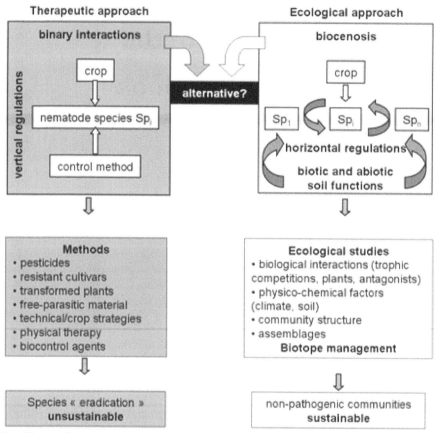

Fig. 17.1. Nematode management, from therapeutic to ecological approaches (Biocenosis- the interacting organisms **living together in a** habitat, **Biotope - habitat).**

Specific and functional evolution and rapid multiplication of plant-parasitic nematodes are enhanced by agriculture intensification and by environment adaptation. In fact, plant-parasitic nematodes are only considered as predators, and not as one of the soil components. Up-to-date, all control strategies developed in agriculture are focused on the eradication of target species. They induce biotic gaps, community rearrangements, insurgence of virulent races, increased aggressiveness of minor species, *etc*. Then, the "soil cleaning" strategy appears to be not sustainable.

The development of sustainable management strategies must move from the "therapeutic approach", mostly adopted in research programs carried out in the world, to the "ecological approach". This approach will seek for information and knowledge about biotic trade-offs in ecosystems, in order to introduce them in agro-systems (resilience). That questioning can be declined as follows:

» Why seek with plant-parasitic nematode eradication?

» Can agronomic problems be solved by agronomic strategies only?

» Can nematode diversity in communities be considered as an auxiliary for nematode management?

Effector biogenesis has not been studied for 30 years because the system was intractable. With recent development in functional genetics, the E-biogenesis is now well placed to dissect this strategically important knowledge gap. *E-biogenesis* aims to investigate methods of blocking crop parasitism by attacking the biogenesis machinery that produces the molecules that cause disease. The vision of the *E-biogenesis* is centred on the ability of nematodes to produce molecules that cause disease; so called "effectors". Blocking effectors blocks parasitism. However, blocking individual effectors is insufficient. Rather, there is a need to attack the features that unite most effectors; their biogenesis machinery.

This research will highlight a multiplicity of attractive targets for the biotechnological control of globally important plant pests. To achieve this, there is a need to open a whole new area of effector study, contribute new fundamental knowledge of strategic relevance, and provide a platform for a paradigm shift from individual effectors to a holistic view of effector biogenesis. Surely, effector biogenesis (*E-biogenesis*) is the Achilles heel of the nematode. Recently (2020), a project on *E-biogenesis has been rgranted by European Research Council to Dr.* Sebastian Eves-van den Akker, Crop Science Centre, University of Cambridge.

17.3. PROPOSED APPROACHES

The nematode diversity in communities would represent the central object of researches in the future. In that way, diversity and population levels in communities can be considered as agro- and ecosystem indicators. They can thus inform on environment disturbances and on the aptitude of environment to facilitate or not epidemic phenomena, on "buffer" and "recovery" capacities, or "plug" of soils (resistance and resilience). The comparative studies on environments with contrasted characteristics or with different adaptation levels will necessarily bring information for understanding and management of more or less intensive agro-systems or weakened environments.

Three types of environments can be studied which include:

» Organic agriculture

» Ecosystems

» Land use changes

17.3.1. Organic agriculture

In organic agriculture, the management of plant-parasitic nematodes implies crop diversification, rotations with non-host or antagonistic plants, amendments with green manures, and biofumigation methods. All these methods enhance biodiversity in soils, as a source of significant biological competitions. Organic agriculture makes it possible to analyze consequences of methods targeting specifically "major" species on the whole nematode communities, without skews induced by chemical treatments.

17.3.2. Ecosystems

These systems are especially fit for studies on plant-nematode trade-offs. They include many regulations as the "horizontal" biotic regulations defined by the interspecific competitions in the communities (Van der Putten *et al.*, 1998). These competitions can be of trophic nature (exploitation of the same resource), of demographic nature (multiplication rates and life traits of the different species), and of ecological nature (habitats). These studies are essential in order to determine the global pathogenic effect of nematode communities, and the respective role of the species in them. They also include "vertical" biotic regulations defined by crop and soil (microbial antagonists) constraints on the species within communities. Obviously, as plant-parasitic nematodes are obligate parasites, the plant plays a fundamental role in the nematode community structure, which depends on both plant susceptibility to different species and on their pathogenicity. Because of their specificities, microbial antagonists also have a marked impact on community structures (De Rooij *et al.*, 1995). Eventually, abiotic regulations (soil physicochemical factors and functions) also affect the space-time structure of nematode communities (Cadet *et al.*, 1994; Cadet and Thioulouse, 1998).

17.3.3. Land use changes

These situations make it possible to study ecological determinants of nematode community structures. They also are informative on the original nematode structure, before land use changes. Crop ages give essential elements for understanding space and time development of nematode communities.

17.4. FUTURE OUTLOOK

Current research for nematode control is focusing on the development of effective cultural practices such as those traditionally practiced before the advent of broad-spectrum nematicides. In looking to the future, what are some of the things that we need to do from a realistic and practical stand point of view?

Through surveys and identifications, determine more accurately the numbers and kinds of nematodes which actually exist and pinpoint their distribution. Success of crop rotations and resistant cultivars depend upon knowing what nematodes occur in which areas and fields.

With regard to viruses transmitted by soil-borne nematodes, it is advisable to explore the possibility of finding dorylaimid nematodes from the rhizosphere and their role in transmitting soil-borne viruses mainly in crops like fruits, vegetables and ornamentals.

We should continue research on nematode management to include efforts to obtain (i) nematicides which are more effective, can be used at lower dosage, and cause less pollution; (ii) resistant cultivars for major crops and at least for the most important nematode species; and knowledge necessary to utilize the benefits and capabilities offered by integrated nematode management systems.

There is a need to develop a better understanding of the biological aspects of all our more important plant nematodes. This would include the molecular genetics of closely related species and races; host-parasite relationships and histochemical details which may provide clues to the nature of plant resistance to nematodes; ecological factors, such as temperature, moisture and soil types which may influence nematode development and survival which in turn affect control measures; and make full use of current biotechnologies as appropriate in this and other nematology research.

There is a need for development of cropping systems based nematode management technologies. Use of green manuring plants such as *Crotalaria spectabilis* prevents the nematode from reproducing. Coastal Bermuda grass *(Cynodon dactylon)* incorporated before planting tobacco or vegetable transplants protects against root-knot nematodes (*Meloidogyne* spp.) (Burton and Johnson, 1987). These plants could be even more effective if they could be genetically engineered to produce nematode attractants or pheromones.

Another area of concern is to advance biological control to its fullest potential, utilizing biological control agents plus organic amendments, sewage, sludge, *etc*. This would include cultural methods for biocontrol parasites, understand their nature, make them practical, and genetically modify the biocontrol agents to improve their effectiveness.

The present research thrust on biological control of nematode pests need to be intensified. More basic studies in biological control and interactions in the rhizosphere are required. Improved assay techniques for assessing the biological antagonism coefficients of various soils must be developed.

Many fungal and bacterial bioagents have been found effective under *in vitro* conditions. Efforts should be made to test them under field conditions. Further, our efforts should be directed towards developing mass production technologies for effective bioagents and easy field application techniques.

In recent years, it has been observed that the combination of fungal bioagents perform better in reducing nematode population and increasing the plant health. Thus the combination of highly toxic fungus, *Aspergillus niger* (killed most of the infective second stage juveniles) and an egg parasite, *Cladosporium oxysporum* (invaded and killed the eggs in egg sac) both at half the doses reduced significantly more *M. incognita* population and exhibited better plant growth than when either of the fungal bioagents in brinjal (Goswami and Singh, 2002).

Another promising area of research in nematode management will be to identify and develop new generation chemicals which may not kill the nematodes but either change their behavior or delay the development and force them into the inactive stages or alert the plant to defend the nematode attack.

The selection and development of varieties for resistance and tolerance to nematode stress will continue. This may involve incorporation of appropriate genetic material into varieties already selected for production, economic, and marketing qualities. It is still important to develop biological and chemical nematicides that are systemic, easily associated with the root system, target organism specific, or a combination of these factors. These pesticides will allow flexibility in management decisions and compensation for management errors that have promoted or amplified nematode stress problems in a particular production system.

Phytoalexins effectively induce plant resistance mechanisms to nematodes, particularly to sedentary ones. However, the mechanism by which phytoalexins are accumulated in plants is not known. Further research in accurately elucidating the effect of potential phytoalexins on the nematodes is urgently needed.

Although valuable improvement has been obtained in development of nematode resistant horticultural crop cultivars, there is a need for breeding varieties with combined resistance to different nematode species and to other organisms such as soil-borne fungi and bacteria which cause wilt and root rot disease complexes.

The development of transgenic plants with durable nematode resistance keeping pace with the evolution of resistance breaking populations should be considered. Biotechnological applications including RNAi or miRNA represent a potential breakthrough in the application of functional genomics for plant nematode control. Here, a comparison is made between some old and modern nematode management practices but recent data shows that application of RNAi or miRNA has a better option of nematode control in some crop plants.

Once the INM technologies are developed, it is necessary that they should be put to practice under field conditions. The implementation of these programs will require scientists/trained technicians and efficient extension media. In this direction, National Institutes and SAU's can play an important role by holding training courses and carrying out problem oriented researches. Since integrated approach may involve more than one discipline, it is therefore, essential that professional scientists must collaborate in its planning and execution.

Our future research programs need to be more based towards development of sustainable and non-chemical methods for the management of nematode pests. We have made much progress in nematology; but we still have a long way to go in the battle against these destructive and insidious plant nematodes. Through our continued dedication and strong research efforts, greater knowledge and better understanding of nematodes will be achieved. This will result in further reductions in losses caused by nematodes and a significant increase of food and fiber in our countries.

17.5. CONCLUSION

Plant-parasitic nematodes cause substantial damage to major crops throughout the world, including vegetables, fruits, and grain crops. These may become a major threat to the agricultural production system worldwide if management fails. Some important genera of plant-parasitic nematodes such as *Meloidogyne* spp., *Heterodera* spp., and *Pratylenchus* spp. have been ranked uppermost in the list of the most economically and scientifically significant species of nematodes due to their complicated relationship with the host plants, wide host range, and the level of damage due to infection in crops. The economic importance and diagnostic methods of plant-parasitic nematodes, including a comprehensive account of existing strategies used for their management ranging from conventional to modern techniques needs to be emphasized. Further, obstacles encountered in parasitic nematode diagnosis by classical morphology-based methods have been resolved by the adoption of novel molecular techniques, which are rapid, precise, and cost-effective. As far as the existing cultural management techniques are concerned, crop rotation with non-host crops can suppress a wide range of nematode species effectively, followed by the use of organic soil amendments. Nematicide application is effective when speedy control of nematodes is required; however, the use is reappraised due to environmental concerns. Biological control of nematodes by fungi and bacteria is highly favored due to its environmentally friendly nature. In addition, bio-pesticides are becoming a promising option for the management of plant-parasitic nematodes. In conclusion, a sustainable management of plant-parasitic nematodes is feasible when two or more compatible tactics are applied concurrently while appraising environmental protection.

Considering the importance of economic losses caused by nematodes and the fact that the restrictions governing the use of chemical nematicides are elevating, it is clear that there is need for the development of new environmentally benign strategies. It is also crucial to continue improving the current green methods in search of making them more efficient. For example, some questions to be addressed for antagonistic fungi/bacteria which include what are the optimum rate, timing, frequency, and method of application for biocontrol agents especially under field conditions? In addition, agricultural industry and scientists should focus on keeping these developed and/or optimized methods economically advantageous, so that they can be adopted by growers on different farming scales. Biocontrol of nematodes has been around for decades, but it is still capable of achieving much more attention and better results as new species are identified, characterized and evaluated for their efficacy against nematodes.

To date, environmentally benign methods are mostly not fully competent on their own with the traditional chemical practices in terms of protecting plants against nematodes. Therefore, it is critical to consider the development and improvement of multidisciplinary management strategies for nematodes such as combining microbial strategies using both bacterial and fungal agents with other cultural control practices or host resistance. Both biocontrol and application of soil amendments have been studied to some extent against PPNs, but with the recent advances in technology there is room for deeper studies on how these two strategies can be synergistically used. For example, studies on how the application of certain amendments may influence the soil microbiome in relation to nematode inhibition. More tools such as Ozonated water (O_3wat) are also becoming available which may be incorporated into the multi-aspect strategies developed.

Nematode community structures are described by descriptive ecology, but the underlying rules and mechanisms are not well known. It is thus necessary to support a cognitive approach to study ecology of nematode communities which will aim to include/understand assemblage rules, their impacts on populations in communities (life trait evolution, adaptation), and the consequences on system functions and management.

Crop practices favor monocultures, which reduce soil biodiversity. That supports proliferation of nematodes. Imbalances induced by these situations strengthened by large inputs, would solve economic questions only. Such strategies delay the effective control of nematodes. On the other hand, taking into account the whole nematode diversity, according to an ecological approach, makes it possible to consider more suitable strategies for sustainable management of these parasites in agriculture, based on an auto-regulation of the global pathogenic effect of the nematode communities (Fig. 17.2). It is consequently critical to focus control practices on agronomic methods because they involve a great impoverishment of the inputs generated by the agro-industrial research (pesticides, biopesticides, and resistant seeds), conferring an acute brittleness in terms of sustainability at the same time.

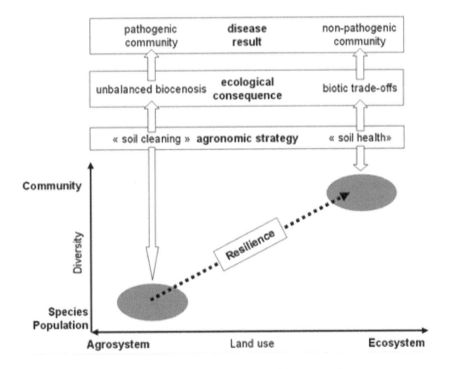

Fig. 17.2. The agronomic soil health approach

Integrated nematode management should hence flank the development of nematodes more than killing them. We must be convinced that both biological balances in soil and environmental factors are fundamental and non-avoiding helps for plant-parasitic nematode management. Every strategy processed in order to eradicate plant parasites would hence invariably prove to be unsustainable.

Biotechnology-, molecular biology-, and nanotechnology-based approaches have added a new dimension to nematode disease diagnosis and management. Identification of genes that reduce nematode's ability to reproduce has allowed the breeding of nematode-resistant plants. Marker-assisted selection, genetic engineering, and RNA interference to confer resistance in crop plants, nematode suppression using host plant proteinase inhibitors, and genome-editing technologies have helped tremendously in developing management strategies for plant-parasitic nematodes.

In conclusion, future studies should focus on environmentally benign approaches which are based on multidisciplinary strategies that can fill the gaps of single sided management methods. Such approaches will also reduce the chance of resistance as the complexity of different nematode management components would make resistance highly improbable. Whatever strategy is devised, future attempts should focus on important factors such as synergism between nematode antagonists, environmental conditions, sustainability, studying the effect of new treatments on non-target organisms, and association of individual plants with nematode antagonists of interest. In summing up, a sustainable management of plant-parasitic nematodes is feasible when two or more compatible tactics are applied concurrently while appraising environmental protection.

References

Abate, T. (ed.). 2012. *Four Seasons of Learning and Engaging Small Farmers: Progress of Phase 1*. International Crops Research Institute for the Semi-Arid-Tropics, PO Box 39063, Nairobi, Kenya, 258 pp.

Abd El Moneem, K.M.H., Fawaz, S.B.M., Saeed, F.A. and El Shehaby, A.I. 2005. Effect of clove size and certain micronutrients on Fusarium & basal rot of garlic. *Assiut J. Agril. Sciences* 36: 163-175.

Abdallah, 1998. Improving vegetable transplants using soil solarization. II. Onion (*Allium cepa*). *Ann. of Agril Sci., Special Issue* 3: 831-843.

Abdel A.S.M and Haroun, M.S. 1990. Efficiency of some herbicides on weed control and yield of onion (*Allium cepa* L.). *Egyptian J. Agron*. 15: 35-44.

Acamovic, T, Sinurat, A, Natarajan, A. *et al*. 2005. Poultry. In E. Owen, A. Kitalyi, N. Jayasuriya and T. Smith (eds.) *Livestock and Wealth Creation: Improving The Husbandry of Animals Kept by Resource-Poor People in Developing Countries*. Nottingham University Press, Nottingham, UK.

Adesogan, A.T., Havelaar, A.H., McKune, S.L. *et al*. 2020. Animal source foods: Sustainability problem or malnutrition and sustainability solution? Perspective matters. *Global Food Security* 25:100325.

Aghora, T.S. and Pathak, C.S. 1991. Heterosis and combining ability in a line x tester cross of onion (*Allium cepa* L.). *Veg. Sci*. 18: 53-58.

Agricultural Statistics at a Glance. 2015. Government of India Ministry of Agriculture & Farmers Welfare Department of Agriculture, Cooperation & Farmers Welfare, Directorate of Economics and Statistics, New Delhi.

Agriwatch. 2018. *Cotton News: USDA Estimates 2018-19 India's Cotton Output Down by 2%*. Accessed February 21, 2019. http://www.agriwatch.com/newsdetails.php? USDA -estimates-2018-19-Indias-Cotton-output-down-by-2%&st=NEWS&commodityid=24 &sid=471851.

Ahir, R.P. and Maharishi R.P. 2008. Effect of pre-harvest application of fungicides and biocontrol agent on black mold (*Aspergillus niger*) of onion in storage. *Ind. Phytopath*. 61: 130–131.

Ahmad, S. and Karimullah. 1998. Relevance of management practices in downy mildew in onion. *Sarhad J. Agri*. 14: 161-162.

Ameta, O.P., Sumeriya, H.K., Mahla, M. and Vyas, A.K. 2001. Yellow revolution: past trends and future potential. *Intensive Agric* 39(5–6): 3–6.

Annapurna, K., Bojappa, K.M. and Bhargava, B.S. 1988. Leaf sampling guide for sapota (*Manilkara achras* M. Foseberg) cv. Cricket Ball. *Crop Res*. 1: 69-75.

Anon. 1994. "Notices." *Federal Register 59, No. 99 (1994)*. http://www.aphis.usda.gov/brs/aphisdocs2/93_25801p_com.pdf.

Anon. 2011. "How'd We 'Make' a Non-browning Apple?" *Arctic Apples* (blog), December 7, 2011. Accessed August 11, 2015, http://www.arcticapples.com/blog/julia/how-did-we-make-nonbrowning-apple.

Anon. 2012. *Del Monte Inquiry Letter to APHIS BRS*, July 30, 2012. Accessed August 11, 2015, http://www.aphis.usda.gov/biotechnology/downloads/reg_loi/del_monte_inquiry_letter.pdf;

Anon. 2013. "Simplot Says It Has Made a Better Potato." *Idaho Statesman*, May 9, 2013. Accessed July 31, 2015, http://www.idahostatesman.com/2013/05/09/2569097/a-better-potato.html.

Anon. 2014a. *"Anti-GMO Activists in Bangladesh Tell Lies to Farmers and the Media,"* April 8, 2014. Accessed July 31, 2015, http://btbrinjal.tumblr.com/post/82090416816/anti-gmo-activists-in-bangladesh-tell-lies-to;

Anon. 2014b. "Super-rice Defies Triple Whammy of Stresses." *New Scientist*, February 28, 2014. Accessed July 31, 2015, https://www.newscientist.com/article/dn25147-super-rice-defies-triple-whammy-of-stresses/.

Anon. 2015a. *"Roundup Ready Soybean Patent Expiration."* Monsanto. Accessed July 31, 2015, http://www.monsanto.com/newsviews/pages/roundup-ready-patent-expiration. aspx.

Anon. 2015b. "Press Release: Non-browning Arctic® Apples to Be Granted Approval." *Arctic Apples* (blog), February 13, 2015. Accessed August 11, 2015, http:// www.arcticapples.com/blog/joel/press-release-nonbrowning-arctic%C2%AE-apples-be-granted-approval.

Anon. 2015c. "Press Release: Arctic® Apples Receive Canadian Approval." *Arctic Apples* (blog), March 20, 2015. Accessed August 11, 2015, http://www.arcticapples. com/ blog/joel/press-release-arctic%C2%AE-apples-receive-canadian-approval.

Anon. 2015d. Haro von Mogel, K. 2015. "*Q&A with Haven Baker on Simplot's Innate™ Potatoes.*" Biology Fortified, May 8, 2013. Accessed July 31, 2015, http://www. biofortified.org/2013/05/qa-with-haven-baker-innate-potatoes/.

Anon. 2015e. *"Patent Application Title: Pineapple Plant Named Rose (EF2-114)."* PatentDocs. Accessed August 11, 2015, http://www.faqs.org/patents/app/ 20130326768.

Anon. 2015f. *"Myths & Facts,"* *AquaBounty Technologies.* Accessed July 31, 2015, http:// aquabounty. com/press-room/myths-facts/.

Anon. 2015g. *"Who Invented Golden Rice and How Did the Project Start?"* Golden Rice Project. Accessed August 11, 2015, http://www.goldenrice.org/Content3-Why/why3_ FAQ. php#Inventors.

Anon. 2015h. "Cotton Crop Scraps Become Healthy Snacks." *Sciencemuseum*, November 24, 2006. Accessed August 11, 2015, http://www.sciencemuseum.org.uk/antenna/non-toxiccotton/.

Anon. 2018. *Horticultural Statistics at a Glance 2018*. http://agricoop.nic.In/sites/ default/ files/.

Arya, P.S. and Bakshi, B.R. 1999. Onion based cropping systems studies under mid-hill conditions of Himachal Pradesh. *Adv. Horti. Forestry* 6: 79-85.

Aulakh, M.S., Sidu, B.S., Arora, B.R. and Singh, B. 1985. Content and uptake of nutrients by pulses and oilseed crops. *Indian J Ecol* 12(2): 238–242.

Ayala-Zavala, J.F., Vega-Vega, V., Rosas-Domínguez, C. *et al.* 2011. Agro-industrial potential of exotic fruit by-products as a source of food additives. *Food Research International* 44(7): 1866- 1874.

Ayyappan, S. 2012. Indian fisheries: Issues and the way forward. *National Academy of Science Letters* 35: 1-6.

Ayyappan, S., Jena, J.K., Gopalakrishnan, A. and Pandey, A.K. 2013. *Handbook of Fisheries and Aquaculture.* Directorate of Knowledge Management in Agriculture, Indian Council of Agricultural Research, New Delhi, India, 1116 pp.

BAHS (Basic Animal Husbandry Statistics). 2019. Department of Animal Husbandry, Dairying and Fisheries. Ministry of Agriculture, Government of India, Available from, http://dadf.gov.in/sites/default/filess/BAHS%20%28Basic%20Animal%20 Husbandry% 20Statistics-2019%29.pdf.

Bajaj, K.L., Kaur, G. and Singh, T. 1979. Lachrymatory factor and other chemical constituents of some varieties of onion. *Journal of Plant Food* 3: 119-203.

Balasubramayam, V.R.A., Dhake, V. and Moitra, P. 2000. Micro Irrigation and fertigation of V-12 Onion (Abstr.). *International Conference on Micro and Sprinkler Irrigation Systems*, Jalgaon, p.77.

Barnwal M.K. and Prasad, S.M. 2005. Influence of date of sowing on Stemphylium blight disease. *J. Res., Birsa Agriculture University* 17: 63-67.

Barnwal, M.K., Prasad, S.M. and Kumar, S. 2006. Cost effective fungicidal management of Stemphylium blight of onion. *J. Res., Birsa Agricultural University* 18: 153-155.

BBC News. 2013. *"Uganda's Genetically Modified Golden Bananas."* BBC News, March 27, 2013. Accessed August 11, 2015, http://www.bbc.com/news/world-africa-21945311.

Beintema, N., Stads, G.J., Fuglie, K. and Heisey, P. 2012. *ASTI Global Assessment of Agricultural R&D Spending.* International Food Policy Research Institute, Washington D.C., U.S.A.

Belton, B., Padiyar, A., Ravibabu, G. and Rao, K.G. 2017. Boom and bust in Andhra Pradesh: Development and transformation in India's domestic aquaculture value chain. *Aquaculture* 470: 196-206.

Bender, A. 1992. *Meat and Meat Products in Human Nutrition in Developing Countries.* FAO Food and Nutrition Paper #53. *Food Policy and Nutrition Division of FAO* 2: 1–88.

Bender, D.A. 1993. Onions. In: W.F. Bennett (ed.) *Nutrient Deficiencies and Toxicities in Crop Plants*, pp. 131–135. APS Press, St. Paul, Minnesota.

Bhambal, S.B. 1987. *Effect of Foliar Application of Micronutrients on Growth, Yield, Fruit Quality and Leaf Nutrient Status of Pomegranate* (Punica granatum *L.*) cv. *Ganesh.* M. Sc. (Ag.) thesis, Mahatma Phule Agri. Univ., Rahuri, Maharashtra.

Bhargava, B.S. and Chadha, K.L. 1993. Leaf nutrient guide for fruit crops. In K.L. Chadha and O.P. Pareek (eds.) *Advances in Horticulture*, Vol. 2, pp. 973-1029. Malhotra Publishing House, New Delhi.

Bhargava, B.S., Raturi, G.B. and Hiwale, S.S. 1990. Leaf sampling technique in ber (*Zizyphus mauritiana* Limk) for nutritional diagnosis. *Singapore J. Prim. Ind.* 18: 85.

Bhonde S.R., Sharam, S.B. and Chougale, A.B.1997. Effect of biofertilzer in combination with nitrogen through organic sources on yield and quality of onion. *NHRDF News Lett.* 17(2): 1-3.

Biji, K.B., Ravishankar, C.N., Mohan, C.O. and Srinivasa Gopal, T.K. 2015. Smart packaging systems for food applications: A review. *Food Sci. Technol.* 52(10): 6125-6135.

Birthal, P.S., Joshi, P.K., Roy, D. and Thorat, A. 2007. *Diversification in Indian Agriculture towards High-Value Crops: The Role of smallholders.* IFPRI Discussion Paper 00727, International Food Policy Research Institute, Washington, D.C.

Birthal, P.S., Khan, M.T., Negi, D.S. and Agarwal, S. 2014b. Impact of climate change on yields of major food crops in India: Implications for Food Security. *Agricultural Economics Research Review* 27(2): 145-155.

Birthal, P.S., Kumar, S., Negi, D.S. and Roy, D. 2015c. The impact of information on returns from farming: evidence from a nationally representative farm survey in India. *Agricultural Economics* 46:1-13.

Birthal, P.S. and Negi, D.S. 2012. Livestock for higher, sustainable and inclusive agricultural growth. *Economic and Political Weekly* 47(26-27): 89-99.

Birthal, P.S., Negi, D.S., Khan, M.D. and Agarwal, S. 2015b. Is Indian agriculture becoming resilient to droughts?. *Food Policy* 56: 1-12.

Birthal, P.S., Negi, D.S., Kumar, S. *et al.* 2014a. How sensitive is Indian agriculture to climate change?. *Indian Journal of Agricultural Economics* 69(4): 474-487.

Birthal, P.S., Roy, D. and Negi, D.S. 2015d. Assessing the impact of crop diversification on farm poverty in India. *World Development* 72: 70-92.

Blaise, C.D., Ravindran, C.D. and Singh, J.V. 2006. Trends and stability analyses to interpret results of long-term effects of application of fertilizers and manure to rainfed cotton. *Journal of Agronomy and Crop Science* 192: 319-330.

Blaise, D. 2006. Balanced fertilization for high yield and quality cotton In: D.K. Benbi, M.S. Brar, and S.K. Bansal (eds.) *Balanced Fertilization for Sustaining Crop Productivity*, pp. 255-271. Proc. Intl. Symp. PAU, Ludhiana, IPI.

Blaise, D. 2011. Tillage and green manure effects on Bt transgenic cotton (*Gossypium hirsutum* L.) hybrid grown on rainfed vertisols of central India. *Soil Tillage Res.* 114: 86-96.

Blaise, D. and Prasad, R. 2005. Integrated plant nutrient supply: An approach to sustained cotton production. *Indian J. Fert.* 1: 37-46.

Blaise, D. and Ravindran, C.D. 2003. Influence of tillage and residue management on growth and yield of cotton grown on a Vertisol over 5 years in a semi-arid region of India. *Soil and Tillage Research* 70: 163-173.

Borlaug, N.E. 1972. "Nobel Lecture, December 11, 1970." In *Nobel Lectures, Peace 1951–1970*, edited by Frederick W. Haberman. Elsevier Publishing Company, Amsterdam, 1972. http://www. nobelprize.org/nobel_prizes/peace/laureates/1970/borlaug-lecture. html.

Borlaug, N.E. 2000. Ending world hunger: The promise of biotechnology and the threat of antiscience zealotry. *Plant Physiology* 124(2): 487-490.

Bowonder, B. 1979. Impact analysis of the green revolution in India. *Technol Forecast Soc Chang.* 15: 297–313.

Brar, K.S., Sidhu, A.S. and Chadha, M.L. 1993. Screening onion varieties for resistance to *Thrips tabaci* Lind. and *Helicoverpa armigera* (Hubner). *J. Insect Sci.* 6: 123-124.

Brown, L.R., Christopher F., Hilary, F. *et al.* (eds.). 2001. *State of the World 2001: A World-Watch Institute Report on Progress Toward a Sustainable Society.* Norton, New York.

Carpenter, J. 2014. "*How Many Pounds of GM Foods Are Produced Each Year in the U.S.A.*" GMO Answers, May 8, 2014. Accessed July 31, 2015, https://gmoanswers.com/ ask/how-many-pounds-gm-foods-are-produced-each-year-usa.

Castle, L.A., Wu, G. and McElroy, D. 2006. Agricultural input traits: Past, present and future. *Curr. Opin. Biotechnol.* 17: 105-112.

Chand, R. 2016. "*Doubling Farmers' Income: Strategy and Prospects*". Presidential Address delivered at 76th Annual Conference of Indian Society of Agricultural Economics, Assam Agricultural University, Jorhat, Assam.

Chand, R., Kumar, P. and Kumar, S. 2012. Total factor productivity and contribution of research investment to agricultural growth in India. *Agricultural Economics Research Review* 25(2): 181-194.

Charles, D. 2015. "*A Top Weed Killer Could Cause Cancer. Should We Be Scared?*," March 24, 2015. NPR-*The Salt* (blog). http://www.npr.org/sections/thesalt/2015/03/24/ 394912399/a-top-weedkiiller-probably-causes-cancer-should-we-be-scared.

Chaturvedi, S.K. 2009. Pulses research and development in achieving millennium development goals. National symposium on "*Achieving Millennium Development Goals: Problems and Prospects.*" Bundelkhand University, Jhansi, India, pp. 1-5.

Chel, A., Nayak, J.K. and Kaushik, G. 2008. Energy conservation in honey storage building using Trombe wall. *Energy Build*. 40: 1643–1650.

Chel, A. and Tiwari, G.N. 2010. Stand-alone photovoltaic (PV) integrated with earth to air heat exchanger (EAHE) for space heating/cooling of adobe house in New Delhi (India). *Energy Convers. Manage.* 51: 393–409.

Chopade, S.O, Bansode, P.N and Hiwase, S.S. 1998. Studies on fertilizer and water management to onion. *PKV Res. J.* 22: 44-46.

Chopra, S. 2010. Horticultural interventions for food security challenges. In: *Souvenir of the Fourth Indian Horticulture Congress*, New Delhi.

Christopher, B., Barange, M., Subasinghe, R. *et al.* 2015. Feeding a billion by 2050 - Putting fish back on the menu. *Food Security* 7: 261-274.

CMFRI. 2020. *Marine Fish Landings in India 2019*. Technical Report, CMFRI Booklet Series No. 24/2020. ICAR-Central Marine Fisheries Research Institute, Kochi, India, 15 pp.

Cranshaw, W.S. 2015. "*Bacillus thuringiensis.*" Colorado State University. Accessed July 31, 2015, http://www.ext.colostate.edu/pubs/insect/05556.html.

DAC&FW. 2017b. *Horticultural Statistics at a Glance 2017*. Ministry of Agriculture & Farmers Welfare, Government of India, New Delhi.

Danforth Center. 2015. "*Virus Resistant Cassava for Africa.*" Danforth Center. Accessed August 12, 2015, http://www.danforthcenter.org/docs/default-source/newsmedia/ infographics/infographic-horozontal.pdf?sfvrsn=0.

Darshan Singh, D., Sidhu, A.S., Thakur, J.C. and Singh, D. 1986. Relative resistance of onion and garlic cultivars to *Thrips tabaci* Lind. *J. Res. Punjab Agril. Uni.* 23: 424-427.

Dastagiri M.B. 2004. *Demand and Supply Projections for Livestock Products in India*. Policy Paper 21.

Datt, G. and Ravallion, M. 1998. Farm productivity and rural poverty in India. *Journal of Development Studies* 34(4): 62-85.

Datt, G., Ravallion, M. and Murgai, R. 2016. *Growth, Urbanization and Poverty Reduction in India*, Policy Research Working Paper 7568, The World Bank, Washington D.C., U.S.A.

Datta, A. 2013. Genetic engineering for improving quality and productivity of crops. *Agriculture & Food Security* 2: 15.

Dawe, D. 1998. Re-energizing the green revolution in rice. *American Journal of Agricultural Economics* 80: 948–953.

De Janvry, A. and Subbarao, K. 1986. *Agricultural Price Policy and Income Distribution in India*. Studies in Economic Planning, Oxford University Press, New Delhi.

Debajit S., Akhtar, M.S., Pandey, N.N. *et al.* 2011. *Nutrient Profile and Health Benefits of Cold-Water Fishes*, Bulletin No. 17. ICAR Directorate of Coldwater Fisheries Research, Bhimtal, Uttarakhand, India, 40 pp.

Delgado, C. 2003. Rising consumption of meat and milk in developing countries has created a new food revolution. *Journal of Nutrition* 133 (11, sup 2): 3907S–3910S.

Delgado, C., Rosegrant, M. and Meijer, S. 2001. Livestock to 2020: the revolution continues. *International Agricultural Trade Research Consortium. http://www.iatrcweb.org/ oldiatrc/Papers/ Delgado.pdf* (Accessed October 22, 2002).

Department of Animal Husbandry, Dairying and Fisheries. 2017. *Basic Animal Husbandry and Fisheries Statistics*. Government of India, New Delhi. Accessed September 20, 2018. http://www.dahd. nic.in/sites/default/filess/Tables%20of%20BAH%26amp% 3BFS% 202017%20% 281% 29.pdf.

Department of Animal Husbandry, Dairying and Fisheries. 2017. *National Action Plan for Egg and Poultry—For Doubling Farmers' Income by 2022*. Ministry of Agriculture and Farmers Welfare, Government of India, New Delhi.

DES. 2013. Agricultural statistics at a glance 2013. Directorate of Economics and Statistics. Ministry of Agriculture, Government of India. http://www.nhrdf.com/contentPage. asp?sub_ section_ code=104.

Desh, B.B. 2002. *Status of S in Groundnut Growing Red and Laterite Soils of Orissa and its Integrated Management in Groundnut-Sesame and Groundnut-Finger Millet System*. Ph.D. Dissertation, Orissa University of Agriculture and Technology, Bhubaneswar, India.

Devulkar, N.G., Bhanderi, D.R., More, S.J. and Jethava, B.A. 2015. Optimization of yield and growth in onion through spacing and time of planting. *Green Farming* 2: 305-307.

Directorate of Cotton Development. 2017. *Status Paper of Indian Cotton*. Ministry of Agriculture and Farmers Welfare, Government of India, Nagpur.

Discovery. 2013. "Bananas Get Pepper Power." *Discovery*, February 11, 2013. Accessed July 31, 2015, http://news.discovery.com/earth/bananas-peppers-genes.htm.

Djurfeldt, G. and Jirström, M. 2005. The puzzle of the policy shift—The early green revolution in India, Indonesia, and the Philippines. In Göran Djurfeldt, Hans Holmén, Magnus Jirström, and Ron Larsson (eds.) *The African Food Crisis: Lessons from The Asian Green Revolution*. CABI, Wallingford.

Dodamani, B.M., Hosmani, M.M. and Hunshal, C.S. 1993. Management of chilli + cotton + onion intercropping systems for higher returns. *Farming Systems* 9: 52-55.

DOGR. 2011. *Annual report 2010-11*. Directorate of Onion and Garlic Research, Rajgurunagar, Pune, India. 80p.

DOGR. 2012. *Annual report 2011-12*. Directorate of Onion and Garlic Research, Rajgurunagar, Pune, India. 72p.

DOGR. 2013. *Annual report 2012-13*. Directorate of Onion and Garlic Research, Rajgurunagar, Pune, India. 92 pp.

Dow AgroSciences. 2014. *"USDA Allows Commercialization of Dow Agrosciences' Enlist™ Corn, Soybean Traits: Farmer Voice Supports USDA's Action on Enlist."* Dow AgroSciences, September 17, 2014. Accessed July 31, 2015, http://newsroom.dowagro. com/press-release/usda-allows-commercialization-dow-agrosciences-enlist-corn-soybean -traits.

Dumas, A., Dijkstra, J. and France, J. 2008. Mathematical modelling in animal nutrition: A centenary review. *J. Agric. Sci.* 146: 123–142.

Economist. 2013. "Genetically Modified Trees: Into the Wildwood." *Economist*, May 4, 2013. Accessed August 11, 2015, http://www.economist.com/news/science-and-technology/21577033-gm-species-may-soon-be-liberated-deliberately-wildwood.

Edwards, P., Zang, W., Belton, B. and Little, D.C. 2019. Misunderstandings, myths and mantra in aquaculture: Its contribution to world food supplies has been systematically over reported. *Mar. Policy* 106: 103547.

Elangovan, M., Suthanthirapandian, I.R and Sayed, S. 1996. Intercropping of onion in chilli. *Annals of Agricultural Science* 34: 839-857.

Enting, H., Kooij, D., Dijkhiuzen, A.A. *et al.* 1997. Economic losses due to clinical lameness in dairy cattle. *Livestock Products Science* 49: 259-267.

Evenson, R.E., Carl, P. and Mark, W.R. 1998. *Agricultural Research and Productivity Growth in India*, Vol.109. Intl. Food Policy Res. Inst., Washington, DC, USA.

Evenson, R.E., Pray, C.E. and Rosegrant, M.W. 1999. *Agricultural Research and Productivity Growth in India*. Research Report 109, International Food Policy Research Institute, Washington, D.C.

FAO. 2007. Poultry. *Proceedings of The International Conference Poultry in The Twenty-First Century: Avian Influenza And Beyond*, Bangkok, Thailand.

FAO. 2012. *World Agriculture Towards 2030/2050: The 2012 Revision*. ESA Working paper No. 12-03.

FAO. 2014. *Innovation in Family Farming*. Food and Agriculture Organization of United Nations, Rome.

FAO. 2015a. *"Why Is Provitamin A Important for Health?"* Golden Rice Project. Accessed August 11, 2015, http://www.goldenrice.org/Content3-Why/why3_FAQ.php# Inventors.

FAO. 2015b. *Myanmar Floods Deal Major Blow to Country's Agriculture*. Food and Agriculture Organization (FAO), Rome.

FAO. 2018. *Transforming the Livestock Sector through the SDGs*. FAO, Rome. http:// www.fao.org/ 3/ CA1177EN/ca1177en.pdf.

FAO. 2020. *The State of World Fisheries and Aquaculture*. Food and Agriculture Organisation of the United Nations, Rome, Italy, 244 pp.

Farghali, M.A. and Zeid, M.I.A. 1995. Phosphorus fertilization and plant population effects on onion grown in different soils. *Assiut J. Agric. Sci.* 26(4): 187-203.

Ferguson, E.L., Gibson, R.S., Opare-Obisau, C. *et al.* 1993. The zinc nutrition of preschool children living in two African countries. *J Nutr* 123: 1487–1496.

Fournier, F., Guy B., and Robin S. 1995. Effect of *Thrips tabaci* (Thysanoptera: Thripidae) on yellow onion yields and economic thresholds for its management. *Entom. Soc. Amer.* 88: 1401-1407.

GAIN Report. 2008. *India Livestock and Products Annual 2008* (IN8098). ThermoFisher Scientific.

Ghosh, B., Bose, T.K. and Mitra, S.K. 1986. Chemical induction of flowering and control of fruit drop in litchi (*Litchi chinensis* Sonn.). *Proc. 22nd Int. Hort. Cong.*, California, Abstr. 1189.

Gibson, R.S.1994. Content and bioavailability of trace elements in vegetarian diets. *Am J Clin Nutr* 59: 1223S–1232S.

Giger, E., Prem, R. and Leen, M. 2009. Increase of agricultural production based on genetically modified food to meet population growth demands. *School of Doctoral Studies (European Union) Journal* 1: 98-124.

Gillis, J. 2009. "Norman Borlaug, Plant Scientist Who Fought Famine, Dies at 95." *New York Times*, September 13, 2009. http://www.nytimes.com/2009/09/14/business/energy-environment/14borlaug.html?pagewanted=all.

GOI. 2007. *Agriculture Strategy for Eleventh Plan: Some Critical Issues.* Planning commission, Govt. of India, New Delhi.

GOI. 2007. *Report of the Working Group on Horticulture, Plantation Crops and Organic Farming for the XI Five Year Plan (2007-12).* Planning Commission, Govt. of India, New Delhi.

GOI. 2009. *Report of the Task Force on Irrigation.* Planning Commission of India, Government of India, New Delhi.

GOI. 2014. *Mission for Integrated Development of Horticulture: Operational Guidelines.* Horticulture Mission, Govt. of India, New Delhi.

GoI. 2018. *Handbook on Fisheries Statistics* 2018. Department of Animal husbandry, Dairying and Fisheries, Ministry of Agriculture, Government of India, Krishi Bhavan, New Delhi.

GoI. 2020b. *PIB Release on New Schemes for Fisheries.* Ministry of Fisheries, Animal Husbandry and Dairying, Govt. of India, New Delhi, India.

Gollin, D., M. Morris, M.W. and Byerlee, D. 2005. Technology adoption in intensive post-green revolution systems. *American Journal of Agricultural Economics* 87(5): 1310–1316.

Gonsalves, D. 2015. "Transgenic Papaya in Hawaii and Beyond." *AgBioForum* 7(1–2) (2004). Accessed July 31, 2015, http://www.agbioforum.org/v7n12/v7n12a07-gonsalves.htm;

Gonsalves, D. *et al.* 2015. "*Papaya Ringspot Virus.*" American Phytopathological Society. Accessed July 31, 2015, http://www.apsnet.org/edcenter/intropp/lessons/viruses/Pages/ PapayaRingspotvirus.aspx.

Gopal, J. 2014. Pre-and post-harvest losses in onion. *Proceedings National Conference on Pre-/Post -Harvest Losses & Value Addition in Vegetables*, IIVR, Varanasi, pp. 25-29.

Gulati, A. and Juneja, R. 2018. *From Plate to Plough: Timidity and Technology.* The Indian Express. Accessed December 6, 2018. https://indianexpress.com/article/opinion/ columns/indian-cotton-growers-farmers-cotton-crop-cotton-farming-narendra-modi-govt-5292899/.

Gulati, A. and Verma, S. 2016. *From Plate to Plough: A Clear Trend towards Non-Vegetarianism in India. The Indian Express.* Accessed September 15, 2018. https:// indianexpress.com/article/opinion/ columns/india-diet-indian-palate-non-vegetarian-vegetarianism-3099363/.

Gupta, R.B.L. and Pathak, V.K. 1987. Management of purple blotch *Alternaria porri* (Ellis:) Clif. of onion by summer ploughing and alteration of date of sowing. *Z. Microbiol.* 142: 163-166.

Gupta, R.P., Srivastava, P.K. and Pandey, U.B. 1986. Control of purple blotch of onion seed crop. *Indian Phytopath*. 39: 303-304.

Gupta, R.P., Srivastava, P.K. and Sharma, R.C. 1996a. Chemical control of purple blotch and Stemphylium blight diseases of onion. *NHRDF News Letter* 16: 14-16.

Gupta, R.P., Srivastava, P.K. and Sharma, R.C. 1996b. Effect of foliar spray of different fungicides on the control of Stemphylium blight disease and yield of onion bulb. *NHRDF News Letter* 16: 13-14.

Gupta, S.C. and Gangwar, S. 2012. Effect of molybdenum, iron and microbial inoculants on symbiotic traits, nutrient uptake and yield of chickpea. *Journal of Food Legumes* 25(1): 45-49.

Gustav R., Anne H., Thomas F. *et al*. 2008. Potentials and prospects for renewable energies at global scale. *Energy Policy* 36: 4048–4056.

Hallikeri, S.S., Halemani, H.L., Patil, B.C. and Nanadagavi, R.A. 2011. Influence of nitrogen management on expression of cry protein in Bt-cotton (*Gossypium hirsutum*). *Indian Journal of Agronomy* 56: 62-67.

Hamilton, B.K., Yoo, K. and MPike, L. 1998. Changes in pungency of onions by soil type, sulphur nutrition and bulb maturity. *Scientia Horticulturae* 74(4): 249-256.

Hanumashetti, S.I., Rao, M.M. and Bankapur, V.M. 1981. Preliminary studies on the effect of growth regulators and chemicals on the improvement of colouration in Gulabi grapes. *Curr. Res.* 10: 45-46.

Hayes, B.J., Bowman, P.J., Chamberlain, A.J. and Goddard, M.E. 2009. Genomic selection in dairy cattle: Progress and challenges. *J. Dairy Sci.* 92: 433–443.

Hazell, P.B.R. 2003. Green revolution, curse or blessing? In Joel Mokyr (ed.) *The Oxford Encyclopedia of Economic History*. Oxford University Press, Oxford, UK.

Hegde, D.M. and Sudhakara Babu, S.N. 2009. Declining factor productivity and improving nutrient use efficiency in oilseeds. *Indian J Agron* 54(1): 1–8.

Helikson, H.J., Haman, D.Z. and Baird, C.D. 1991. *Pumping Water for Irrigation Using Solar Energy*. University of Florida, Florida Cooperation Extension Services, Institute of Food and Agriculture Sciences, Fact sheet EES-63, USA.

Hiloidhari, M., Das, D. and Baruah, D.C. 2014. Bioenergy potential from crop residue biomass in India. *Renewable and Sustainable Energy Reviews* 32: 504-512.

Hisham El-Osta, S. and Mitchell, J.M. 2000. Technology adoption and its impact on production performance of dairy operations. *Review of Agricultural Economics* 22(2): 477-498.

Hokkanen, H.M.T. 1998. Ecological impact of transgenic, insect resistant crops. In *Génie Génétique*: perspectives, inconnues et risques. Les applications en agriculture et dans l′alimentation. Colloque international, Groupe des Verts au Parlement Européen Bruxelles, p. 8.

IARI. 2010. *Annual Report 2011-12*. Indian Agricultural Research Institute, New Delhi, India. 200 pp.

Ibrahim, S.T., Khalil, H.E and Kamel, A.S. 2005. Growth and productivity of sugar beet, onion and garlic grown alone and associations under different inter and intraspacing. *Annals of Agricultural Science* 43 :497-516.

IIPR. 2011. *Vision 2030*. Indian Institute of Pulses Research, Kanpur.

Indira, A., Bhagavan, M.R. and Virgin, I. 2005. *Agricultural Biotechnology and Biosafety in India: Expectations, Outcomes and Lessons*. Stockholms Environment Institute, Stockholm.

IPCC (Intergovernmental Panel on Climate Change). 2007. *Climate Change 2007: Impacts, Adaptation and Vulnerability. Summary for Policy Makers*. See http://www.ipcc. ch/publications_and_data/ ar4/wg2/en/spm.html.

IRRI. 2015. "Why Is Golden Rice Needed in the Philippines Since Vitamin A deficiency Is Already Decreasing?" IRRI. Accessed August 11, 2015, http://irri.org/index.php? option =com_k2&view=item&id=12352&lang=en.

IRRI. 2015a. "*About Golden Rice*." IRRI. Accessed August 11, 2015, http://irri.org/index. php?option=com_k2&view=item&id=10202&lang=en.

IRRI. 2015b. "*Does Golden Rice Contain Daffodil Genes?*" IRRI. Accessed August 11, 2015, http://irri. org/golden-rice/faqs/does-golden-rice-contain-daffodil-genes.

ISAAA. 2008. *Global Status of Commercialized Biotech/GM Crops: 2008 The First Thirteen Years, 1996 to 2008*. ISAAA Brief 39-2008: Executive Summary. Ithaca, NY.

ISAAA. 2015. "*Pocket K No. 10: Herbicide Tolerance Technology: Glyphosate and Glufosinate*." ISAAA. Accessed July 31, 2015, https://isaaa.org/resources/ publications/pocketk/10/default.asp.

ISAAA. 2016. *Global Status of Commercialized Biotech/GM Crops: 2016*. ISAAA *Brief* No. 52, ISAAA, Ithaca, NY.

Islam, M.K., Alam, M.F. and Islam, A.K.M.R. 2007. Growth and yield response of onion (*Allium cepa* L.) genotypes to different levels of fertilizers. *Bangladesh Journal of Botany* 36(1): 33-38.

Jabeda, A., Wagner, S., McCracken, J. *et al.* 2012. Targeted microRNA expression in dairy cattle directs production of β-lactoglobulin-free, high-casein milk. *PNAS* 109(42): 16811-16816.

Jalota, S.K., Buttar, G.S., Sood, A. *et al.* 2008. Effects of sowing date, tillage and residue management on productivity of cotton (*Gossypium hirsutum* L.)–wheat (*Triticum aestivum* L.) system in northwest India. *Soil and Tillage Research* 99: 76-83.

Jat, M.L., Gathala, M.K., Ladha, J.K. *et al.* 2009. Evaluation of precision land levelling and double zero-till systems in the rice–wheat rotation: water use, productivity, profitability and soil physical properties. *Soil and Tillage Research* 105(1): 112-121.

Jayasankar, P. 2018. Present status of freshwater aquaculture in India-A review. *Indian J. Fish.* 65(4): 157-165.

Jena, D., Sahoo, R., Sarang, D.R. and Singh, M.V. 2006. Effect of different sources and levels of S on yield and nutrient uptake by groundnut-rice cropping system in an Inceptisol of Orissa. *J Indian Soc Soil Sci* 54(1): 126–129.

Jha, G.K., Pal, S. and Singh, A. 2012. Changing energy-use pattern and the demand projections for Indian agriculture. *Agricultural Economics Research Review* 25(1): 61-68.

Jha, S.N., Vishwakarma, R.K., Ahmad, T. *et al.* 2015. *Assessment of Quantitative Harvest and Post-Harvest Losses of Major Crops and Commodity in India*. ICAR—All-India Co-ordinated Research Project on Post-Harvest Technology, ICAR-CIPHET.

John Innes Centre. 2014. "*Bumper Harvest for GM Purple Tomatoes*." John Innes Centre, January 25, 2014. https://www.jic.ac.uk/ news/2014/01/gm-purple-tomatoes/.

Johnson, M., Hazell, P.B.R. and Gulati, A. 2003. The role of intermediate factor markets in Asia's green revolution: Lessons for Africa? *American Journal of Agricultural Economics* 85(5): 1211–1216.

Joshi, P.K. and Kumar, P. 2016. Food demand and supply projections for India. In F. Brower and P.K. Joshi (eds.) *International Trade and Food Security: The Future of Indian Agriculture*. CAB International, Wallingford, U.K.

Kaladharan, P., Johnson, B., Nazar, A.K. *et al.* 2019. Perspective plan of ICAR-CMFRI for promoting seaweed mariculture in India. *Mar. Fish. Inf. Serv. T&E Ser.* 240: 17-22.

Kalra, C.L., Beerh, J.K., Manan, J.K. *et al.* 1986. Studies on influence of cultivars on the quality of dehydrated onion (*Allium cepa* L.). *Indian Food Packer* 40: 20-27.

Kalt, W. 2002. Health functional phytochemicals of fruits. *Horticulture Review* 27: 269-315.

Karanja, F., Gilmour, D. and Fraser, I. 2012. Dairy productivity growth, efficiency change and technological progress in Victoria, Paper presented at *Annual Conference of Australian Agricultural and Resource Economics Society*, Fremantle, Western Australia.

Katiha, P.K., Jena, J.K., Pillai, N.G.K. *et al.* 2005. Inland aquaculture in India: Past trend, present status and future prospects. *Aquac. Econ. Manage.* 9: 237-264.

Kaur, G., Brar, Y.S. and Kothari, D.P. 2017. Potential of livestock generated biomass: Untapped energy source in India. *Energies* 10: 3-15.

Kenna, J.P., Gillett, W.B., Power, I.T. and Halcrow, S.W. 1985. *Handbook on Solar Water Pumping*. World Bank, Washington, DC and IT Publications, London.

Khan, A.A., Jilani, G., Akhtar, M.S. *et al.* 2009. Phosphorous solubilizing bacteria: Occurrence, mechanisms and their role in crop production. *Journal of Agriculture and Biological Sciences* 1: 48-58.

Khan, M.I., Shah, M.H., Raja, W. and Teeli, N.A. 2006. Effect of intercropping on the soil fertility and economics of sunflower and companion legumes under temperate conditions of Kashmir. *Environment Ecology* 245(1): 171–173.

Khanduja, S.D. and Garg, Y.K. 1984. Macro-nutrient element composition of leaves from jujube (*Zizyphus mauritiana* Lamk.) tree in north India. *Indian J. Hort.* 41: 22-24.

Khura, T.K., Indra-Mani and Srivastava, A.P. 2011. Design and development of tractor drawn onion (*Allium cepa*) harvester. *Indian Journal of Agricultural Science* 81: 528-532.

Khurana, S.C and Bhatia, A.K. 1991. Intercropping of onion and fennel with potato. Indian *Journal of Weed Science* 23: 64-66.

King, D.A., Peckham, C., Waage, J.K. *et al.* 2006. Infectious diseases: Preparing for the future. *Science* 313: 1392–1393.

King, S.L., Kratochvil, J.A. and Boyson, W.E. 2001. *Stabilization and Performance Characteristics of Commercial Amorphous-Silicon PV Modules*. Sandia National Laboratories. http://photovoltaics. sandia.gov/docs/PDF/kingkrat.pdf.

Knapp, S. 2008. Potatoes and poverty. *Nature* 455: 170–171

Kothari, S.K., Singh U.B., Sushil Kumar and Kumar S. 2000. Inter-cropping of onion in menthol mint for higher profit under subtropical conditions of north Indian plains. *Journal of Medicinal Aromatic Plant Science* 22: 213-218.

Kranthi, K.R., Naidu, S., Dhawad, C.S. *et al.* 2005. Temporal and intra-plant variability of cry1Ac expression in Bt cotton and its influence on the survival of the cotton bollworm, *Helicoverpa armigera* (Noctuidae: Lepidoptera). *Current Science* 89: 291-297.

Krishna, V., Erenstein, O., Sadashivappa P. and Vivek, B.S. 2014. Potential economic impact of biofortified maize in the Indian poultry sector. *International Food and Agribusiness Management Review* 17(4): 111-140.

Kumar, V., Patil, R.G. and Patel, J.G. 2011. Efficient water management technology for sustainable cotton production in central India. In: K.R. Kranthi, M.V. Venugopalan, R.H. Balasubrahmanya, S. Kranthi, S.B. Singh and D. Blaise (eds.) *World Cotton Research Conference -5 Book of Papers*, pp. 376-385. Excel Publishers, New Delhi.

Kupferschmidt, K. 2013. "Activists Destroy 'Golden Rice' Field Trial." *Science Magazine*, August 9, 2013. Accessed August 11, 2015. *http://news.sciencemag.org/asiapacific/ 2013/08/ activists-destroy-golden-rice-field-trial.*

Lal, K. 2000. Foot and mouth disease: Present status and future strategies for control in India. *Indian Farming* 50: 28-31.

Lal, K.K. and Jena, J.K. 2019. Fish genetic resources - India. In: R.K. Tyagi, D.H.N. Munasinghe, K.H.M.A. Deepananda, F. Niranjan and R.K. Khetarpal (eds.) *Regional Workshop on Underutilized Fish and Marine Genetic Resources and their Amelioration - Proceedings and Recommendations.* Asia-Pacific Association of Agricultural Research Institutions (APAARI), Bangkok, Thailand, 55 pp.

Lambert, A.D., Smith, J.P. and Dodds, K.L. 1991. Shelf life extension and microbiological safety of fresh meat - A review. *Food Microbiology* 8: 267-297.

Landes, M.R. 2010. *Growth and Equity Effects of Agricultural Marketing Efficiency Gains in India.* DIANE Publishing.

Layrisse, M., Martinez-Torres, C., Mendez-Costellaro, H. *et al.* 1990. Relationship between iron bioavailability from diets and the prevalence of iron deficiency. *Food and Nutr. Bull.* 12: 301–309.

Leaf, M.J. 1984. *Song of Hope: The Green Revolution in a Punjab Village.* Rutgers University Press, New Brunswick, NJ.

Leakey, R. and Kranjac-Berisavljevic, G. 2009. Impacts of AKST (Agricultural Knowledge Science and Technology) on development and sustainability goals. In B.D. McIntyre, H.R. Herren, J. Wakhungu and R.T. Watson (eds.) *Agriculture at A Crossroads*, pp. 145–253. Island Press, Washington, DC.

Lim, X.Z. 2014. "*Is Glyphosate, Used with Some GM Crops, Dangerously Toxic to Humans?*," April 30, 2014. Accessed August 8, 2015, http://www. Geneticliteracy project.org/2014/04/30/is-glyphosate-used-with-some-gm-crops-dangerously-toxic-to-humans/.

Lipton, M.L. 1985. Research and design of a policy frame in agriculture. In T. Rose (ed.) *Crisis and Recovery in Sub-Saharan Africa*. Development Center, OECD, Paris.

Lopez-Pereira, M.A. 1993. Economics of quality protein maize as a feedstuff. *Agribusiness* 9 (6): 557-568.

Lu, F.M. 1990. Colour preference and using silver mulches to control onion thrips, *Thrips tabaci* Lindeman. *Chinese J. Entom.* 10: 337-342.

Mahajan V., Lawande, K.E., Krishnaprasad, V.S.R. and Srinivas, P.S. 2011. Bhima Shubra and Bhima Shweta - new white onion varieties for different seasons. In *Souvenir & Abstract: National Symposium on Alliums: Current Scenario and Emerging Trends*, p. 162.

Maini, S.B., Diwan, B. and Anand, J.C. 1984. Storage behaviour and drying characteristics of commercial cultivars of onion. *Journal of Food Science and Technology* 21: 417-419.

Mani, V.P., Chauhan, V.S., Joshi, H.C. and Tandon, J.P. 1999. Exploiting gene effects for improving bulb yields in onion. *Ind. J. Genet. Pl. Breed.* 59: 511-514.

Manjula, K. and Saravanan, G. 2015. Poultry industry in India under globalised environment— opportunities and challenges. *International Journal of Scientific Research* 4(8): 391-393.

Marikhur, R.K., Dhar, A.K. and Kaw, M.R. 1977. Downy mildew of *Allium cepa* and its control with fungicides in Kashmir Valley. *Ind. Phytopath.* 30: 576-577.

Mathur, K., Sharma, S.N. and Sain, R.S. 2006. Onion variety RO-59 has higher yield and resistance to purple blotch and Stemphylicum blight. *J. Mycol. Pl. Pathol.* 36: 49-51.

Maxham, A. 2014. "Masked Eggplant Thugs Plant a Field of Lies." *Voices for Reason* (blog), April 17, 2014. https://ari.aynrand.org/blog/2014/04/17/masked-eggplant-thugs-plant-a-field-of-lies.

McIntire, J., Bourzat, D. and Pingali, P. 1992. *Crop-Livestock Interactions in sub-Saharan Africa.* World Bank, Washington, DC.

Meenakshi, J.V. and Banerji, A. 2005. The unsupportable support price: An analysis of collusion and government intervention in paddy auction markets in north India. *Journal of Development Economics* 76(2): 377-403.

Mehta, R. 2003. *The WTO and the Indian Poultry Sector.* Asia Pacific School of Economics and Government, The Australian Natl. Univ. http://aspem.anu.edu.au.

Mehta, R. and Nambiar, R.G. 2007. *The Poultry Industry in India.* The Food and Agriculture Organization (FAO), Bangkok.

Mitra, J., Shrivastava, S.L. and Rao, P.S. 2012. Onion dehydration: A review. *Journal of Food Science and Technology* 49: 267-277.

Mohammad, M.J. and Zuraiqi S. 2003. Enhancement of yield and nitrogen and water use efficiencies by nitrogen drip fertigation of garlic. *J. Pl. Nutr.* 26: 1749- 1766.

Mohanty, B., Vivekanandan, E., Mohanty, S. *et al.* 2017. The impact of climate change on marine and Inland fisheries and aquaculture in India. In: B.F. Philips and M. Perez-Ramirez (eds.) *Climate Change Impacts on Fisheries and Aquaculture: A Global Analysis*, Vol. 2, pp. 569-602. Wiley-Blackwell, New Jersey, USA.

Mollah., M.R.A., Rahman., S.M.L., Khalequzzaman., K.M. *et al.* 2007. Performance of intercropping groundnut with garlic and onion. *International Journal of Sustainable Crop Production* 2: 31-33.

Moloney, C. 2016. India's major agricultural produce losses estimated at Rs 92,000 crore. *Business News*, August 11, 2016.

Monsanto. 2015. "*YieldGard Biotech Maize: Increasing Yields Protection Against the Maize Stalk Borer Through Biotechnology.*" Monsanto. Accessed July 31, 2015, http://www. monsantoafrica.com/products/farmersguides/yieldgard.asp.

Mottet, A., Haan, C., Falcucci, A. *et al.* 2017. Livestock: On our plates or eating at our table? A new analysis of the feed/food debate. *Global Food Security* 14: 1-8.

Munilkumar, S. and Nandeesha, M.C. 2007. Aquaculture practices in Northeast India: Current status and future directions. *Fish Physiol. Biochem.* 33(4): 399-412.

Murkute, A.A. and Gopal, J. 2013. Taming the glut. *Agriculture Today* 16:28-30.

Murphy, S.P., Beaton, G.H., Calloway, D.H. 1992. Estimated mineral intakes of toddlers: Predicted prevalence of inadequacy in village populations in Egypt, Kenya, and Mexico. *Am J Clin Nutr* 56: 565–572.

Nagalaxmi, K., Annap, P., Venkatshwarlu, G. *et al.* 2015. Mislabelling in Indian seafood: An investigation using DNA barcoding. *Food Control* 59: 196-200. DOI:10.1016/j. foodcont. 2015.05.018.

Naik, R,, Annamali, SJ.K. and Ambrose, D.C.P. 2007. Development of batch type multiplier onion peeler. *Proceedings of the International Agricultural Engineering Conference on Cutting-edge Technologies and Innovations on Sustainable Resources for World Food Sufficiency.* Bangkok, Thailand.

Nair, L. 2014. Emerging trends in Indian aquaculture. *J. Aquat. Biol. Fish.* 2(1): 1-5.

Nalayini, P., Raj, S.P. and Sankaranarayana, K. 2011. Growth and yield performance of cotton (*Gossypium hirsutum*) expressing *Bacillus thuringiensis* var: *kurstaki* as influenced by polyethylene mulching and planting techniques. *Indian Journal of Agricultural Sciences* 81: 55-59.

Namukwaya, B., Tripathi, L., Tripathi, J.N. *et al.* 2012. Transgenic banana expressing *Pflp* gene confers enhanced resistance to *Xanthomonas* wilt disease. *Transgenic Research* 21(4): 855-862.

Narayanamoorthy, A. 2008. Drip irrigation and rainfed crop cultivation nexus: The case of cotton crop. *Indian Journal of Agricultural Economics* 63: 487-501.

NARL. 2015. *"About the National Agricultural Research Laboratories."* NARL. Accessed August 11, 2015, http://www.narl.go.ug/.

NDDB. 2015. *Handbook of Good Animal Husbandry Practices [Internet].* Available from: www. dairyknowledge. in/article/handbook-good-dairyhusbandry-practices [Accessed: May 15, 2021].

Negi, D.S., Birthal, P.S., Roy, D. and Khan, M.T. 2018. Farmers' choice of market channels and producer prices in India: Role of transportation and communication networks. *Food Policy* 81(C): 106-121.

New Scientist. 2006. "Edible Cotton Breakthrough May Help Feed the World." *New Scientist,* November 20, 2006. Accessed August 11, 2015, https://www.newscientist.com/ article/dn10612-edible-cotton-breakthrough-may-help-feed-the-world/.

New, M.B., Valenti, W.C., Jidwal, J.H. *et al.* 2010. *Freshwater Prawns: Biology and Farming.* Wiley-Blackwell, New Jersey, USA, 531 pp.

NFDB. 2019. *Aquaculture Technologies Implemented by NFDB.* National Fisheries Development Board, Department of Fisheries Ministry of Fisheries, Animal Husbandry and Dairying, Govt. of India, Hyderabad, India, 58 pp.

Nielsen, R.L. 2015. *"A Compendium of Biotech Corn Traits."* Purdue University, May 2010. Accessed July 31, 2015, http://www.kingcorn.org/news/timeless/BiotechTraits.html.

Nimbalkar, V., Verma, H., Singh, J. and Kansal, S. 2020. Awareness and adoption level of subclinical mastitis diagnosis among dairy farmers of Punjab, India. *Turkish Journal of Veterinary and Animal Sciences* 44: 845-852.

NRCOG. 2004. *Annual Report 2003-04.* National Research Centre on Onion and Garlic, Rajgurunagar, Pune. 52pp.

NREL. 1997. *Twenty Years of Clean Energy.* National Renewable Energy Laboratory, Washington D.C.

OECD/FAO. 2020. Dairy and dairy products. In *OECD-FAO Agricultural Outlook 2020-2029*, OECD Publishing, Paris.

Osborn, D.E. 2003. *Overview of Amorphous Silicon (a-Si) Photovoltaic Installations at SMUD*. ASES 2003: America's Secure Energy, Austin, TX, 8 pp.

Palanisami, K., Mohan, K., Kakumanu, K.R. and Raman, S. 2011. Spread and economics of microirrigation in India: Evidence from nine states. *Economic and Political Weekly* 46(26-27): 81-86.

Palanisami, K. and Raman, S. 2012. *Potential and Challenges in Up-scaling Micro-irrigation in India Experiences from Nine States*. Water Policy Research HIGHLIGHT IWMI-TATA Water Policy Programme.

Palti, J. 1989. Epidemiology, production and control of downy mildew of onion caused by *Perasospora destructor*. *Phytoparasitica* 17: 1.

Pandotra, V.R. 1965. Purple blotch disease of onion in Punjab II: Studies on the life history, viability and infectivity of the causal organism *Alternaria porri*. *Proc. Ind. Acad. Sci. Sec. B* 61: 326-330.

Panwar, A.S., Singh, N.P., Munda, G.C. and Patel, D.P. 2001. Groundnut – production technology for hill region. *Intensive Agric* 39(5–6): 7–9.

Pasricha, N.S. and Tandon, H.L.S. 1993. Fertilizer management in oilseeds. In: H.L.S. Tandon (ed.) *Fertilizer Management in Commercial Crops*, pp. 65–66. Fertilizer Development and Consultation Organisation, New Delhi, India.

Pathak, C.S., Singh, D.P., Deshpande, A.A. and Sreedhar, T.S. 1986. Sources of resistance to pruple blotch in onion. *Veg. Sci.* 13: 300-303.

Patton, L. 2015. "McDonald's Pursuit of the Perfect French Fry." *Bloomberg Business*, April 19, 2012. Accessed July 31, 2015, http://www.bloomberg.com/bw/articles/2012-04-19/mcdonalds-pursuit-of-the-perfect-french-fry.

Pawar, D.D., Dingre, S.K., Bhakre, B.D. and Surve, U.S. 2013. Nutrient and water use by Bt cotton (*Gossypium hirsutum*) under drip fertigation. *Indian Journal of Agronomy* 58: 237-242.

Perkins, J.H. 1997. *Geopolitics and the Green Revolution: Wheat, Genes, and the Cold War*. Oxford University Press, New York.

Perkowski, M. 2013. "*Del Monte Gets OK to Import Biotech Pineapple*" Capital Press, April 25, 2013 (updated May 23, 2013). Accessed August 11, 2015, http://www. capitalpress. com/content/mp-transgenic-pineapple-041613.

Perry, B. and Sones, K. 2009. *Global Livestock Disease Dynamics Over the Last Quarter Century: Drivers, Impacts and Implications*. FAO, Rome, Italy. (Background paper for the SOFA 2009).

Peter, K. 1999. *Informatics on Turmeric and Ginger*. Indian Inst. of Spices Research, Calicut, Kerala, India.

Peter, K.V. 1999. Spices research and development – An updated review. *National Seminar on Sustainable Horticultural Production in Tribal Regions*, Central Hort. Expt. Stn., Ranchi, pp. 48-54.

Pingali, P.L. and Raney, T. 2005. *From the Green Revolution to the Gene Revolution: How will the Poor Fare?* ESA Working Paper No. 05-09.

Potrykus, I. 2015. "The 'Golden Rice' Tale." *AgBioWorld*. Accessed August 11, 2015, http://www.agbioworld.org/biotech-info/topics/goldenrice/tale.html.

Prasanna, B.M., Vasal, S.K., Kassahun, B. and Singh, N.N. 2001. Quality protein maize. *Current Science* 81(10): 1308-1319.

Press Information Bureau (PIB). 2017. *India Becomes Second Largest Fish Producing Country in the World.* Ministry of Agriculture and Farmers Welfare, Government of India, New Delhi. Accessed May 13, 2018. http://www.pib.nic.in/newsite/mbErel. aspx?relid= 173699.

Raheja, P.C. 1973. *Mixed cropping.* Indian Council of Agricultural Research, New Delhi, India, Technical Bulletin (Agric.) No. 42, pp 24–26.

Rahman, M.A., Chiranjeevi, C.H. and Reddy, I.P. 2000. Management of leaf blight disease of onion. *Proc. National Symposium on Onion Garlic Production and Post-Harvest Management. Challenges and Strategies*, Nasik (India), pp. 147-149.

Rajkumar, U., Rama Rao, S.V. and Sharma, R.P. 2010. Backyard poultry farming-changing the face of rural and tribal livelihoods. *Indian Farming* 59: 20-24.

Ramaswamy, N. 1971. *Studies on the Effect of Nitrogen on the Growth and Development of 'Robusta'* (Musa cavendishi *L.*). M. Sc. (Ag.) thesis, Annamalai Univ., Annamalinagar.

Rana, M.K. 2010. Fruits and Vegetables: A potential source of non-nutrients bioactive substances (Functional foods). *Processed Food Industries* 27: 26-33.

Rani, V. and Srivastava, A.P. 2012. Design and development of onion detopper. *AMA Agriculture Mechanization in Asia, Africa and Latin America* 43: 69-73.

Ranjan, R., Megaranjan, S., Xavier, B. *et al.* 2018. Broodstock development, induced breeding and larval rearing of Indian pompano, *Trichinotus mookalee* (Civier 1832) - A new candidate species for aquaculture. *Aquaculture* 495: 265-272.

Rao, M.M. 1997. Studies on the improvement of finger size in Munavalli (Musa AAB) banana. In M.M. Rao and G.S. Sulikeri (eds.) *Research and Development in Fruit Crops in North Karnataka*, pp. 59-61. Univ. of Agri. Sci., Dharwad, Karnataka.

Rao, M.M., Narasimhan, P., Nagaraja, N. and Anandaswamy, B. 1968. Effect of naphthalene acetic acid and p-chlorophenoxy acetic acid on control of berry drop in Anab-e-Shahi grape. *J. Food Sci. Tech.* 5: 127-128.

Rathod, P. and Chander, M. 2016. Adoption status and factors influencing adoption of livestock vaccination in India: An application of multinomial logit model. *Indian Journal of Animal Sciences* 86(9): 1061-1067.

Raveloson, C. 1990. Situation et contraintes de l'aviculture villageoise à Madagascar In: *CTA Seminar Proceedings, Smallholder Rural Poultry Production, Thessaloniki, Greece* 2: 135-138.

Reddy, K.C. and Reddy, K.M. 2005. Differential levels of vermicompost and nitrogen on growth and yield in onion (*Allium cepa* L.) radish (*Raphanus sativus* L.) cropping system. *J. Res., ANGRAU* 33:11-17.

Remiro, D. and Kirmati, H. 1975. Control of seven curls disease of onion with benomyl. *Sum. Phytopathology*, pp. 51-54.

Reuther, W., Batchelor, H.J. and Webber, H.J. 1962. *Citrus Industry, Vol. II.* Univ. of California Press, Berkeley, California.

Rice, X. 2011. "Ugandan Scientists Grow GM Banana as Disease Threatens Country's Staple

Food." *Guardian*, March 8, 2011. Accessed July 31, 2015, http://www.theguardian.com/world/2011/mar/09/gm-banana-crop-disease-uganda.

Rosen, R.J. 2013. "Genetically Engineering an Icon: Can Biotech Bring the Chestnut Back to America's Forests?" *Atlantic*, May 21, 2013. Accessed August 11, 2015, http://www.theatlantic.com/technology/archive/2013/05/genetically-engineering-an-icon-can-biotech-bring-the-chestnut-back-to-americas-forests/276356/.

Ruiz, R.S. and Escaff, M. 1992. Nutrición y fertilización de la cebolla. Serie La Platina-Instituto de Investigaciones Agropecuarias. *Estación Experimental La Platina (Chile)* 37: 69-73.

Rumpel, K.S and Dysko, J. 2003. Effect of drip irrigation and fertilization timing and rate on yield of onion. *J. Veg. Crop Prod.* 9: 65-73.

Ryan, C. 2015 . "The Dose Makes the Poison." *Cami Ryan* (blog), March 5, 2014. Accessed July 31, 2015, https://doccamiryan.wordpress.com/2014/03/05/the-dose-makes-the-poison/.

Saimbhi, M.S. and Bal, S.S. 1996. Evaluation of different varieties of onion for dehydration. *Punjab Vegetable Grower* 31: 45-46.

Salakinkop, S.R. 2011. Enhancing the productivity of irrigated *Bt* cotton (*Gossypium hirsutum*) by transplanting technique and planting geometry. *Indian Journal of Agricultural Sciences* 81: 150–153.

Sample, I. 2012. "GM Cow Designed to Produce Milk Without an Allergy-causing Protein." *Guardian*, October 1, 2012. http://www.theguardian.com/science/2012/oct/ 01/ gm-cow-milk-alllergy-protein.

Samra, J.S., Thakur, R.S. and Chadha, K.L. 1978. Comparison of some mango cultivars in terms of their macronutrient status in fruiting and non-fruiting terminals. *Indian J. Hort.* 35: 144-187.

Sankar, V., Lawande, K.E. and Tripathi, P.C. 2008. Effect of micro-irrigation practices on growth and yield of garlic. *J. Spices Arom. Crops* 17: 232-234.

Sankar, V, Qureshi, A.A, Tripathi, P.C. and Lawande, K.E., 2005. Production potential and economics of onion based cropping systems under western Maharashtra region (Abstr.). Paper presented on *National Symposium on Current Trends in Onion, Garlic, Chillies and Seed Spices– Production, Marketing and Utilization, (SYMSAC-II)*, Rajgurunagar, Pune, p 79.

Sankar, V., Thangasamy A., and J. Gopal. 2014. *Improved Cultivation Practices for Onion*. Tech Bulletin No. 21, Directorate of Onion and Garlic Research, Rajgurunagar, 23 pp.

Sankar, V., Tripathi, P.C., Qureshi, A. and Lawande, K.E. 2005. Fertigation studies in onion and garlic (Abstr.). *National Symposium on Current Trends in Onion, Garlic, Chillies and Seed Spices– Production, Marketing and Utilization, (SYMSAC-II)*, Rajgurunagar, Pune, pp. 62 & 80.

Sankar, V., Veeraragavathatham, D. and Kannan, M. 2009. Organic farming practices in white onion (*Allium cepa* L.) for the production of export quality bulbs. *Journal of Eco-friendly Agriculture* 4: 17-21.

Sarsavadia, P.N., Sawhney, R.L., Pangavhane, D.R. and Singh S.P. 1999. Drying behaviour of brined onion slices. *Journal o f Food Engineering* 40: 219-226.

Scholten, B. and Basu, P. 2009. White counter-revolution? India's dairy cooperatives in a neoliberal era. *Human Geography* 2 (1): 17–28.

Scott, N.R. 2006. *Impact of Nanoscale Technologies in Animal Management*. Wageningen Academic Publishers, The Netherlands.

Scott, S.J., McLeod, P.J., Montgomery, F.W. and Hander, C.A. 1989. Influence of reflective mulch on incidence of thrips (Thysanoptera: Thripidae: Phlaeothripidae) in stacked tomatoes. *J. Entom. Sci.* 24: 422-427.

Sen, A.K. 1981. *Poverty and Famines: An Essay on Entitlement and Deprivation.* Clarendon, Oxford, UK.

Sentenac, H. 2014. "GMO Salmon May Soon Hit Food Stores, but Will Anyone Buy It?" *FoxNews*, March 11, 2014. Accessed July 31, 2051. http://www.foxnews.com/leisure/ 2015,2014/ 03/11/ gmo-salmon-may-soon-hit-food-stores-but-will-anyone-buy-it/.

Shaji, C., Sajal, K.K. and Vishal, T. 2014. Storm surge studies in the North Indian Ocean: A review. *Indian J. Geo-Mar. Sci.* 43(2): 125-147.

Shantharam, S. 2010. Setback to Bt brinjal will have long-term effect on Indian science and technology. *Curr. Sci.* 98(8): 996-997.

Sharangi, A.B, Pariari, A, Datta, S and Chatterjee, R .2003. Effect of boron and zinc on growth and yield of garlic in New Alluvial Zone of West Bengal. *Crop Research* 25: 83-85.

Sharma, A., Sharma, P., Brar, M.S. and Dhillon, N.S. 2009. Comparative response to sulphur application in raya (*Brassica juncea*) and wheat (*Triticum aestivum*) grown on light textured alluvial soils. *J Indian Soc Soil Sci* 57(1): 62–65.

Sharma, I.M. 1997. Screening of onion varieties/lines against purple blotch caused by *Alternaria porri* under field conditions. *Pl. Dis. Res.* 12: 60-61.

Sharma, O.P., Bantewad, S.D., Patange, N.R. *et al.* 2015. Implementation of integrated pest management in pigeon pea and chickpea pests in major pulse-growing areas of Maharashtra. *Journal of Integrated Pest Management* 15(1): 12.

Sharma, P.K., Kumar, S., Yadav, G.L. *et al.* 2007. Effect of last irrigation and field curing on yield and post-harvest losses of rabi onion (*Allium cepa*). *Annals of Biology* 23: 145-148.

Sharma, R.P. and Chatterjee, R.N. 2009. Backyard poultry farming and rural food security. *Indian Farming* 59: 36-37, 48.

Shende, D.G. 1977. *Effect of N, P and K on Growth, Yield and Quality of Pomegranate* (Punica granatum L.). M. Sc. (Ag.) thesis, Mahatma Phule Agri. Univ., Rahuri, Maharashtra.

Shinoj, P., Gopalakrishnan, A. and Jena, J.K. 2020. *Demographic Change in Marine Fishing Communities in India.* FAO, Bangkok.

Shiva, V. 1991. *The Violence of the Green Revolution: Third World Agriculture, Ecology, and Politics.* Third World Network, Penang, Malaysia.

Shiva, V. 1993. *Monocultures of the Mind: Perspectives on Biodiversity and Biotechnology.* Zed Press, London.

Shrivastava, G.K., Khanna, P., Tomar, H.S. and Tripathi, R.S. 2000. Sorghum cultivation and its production technology for eastern Madhya Pradesh. *Intensive Agric* 38(5–6): 1–5.

Shukla, P.K. and Nayak, S. 2015. Challenges in export of poultry and poultry products. In: Souvenir, *32nd Annual conference of IPSA and National symposium,* College of Avian Sciences and Management, Tiruvazhamkunnu, Palakkad, Kerala, pp. 95-108.

Siddiqui, M.W., Ayala-Zavala, J.F. and Dhua, R.S. 2015. Genotypic variation in tomatoes affecting processing and antioxidant attributes. *Critical Review in Food Science and Nutrition* 55(13): 1819-1835.

Siddiqui, M.W. and Dhua, R.S. 2010. Eating artificially ripened fruits is harmful. *Current Science* 99(12): 1664-1668.

Siddiqui, M.W., Momin, C.M., Acharya, P. *et al.* 2013. Dynamics of changes in bioactive molecules and antioxidant potential of *Capsicum chinense* Jacq. cv Habanero at nine maturity stages. *Acta Physiologea Plantarum* 35 (4): 1141-1148.

Simm, G. 1998. *Genetic Improvement of Cattle and Sheep.* CABI Publishing, Wallingford, UK.

Simm, G., Bünger, L., Villanueva, B. and Hill, W.G. 2004. Limits to yield of farm species: Genetic improvement of livestock. In R. Sylvester-Bradley and J. Wiseman (eds.) *Yields of Farmed Species: Constraints and Opportunities in the 21st Century*, pp. 123–141. Nottingham University Press, Nottingham, UK.

Singh, D. 1996. Comparative study of autumn v/s spring sugarcane crop in different cropping systems. *Indian Sugar* 46 (9): 727-729.

Singh, R. 2015. "*Papaya Ringspot.*" The American Phytopathological Society. Accessed July 31, 2015, http://www.apsnet.org/publications/imageresources/Pages/ fi00157.aspx.

Singh, R., Nandal, T.R., Singh, R. 2002. Studies on weed management in garlic (*Allium sativum* L.). *Ind. J. Weed Sci.* 34: 80-81.

Singh, R.A. 1999. A case study: Farming system in Farrukhabad and Kannauj districts (UP). *Agric Ext Rev* 11(2): 22–28.

Singh, R.B. 2009. Serving farmers to render India prosperous. *Agric Today* 12(2): 24–26.

Singh, R.K., Ghosh, P.K., Bandyopadhyay, K.K. *et al.* 2006. Integrated plant nutrient supply for sustainable production in soybean-based cropping system. *Indian J Fertilizers* 1(11): 25–32.

Sinha, V.R.P., Gupta, M.V., Banerjee, M.K. and Kumar, D. 1973. Composite fish culture in Kalyani. *J. Inland Fish. Soc. India* 5: 201-208.

Smith, P., Martino, D., Cai, Z. *et al.* 2007. Agriculture. In B. Metz, O.R. Davidson, P.R. Bosch, R. Dave and L.A. Meyer (eds.) *Climate Change 2007: Mitigation*, pp. 497-540. Contribution of Working Group III to the Fourth Assessment Report of the Intergovernmental Panel on Climate Change, Cambridge University Press, Cambridge, UK.

Smith, P., Martino, D., Cai, Z. *et al.* 2008. Greenhouse gas mitigation in agriculture. *Phil. Trans. R. Soc. B* 363: 789 –813.

Smith. C.J.S., Watson, C.F., Ray, J. *et al.* 1988. Antisense RNA inhibition of polygalacturonase gene expression in transgenic tomatoes. *Nature* 334: 724-726.

Srihari, D. and Rao, M.M. 1996. Induction of flowering in "off" phase mango trees by soil application of paclobutrazol. *Vth Int. Mango Symp.*, Tel Aviv, Israel.

Srinivas, P.S. and Lawande K.E. 2006. Maize barrier as a cultural method for management of thrips in onion (*Allium cepa* L.). *Ind. J. Agril. Sci.* 76: 167-171.

Srinivas, P.S. and Lawande, K.E. 2007. Seedling root dip method for protecting onion plants from thrips. *Ind. J. Pl. Prot.* 35: 206-209.

Srivastava, P.K. and Pandey, V.B. 1995. *Compendium of Onion Diseases.* Tech. Bull. No.7, NHRDF, Nasik, 26 pp.

Srivastava, P.K., Sharma, R.C. and Gupta, R.P. 1995. Effect of different fungicides on the control of purple blotch and Stemphylium blight diseases in onion and seed crop. *NHRDF Newsletter* 15: 6-9.

Srivastava, P.K., Tiwari, B.K., Srivastava, K.J. and Gupta, R.P. 1996. Chemical control of purple blotch and basal rot diseases in onion bulb crop during Kharif. *NHRDF Newslett.* 16: 7-9.

Srivastava, R., Agarwal, A., Tiwari, R.S. and Kumar, S. 2005. Effect of micronutrients, zinc and boron on yield, quality and storability of garlic (*Allium sativum*). *Ind. J. Agri. Sci.* 75: 157-159.

Staal, S.J., Pratt, A.N. and Jabbar, M. 2008. *Dairy Development for the Resource Poor—Part 1: A Comparison of Dairy Policies and Development in South Asia and East Africa*. PPLPI Working Paper No. 44-1. International Livestock Research Institute, Addis Ababa, Ethiopia.

Steinfeld, H., Gerber, P., Wassenaar, T. *et al.* 2006. *Livestock's Long Shadow: Environmental Issues and Options*. FAO, Rome, Italy.

Stuertz, M. 2002. "Green Giant." *Dallas Observer*, December 5, 2002. http://www. dallasobserver. com/news/green-giant-6389547.

Sugha, S.K., Develash, R.K. and Tyagi, R.D. 1992. Performance of onion genotypes against purple blotch pathogen. *South Ind. Hort.* 40: 297.

Swaminathan, M.S. 2000. For an evergreen revolution. In *The Hindu Survey of Indian Agriculture 2000*, pp. 9-15.

Syda Rao, G., Imelda Joseph, Philopose, K.K. and Suresh Kumar, M. 2014. Cage culture in India. *Aquac. Int.* 22: 961-962.

Tamil Selvan, C., Valliappan, K. and Sundararajan, R. 1990. Studies on the weed control efficiency and residues of oxyfluorfen in onion (*Allium cepa* L.). *Intl. J. Trop. Agri.* 8: 123-128.

Tandon, H.L.S. and Sekhon, G.S. 1988. *Potassium Research and Agricultural Production in India*. FDCO, New Delhi, India, viii + 144 pp.

Taryn, G., Frank, A., James, A. *et al.* 2020. A global blue revolution: Aquaculture growth across regions, species and countries. *Rev. Fish. Sci. Aquac.* 28: 107-116.

Thilakavathy, S. and N. Ramaswamy. 1998. Effect of inorganic and biofertilizer treatments on yield and quality parameters of multiplier onion. *NHRDF Newslett.* 18: 18-22.

Thind, H.S., Buttar, G.S., Singh, A.M. *et al.* 2012. Yield and water use efficiency of hybrid Bt cotton as affected by methods of sowing and rates of nitrogen under surface drip irrigation. *Archives of Agronomy and Soil Science* 58: 199-211.

Thornton, P.K. and Gerber, P. 2010. Climate change and the growth of the livestock sector in developing countries. *Mitigation Adapt. Strateg. Glob. Change* 15: 169 –184.

Thornton, P.K., van de Steeg, J., Notenbaert, A. and Herrero, M. 2009. The impacts of climate change on livestock and livestock systems in developing countries: A review of what we know and what we need to know. *Agric. Syst.* 101: 113 –127.

Toor, S.S., Singh, S., Garcha, A.I.S. *et al.* 2000. Gobhi sarson as an intercrop in "Autumn Sugarcane" for higher returns. *Intensive Agric* 38(5–6): 29–30.

Trindade-Santos, I., Moyes, F. and Magurran, A.E. 2020. Global change in the functional diversity of marine fisheries exploitation over the past 65 years. *Proc. R. Soc. Biol. Sci. Ser. B.* 287: 2020-2089.

Tripathi, L., Mwaka, H., Tripathi, J.N. and Tushemereirwe, W.K. 2010. Expression of sweet pepper *Hrap* gene in banana enhances resistance to *Xanthomonas campestris* pv. *musacearum*. *Molecular plant pathology* 11(6): 721–731.

Tripathi, L., Mwangi, M., Abele, S. and Aritua, V. 2009. *Xanthomonas* wilt: A threat to banana production in East and Central Africa. *Plant Disease* 93(5): 450-451.

Tripathi, P.C. and Lawande, K.E. 2009. A new gadget for onion grading. *AG Journal* 90: 1-4.

Tripathi, P.C., Sankar, V. and Lawande, K.E. 2010. Influence of micro-irrigation methods on growth, yield and storage of rabi onion. *Indian Journal of Horticulture* 67: 61-65.

Tripathi, P.C., Sankar, V., Mahajan, V. and Lawande, K.E. 2011. Response of gamma irradiation on post-harvest losses in some onion varieties. *Indian Journal of Horticulture* 68: 556-560.

University of Illinois. 2011. "*News Release: Stalk Borers and Corn Borers.*" University of Illinois Extension, June 20, 2011. Accessed July 31, 2015, http://web.extension.illinois. edu/state/newsdetail. cfm?NewsID=21044.

Unnikrishnan, A.S., Manimurali, M. and Ramesh Kumar, M.R. 2006. *Sea-Level Changes along the Indian Coast.* National Institute of Oceanography, Goa, India.

Upton, M, 2007. Scale and structures of the poultry sector and factors inducing change: Inter country differences and expected trends. In *Proc. Poultry in the 21ˢᵗ Century: Avian Influenza and Beyond.* International Poultry Conference, Bangkok, pp. 49- 79.

USDA. 2015 . "*Adoption of Genetically Engineered Crops in the U.S.*" USDA. Accessed July 31, 2015, http://www.ers.usda.gov/data-products/adoption-of-genetically-engineered-crops-in-the-us/recent-trends-in-ge-adoption.aspx.

USDA, Foreign Agricultural Service. 2017. *Production, Supply and Distribution Online Data.* Accessed July 9, 2018. https://apps. fas.usda.gov/psdonline/app/index.html#/ app/downloads.

Valenti, W.C., Kimpara, J.M., Preto, B.D.L. and Moraes Valenti, P. 2018. Indicators of sustainability to assess aquaculture systems. *Ecol. Indic.* 88: 402-413.

Vatta, K., Sidhu, R.S., Lall, U. *et al.* 2018. Assessing the economic impact of a low-cost water-saving irrigation technology in Indian Punjab: the Tensiometer. *Water International* 43(2): 305–321.

Vekaria, R.S., Pandya, R.D. and Thumar, D.N. 2000. Knowledge and adoption behaviour of rainfed groundnut growers. *Agric Ext Rev* 12(1): 23–27.

Venugopalan, M.V., Blaise, D., Yadav, M.S. and Deshmukh, R. 2011. Fertilizer response and nutrient management strategies for cotton. *Indian J. Fert.* 7: 82-94.

Verlodt, H. 1990. *Greenhouses in Cyprus, Protected Cultivation in the Mediterranean Climate.* FAO, Rome.

Verma, L.R., Pandey, U.B., Bhonde, S.R. and Srivastava, K.J. 1999. Quality evaluation of different onion varieties for dehydration. *News Letter National Horticultural Research and Development Foundation* 19 (2/3): 1-6.

VIB News. 2015 . "*MON810 Scientific Background Report.*" VIB (report). Accessed July 31, 2015, http:// www.vib.be/en/news/Documents/VIB_Dossier_MON810_ENG.pdf.

Vick, B.D., Neal, B., Clark, R.N. and Holman, A. 2003. *Water Pumping with AC Motors and Thin-film Solar Panels.* ASES Solar 2003: America's Secure Energy, Austin, TX, 6 pp.

Vijayan, K.K. 2019. Domestication and genetic improvement of Indian white shrimp, *Penaeus indicus*: A complimentary native option to exotic pacific white shrimp, *Penaeus vannamei. J. Coast. Res.* 86(spl.): 270-276.

Vinay Singh, J., Bisen R.K., Agrawal H.P. and Singh, V. 1997. A note on weed management in onion. *Veg. Sci.* 24: 157-158.

Voosen, P. 2011. *"Crop Savior Blazes Biotech Trail, but Few Scientists or Companies are Willing to Follow."* New York Times, September 21, 2011. Accessed July 31, 2015, http://www.nytimes.com/gwire/2011/09/21/21greenwire-crop-savior-blazes-biotech-trail-but-few-scien-88379.html.

Wagner, H. 2003. *"Researchers Get to the Root of Cassava's Cyanide-Producing Abilities."* Ohio State University. http://researchnews.osu.edu/archive/cassava.htm;

Walton, D. 2015. *"GMO Myth: Farmers 'Drown' Crops in 'Dangerous' Glyphosate. Fact: They Use Eye Droppers."* Genetic Literacy Project, January 22, 2015. Accessed July 31, 2015, http://www.geneticliteracyproject.org/2015/01/22/gmo-myth-farmers-drown-crops-in-dangerous-glyphosate-fact-they-use-eye-droppers/.

Warade, S.D, Shinde, S.V. and Gaikwad, S.K.1996. Studies on periodical storage losses in onion. *Allium News Letter* 10: 37-41.

Wattiaux, M.A. 2011. *Mastitis: The Disease and its Transmission. Dairy Essentials, Babcock.* Available from: https://www.yumpu.com/en/document/read/29801928/23-15 (Accessed: May 15, 2021).

Whiffen, H.J.H., Haman, D.Z. and Baird, C.D. 1992. Photovoltaic-powered water pumping for small irrigation systems. *Appl. Eng. Agric.* 8: 625–629.

WHO. 2015. *"Micronutrient Deficiencies."* World Health Organization. Accessed July 31, 2015, http://www.who.int/nutrition/topics/vad/en/.

World Bank. 2005. *India's Water Economy: Bracing for a Turbulent Future.* World Bank, Washington, D.C., U.S.A.

World Bank. 2007. *World Development Report 2008: Agriculture for Development.* World Bank, Washington, D.C.

World Bank. 2013. *Turn Down the Heat: Climate Extremes, Regional Impacts, and the Case for Resilience.* The World Bank, Washington, D.C., U.S.A.

WRI (Water Resources Institute). 2015. *India's Growing Water Risks.* Available at: http://www.wri.org/blog/2015/02/3-maps-explain-india%E2%80%99s-growing-waterrisk.

Wu, E.W.K. and Lau, I.P.L. 2008. *The Potential Application of Amorphous Silicon Photovoltaic Technology in Hong Kong.* Hong Kong Electrical and Mechanical Service Department. http://www.emsd.gov.hk/emsd/e_download/wnew/conf_papers/ emsd paper _ inal.pdf (Accessed June 12, 2010).

Yadav, I.C., Devi, N.L., Syed, J.H. *et al.* 2015. Current status of persistent organic pesticides residues in air, water, and soil, and their possible effect on neighboring countries: A comprehensive review of India. *Sci. Total Environ.* 51(1): 123–137.

YouTube. 2015. *"Hawaii Snapshot—a Legendary Scientist and His Papaya Dreams."* YouTube. Accessed August 10, 2015, https://www.youtube.com/watch?t=95&v=fn_ 0KdbTlR8.

Subject Index

A

Aphelenchoides besseyi, 7,49
A. ritzemabosi, 253
Azospirillum lipoferum, 121
Azotobacter chroococcum, 121

B

Bacillus megaterium, 121
B. pumilus, 121
B. subtilis, 121
BINM-Cultural controls, 223
    Altered planting dates, 225
    Crop rotation, 223
    Disease-free planting material, 224
    Genetic diversity, 223
    Intercropping, 224
    Optimum growing conditions, 225
    Mulching, 225
    Multiple cropping, 224
    Sanitation, 224
    Spacing, 224
    Strip cropping, 224
BINM options, 18
    Proactive options, 18
    Reactive options, 18
BINM-Resistant cultivars, 225
    Biotech crops, 226
Biofumigation-Banana nema mangmt., 68
    Burrowing nematode, 68
    Lance nematode, 68
    Reniform nematode, 68
    Spiral nematode, 68
Biofumigation-Citrus nema mangmt., 67
    Citrus nematode, 67
    Root-knot nematode, 68
Biofumigation-Cyst nema mangmt., 67

Potato, 67
Sugar beet, 67
Biofumigation-Grapevine nema mangmt., 68
    Dagger nematode, 68
Biofumigation-Lesion nema mangmt., 68
    Strawberry, 68
    Tobacco, 68
Biofumigation-Non-Brassica crops, 62
    Chinaberry, 63
    Marigold, 63
    Sorghum, 62
    Sudan grass, 62
Biofumigation-Root-knot nema mangmt., 66
    Aster, 67
    Carrot, 66
    Lettuce, 66
    Muskmelon, 66
    Potato, 66
    Tomato, 66
Biotechnological approaches, 17
    Genome editing, 17
    Recombinant DNA technology, 17

C

Ceratocystis fimbriata, 6
Citrus root-knot nema management, 43

D

Ditylenchus dipsaci, 47

F

Fusarium oxysporum f. sp. *dianthi,* 7

G

Globodera pallida, 7
G. rostochiensis, 7
Glomus mosseae, 121
GM crops-Nematicidal proteins, 191

Bt toxins (Cry proteins), 192
Chemodisruptive peptides, 192
Lectins, 191
Plantibodies, 191

GM crops-Other strategies, 194
Camalexin and callose synthesis, 195
Downregulation of genes, 195
Genome editing, 195
Overexpression of genes, 194
Suppression of genes, 194
Upregulation of genes, 195

Grafting-Nematode management, 173
Brinjal, 180
Cucumber, 175
Muskmelon, 174,175
Sweet pepper, 181
Tomato, 176
Watermelon, 174,175

H
Helicotylenchus dihystera, 68, 253
Heterodera avenae, 49
H. glycines, 197
H. oryzae, 203
H. schachtii, 199
Hirschmanniella oryzae, 203
Hoplolaimus indicus, 68

I
INM – Protected cultivation, 242
In nursery beds, 242
In standing crop, 242

INM – Bell pepper protected cultivation, 247
Biological methods, 247
Chemical methods, 247
Cultural methods, 247
Host resistance, 247
Integrated methods, 247
Physical methods, 247

INM – Carnation protected cultivation, 252,253
Biological methods, 252,253
Integrated methods, 252

INM – Crucifers protected cultivation, 250

Host resistance, 251
Physical methods, 251

INM – Cucumber protected cultivation, 248
Biological methods, 248
Chemical methods, 249
Cultural methods, 248
Host resistance, 249
Integrated methods, 250

INM – Gerbera protected cultivation, 252,253
Chemical methods, 252,253
Integrated methods, 252

INM – Gladiolus protected cultivation, 254
Chemical methods, 254
Integrated methods, 254
Physical methods, 254

INM – Lettuce protected cultivation, 251
Chemical methods, 251
Cultural methods, 251
Physical methods, 251

INM – Lilies protected cultivation, 254
Biological methods, 255
Chemical methods, 255
Cultural methods, 255
Physical methods, 255

INM – Strawberry protected cultivation, 257
Chemical methods, 257
Integrated methods, 257

INM – Tomato protected cultivation, 243
Chemical methods, 243
Grafting, 243
Integrated methods, 245
Soil solarization, 243

M
Meloidogyne arenaria, 6
M. enterolobii, 7
M. graminicola, 6
M. hapla, 257
M. incognita, 6,248,252
M. indica, 6
M. javanica, 6,248
Meloidogyne spp., 243, 246, 250, 251, 254
Myrothecium verrucaria, 121

N

Nano-formulations-Field studies, 105

    Turf root-knot nema mangmt., 105

Nano-formulations-Glasshouse studies, 103

    Attraction of larvae to roots, 103
    Effect on plant growth, 104
    Effect on root-knot populations, 105
    Effect on fruit yield, 104
    Larval root penetration, 103

Nano-formulations-Laboratory studies, 101

    Inhibition of egg hatching, 101
    Larval immobility & mortality, 101
    Morphological changes-eggs, 101
    Morphological changes-larvae, 102

New chemical formulations, 15

    Nimitz, 15
    Velum Prime, 15

P

Pasteuria usage, 121
Phytophthora parasitica, 7
Pochonia chlymydosporia, 121
Polianthes tuberosa, 7
Pratylenchus coffeae, 42
P. penetrans, 68
Pseudomonas aureofaciens, 121
P. fluorescens, 121
Purpureocillium lilacinum, 121

R

Radopholus similis, 68, 255
Rotylenchulus reniformis, 68

S

Solarization-Enhancement of beneficials, 36

    Bacteria, 36
    Earthworms, 36
    Fungi, 36

Solarization-Field crops nema. mangmt., 49

    Rice, 49
    Wheat, 49

Solarization-Fruit crops nema. mangmt., 42

    Banana, 42
    Citrus, 43
    Peach, 43
    Strawberry, 43

Solarization-Plantations nema. mangmt., 48

    Betel vine, 48
    Tea, 48

Solarization-Spices nema. mangmt., 48

    Cardamom, 48
    Mustard (Rape seed), 48

Solarization-Vegetables nema. mangmt., 44

    Beans, 47
    Bell pepper, 47
    Brinjal, 46
    Garlic, 47
    Melon, 48
    Onion, 47
    Tomato, 44

Streptomyces avermectinius, 132

T

Trichoderma hamatum, 121
T. harzianum, 121
T. viride, 121
Tylenchulus semipenetrans, 67

X

Xiphinema index, 68